Applications of Lie Groups to Difference Equations

Differential and Integral Equations and Their Applications

A series edited by:
A.D. Polyanin
Institute for Problems in Mechanics, Moscow, Russia

Applications of Lie Groups to Difference Equations

Vladimir Dorodnitsyn

CRC Press
Taylor & Francis Group
Boca Raton   London   New York

CRC Press is an imprint of the
Taylor & Francis Group, an **Informa** business

A CHAPMAN & HALL BOOK

Chapman & Hall/CRC
Taylor & Francis Group
6000 Broken Sound Parkway NW, Suite 300
Boca Raton, FL 33487-2742

First issued in paperback 2017

ISBN 13: 978-1-138-11823-2 (pbk)
ISBN 13: 978-1-4200-8309-5 (hbk)

Visit the Taylor & Francis Web site at
http://www.taylorandfrancis.com

and the CRC Press Web site at
http://www.crcpress.com

Contents

Preface

This book is dedicated to applications of Lie groups to finite-difference equations, meshes, and difference functionals. The interest in continuous symmetries of discrete equations (i.e., Lie transformation groups admitted by such equations) springs from at least two sources. First, discrete equations serve as primary, fundamental mathematical models in physics and mechanics. Cellular automata and neural nets also clearly belong to the realm of discrete models. Their integrability and the existence of exact solutions and conservation laws are undoubtedly related to the presence of continuous symmetries. This raises the question of finding and using the transformation group admitted by a given discrete equation.

Second, modeling a given system of differential equations with the use of difference equations and meshes can also be based on symmetries. It is well known that one and the same system of differential equations can be approximated by infinitely many difference schemes. Hence finite-difference modeling always involves the problem of selecting the schemes that are in some respect advantageous. The selection criteria are often given by fundamental physical principles present in the original model, such as conservation laws, variational principles, the existence of physically meaningful exact solutions, etc. In this connection, qualitative considerations play a significant role in the construction of numerical algorithms, because they permit including the "physical meaning" of the object under study in the numerical method used to analyze the mathematical model. This point of view has led to the development of methods for constructing conservative difference schemes, to the integro-interpolation approach to constructing numerical schemes, to variational methods for constructing schemes, to symplectic numerical methods, etc.

The invariance of differential equations under continuous transformation groups is certainly a fundamental property of these models and reflects the homogeneity and isotropy of space–time, the Galilean principle, and other symmetry properties that are intuitively (or experimentally) taken into account by the creators of physical models. Therefore, it is apparently important in the theory of difference schemes to preserve the symmetry properties when passing to the finite-difference model, thus adequately representing the symmetry of the original differential model. This may serve as the above-mentioned selection criterion.

The author first became acquainted with the idea that the qualitative (physical) characteristics of differential equations should be preserved in difference models at A. A. Samarskii's seminar in the late 1960s,[1] mainly in connection with the construction of conservative and completely conservative difference schemes and variational numerical methods for gasdynamic and magnetohydrodynamic problems. From the theoretical viewpoint, this work of Samarskii and his scientific school was perhaps ahead of their time. (By the way, all their publications at the time were only in Russian.) Back then, the community of mathematicians dealing with qualitative methods of the theory of differential equations did not pay much attention to difference equations. These ideas were widely implemented and developed only a few decades later. Nowadays, methods putting emphasis on the preservation of geometric and other qualitative properties of the solution set of the original differential system (e.g., the symmetry group; variational principles; the existence of first integrals, conservation laws, and exact solutions; symplecticity; and volume preservation) are being intensively developed. In recent years, they have sometimes been combined under the common name of *geometric numerical integration*. This book largely deals with only one aspect of this new research trend, with attention being mainly paid to continuous symmetries of discrete mathematical models.

The theory of continuous transformation groups was first formulated by Sophus Lie when he was devising general integration methods for ordinary differential equations. Further development of group analysis of differential equations and a systematic study of the structure of their solution sets originated in the work of L. V. Ovsyannikov and his scientific school. (This was indeed a second birth of group analysis, at least in Russia.) The publication of Ovsyannikov's papers and books in the USSR in the 1960s led to a boom of studies and publications on the topic. The work of Ovsyannikov, Birkhoff, and their students and successors has made Lie's idea of describing symmetries of differential equations into an independent scientific field. At present, group analysis is a generally recognized method for describing continuous symmetries of differential and integro-differential equations of mathematical physics.

It is very tempting to use the group-theoretic approach when constructing and studying various mathematical models (including difference schemes), because group analysis has powerful infinitesimal criteria for the invariance of objects under study. Thus, the problem of finding a continuous transformation group can be reduced to solving a system of *linear* equations regardless of whether the original model itself is linear or not. To model a physical process with known symmetry, one should finding a set of differential (or, in our case, difference) invariants, and this problem is also linear.

The knowledge of the transformation group admitted by a mathematical model provides significant information about the solution set of the model, because the structure of this group correlates with the algebraic structure of the set of solutions.

---

[1]It was in the 1950s that Tikhonov and Samarskii [135] recognized the importance of the requirement that the scheme be conservative.

The higher the dimension of the admissible group, the wider are the possibilities for its application. Therefore, it is apparently important to preserve the entire symmetry of the original continuous model in its finite-difference analog.

In this book, attention is focused on the problem of *constructing* difference equations and meshes such that the difference model preserves the symmetry of the original continuous model. The introduction of finite-difference variables is rather formal. But one can intuitively rely on the geometric vision of the "difference" space being embedded in the "continuous" space; i.e., continuous transformations act on the entire Euclidean space of appropriate dimension, but we are only interested in countably many of its points. Therefore, two types of variables, continuous and discrete, are used in the mathematical apparatus. The first are used to describe the tangent fields of continuous transformation groups, and the second serve to construct difference forms and equations. As a result, there arises a rather unusual object, an infinitesimal group operator whose action is a continuous differentiation with respect to discrete variables.

Finite-difference operators, in contrast to differential operators, are defined on finite subsets of countable sets of mesh nodes. (The finite set of mesh points on which a difference equation is written is called a *difference stencil*.) Owing to this *nonlocality* of the operators (which physically means that the problem has characteristic length scales), difference operators possess peculiar properties absent in the local differential models. In particular, one distinguishes between "right" and "left" differentiations (and the corresponding shifts), there are uniform and nonuniform meshes, and the difference Leibniz rule has a specific character. This specific character results in the appearance of a peculiar *calculus of infinitesimal transformations of finite-difference variables* considered in Chapter 1.

Note that all issues in this book are considered locally, just as in the classical group analysis: invariance problems for difference equations and difference meshes are studied in a neighborhood of an arbitrary point. However, unlike in the case of differential equations, a "point" in the case of difference equations is a difference stencil, which has a certain geometric structure, so that *the role of transformations of independent variables is of exceptional importance*. It is a distinguishing feature of our approach to the analysis of group properties of difference equations that the transformations of independent variables are included in the class of admissible transformations. Accordingly, the group action generally transforms the difference stencil (and the entire mesh). To pose the invariance problem for a difference model, we suggest *including the mesh equation characterizing the difference stencil geometry in the difference model*. With this approach, it is possible to preserve the symmetry of the original differential model *exactly* rather than in the form of a group isomorphic to the original one. This finally permits constructing difference models completely preserving the symmetry of the original differential equations.

In Chapter 1, we consider the invariance problem for various difference meshes, uniform and nonuniform, orthogonal and nonorthogonal. It is clearly impossible to list all meshes used in practice, but a series of propositions establishing necessary and sufficient conditions for the invariance of several classes of most widely used meshes allow us to carry out a preliminary analysis of the possibilities of difference modeling in specific situations. In the same Chapter 1, we consider the relationship between the operations of discrete and continuous differentiation and also the problems related to changes of variables in the space of difference variables. We construct a difference representation of the (continuous) total differentiation operator with the help of so-called Newton series. Note that this representation is extremely cumbersome. Hence the well-known equivalence of the symmetry group in the form of evolution fields (for which the independent variables are invariants) is no longer attractive for difference equations. While the point symmetry group of a difference equation can be written out in exactly the same form as for a differential equation, the symmetry in the form of evolution fields has an awkward structure (which uses *all points* of the difference mesh), which is unsuitable in practice. Here the specific features of difference models manifest themselves as well.

In Chapter 2, we use the mathematical technique developed in the preceding chapter to study the symmetry properties of finite-difference models, i.e., difference equations considered together with difference meshes. The main theorem of this chapter provides *necessary and sufficient conditions for the invariance of difference equations and meshes*. We also propose a simple algorithm for constructing invariant difference models from a given transformation group, namely, the *method of finite-difference invariants*. In several examples, we construct finite-difference models completely inheriting the symmetry of the original differential equations. We show that the symmetry of difference models permits applying reduction to subgroups just as in the case of differential equations and thus obtaining invariant (exact) solutions of difference schemes. Since the criterion for finite-difference equations to be invariant on the difference mesh also provides *necessary conditions for the invariance of a difference model*, it is possible to *calculate* the symmetry of a given difference equation. We present an example of such calculations of the admissible group, showing the peculiarity of the splitting procedure.

In subsequent chapters, the technique of point transformation groups is used to study invariant properties of difference equations. In Chapter 3, we consider invariant ordinary difference equations. Just as in the case of first-order ODE, the knowledge of the admissible group permits integrating the difference equation. For second-order ODE, we give a complete group classification of difference equations and meshes. It is of interest to note that this classification presents significantly more invariant difference equations than the corresponding list of invariant differential equations.

In Chapter 4, we construct examples of finite-difference models (i.e., difference equations and meshes) completely preserving the symmetry of the original partial differential equations. We note that a majority of the constructed invariant difference schemes are very unusual and rather different from the traditional ones. In particular, the symmetry of most evolution equations can be preserved with the help of moving mesh schemes. For the nonlinear heat equation with a source (which includes the linear equation as a special case), we present a complete list of invariant difference models corresponding to the list of invariant differential equations obtained earlier. Note that we do not discuss any issues concerning the *numerical implementation* of the invariant difference models obtained earlier, and hence the book does not contain specific numerical calculations.

In Chapter 5, we consider combined models, i.e., equations containing both differential and difference variables. As typical examples, we consider delay differential equations and differential-difference equations, where continuous derivatives with respect to time occur together with difference spatial derivatives. The symmetry of such models is described by admitted transformations in the product of the spaces of differential and difference variables, and its analysis is in some sense simpler than that of purely difference models, because the question of the invariance of geometric properties of the space–time mesh does not arise.

It is well known that conservation laws underlie the construction of mathematical models in a majority of cases. The relationship between the conservation laws and the symmetries of the corresponding variational problem is stated in a definitive constructive form in the Noether theorem, which says that if the variational functional is invariant, then the corresponding Euler differential equations are conservative (i.e., the conservation laws are satisfied on their solutions). In Chapter 6, we consider invariant variational problems for difference functionals and present a difference counterpart of E. Noether's construction. Difference variational problems have their own specific features and in general substantially differ from the continuous version. Nevertheless, fully constructive methods for devising invariant schemes and meshes with difference analogs of the conservation laws are also proposed in the difference case. We show that the invariance of the *finite-difference* functional does not automatically imply the invariance of the corresponding Euler equations. We obtain a *new difference equation* (which, in general, does not coincide with the difference Euler equation) on whose solutions the functional proves to be stationary under the group transformations. This equation, which is called the *quasi-extremal equation* (or a local extremal equation), depends on the coordinates of the group operator and, in the case of an invariant functional, has the corresponding conservation law. Thus, if a difference functional is invariant under several subgroups, then, in general, this leads to several difference equations with conservation laws. (Each equation has its own conservation law.) If this set of quasi-extremals has a nonempty set of general solutions, then it is possible to state a theorem completely similar to the Noether theorem for such an intersection of quasi-extremals of the invariant functional. Note that the proposed difference construction becomes the classical Noether theorem in the continuum limit.

In Chapter 7, the relation between symmetries and first integrals for difference Hamiltonian equations is considered. These results are based on results for continuous canonical Hamiltonian equations considered in the Introduction. It is shown that discrete Hamiltonian equations can be obtained by the variational principle from action functionals. Noether-type difference operator identities are developed. On the basis of these identities, a Noether-type theorem for the canonical Hamiltonian equations is stated. The approach based on the symmetries of discrete Hamiltonians provides a simple, clear way to construct first integrals of difference Hamiltonian equations by means of purely algebraic manipulations. It can be used to preserve the structural properties of underlying differential equations under discretization procedure, which is useful for numerical implementation.

In Chapter 8, we construct examples of *exact* schemes, i.e., difference models that have infinite order of approximation. The set of solutions of exact schemes, which coincide with the corresponding solutions of the differential equation at the mesh points, obviously admits the symmetry group of the differential equation. Therefore, the exact scheme (and the mesh) must be invariant and can be constructed from difference invariants. Several examples show that the parametric family of invariant schemes contains exact schemes. Such schemes can be viewed as a discrete representation of the solution set of the corresponding ODE. Thus, the following peculiar *mathematical dualism* arises: one and the same process can be described either by ODE or by an exact difference model.

At present, there is a comprehensive literature concerning applications of Lie groups to differential equations. Moreover, there are excellently written introductory courses, which allow young researchers to assimilate the main ideas of group analysis rather quickly. Nevertheless, to make our presentation closed and self-sufficient, in the Introduction we briefly recall the elementary notions of group analysis of differential equations and introduce some notation that we need in our studies of applications of Lie groups of transformations to difference equations, functionals, and meshes. In addition, we briefly present some results concerning the Lie–Bäcklund groups (or higher-order symmetries) and Noether-type theorems for Lagrangian and Hamiltonian formalisms in the context of differential equations.

Acknowledgments

I am deeply grateful to L. V. Ovsyannikov and A. A. Samarskii, with whom I communicated for many years; everything I have learned from them determined the main content of the results presented here.

I am grateful to S. P. Kurdyumov, Director of the M. V. Keldysh Institute of Applied Mathematics, Russian Academy of Sciences in 1989–1999, for his utmost attention to and support of this work.

My first papers on the construction of invariant difference models were written together with M. Bakirova, an expert in the theory of difference schemes, who passed away prematurely in 1996. I commemorate our collaboration with gratitude.

I would like to thank R. Kozlov and P. Winternitz for long-term collaboration. Several important results presented in this book were obtained together with them.

I thank A. Aksenov, G. Bluman, A. Bobylev, C. Budd, P. Clarkson, G. Elenin, E. Ferapontov, V. Galaktionov, P. Hydon, N. Ibragimov, O. Kaptsov, D. Levi, S. Meleshko, P. Olver, S. Svirshchevskii, and many, many others for fruitful discussions, useful remarks, and comments.

I am grateful to V. E. Nazaikinskii for help with the translation of the manuscript into English and for some useful remarks.

Vladimir Dorodnitsyn
Moscow

Introduction

0.1. Brief Introduction to Lie Group Analysis of Differential Equations

Nowadays, there is a wide literature dealing with applications of Lie groups to differential equations (e.g., see [13–15, 69, 73, 107, 111, 130]).

To make our presentation closed and self-contained, in this introductory section we briefly recall the required elementary notions and introduce the notation used in group analysis of differential equations and needed to study difference equations, functionals, and meshes with the help of Lie transformation groups in subsequent chapters.

In our presentation, we follow the notation and partly the contents of Ovsyannikov's excellent book [114], which has long since become a bibliographical rarity. We also briefly present some results concerning the Lie–Bäcklund group (or higher order symmetries) and Noether-type theorems for the Lagrangian and Hamiltonian formalisms. Theorems are given without proofs but are illustrated by examples.

0.1.1. One-parameter continuous transformation groups

Consider the Euclidean space R^N of points $x = (x^1, x^2, x^3, \dots, x^N)$ in which some smooth transformations T_s, $s = 1, 2, \dots$, taking R^N to itself, $x^* = T_s x \in R^N$, are given.

The action of T_s can be written as the system of relations

$$x^{i*} = f^i{}_s(x) = f^i{}_s(x^1, x^2, \dots, x^N), \qquad i = 1, 2, \dots, N, \quad s = 1, 2, \dots .$$

We assume that the functions $f^i{}_s$ determining this transformation are locally invertible and three times continuously differentiable. The inverse transformation is denoted by T_s^{-1}.

The *product* $T_1 T_2$ *of transformations* T_1 *and* T_2 is understood as the successive application first of T_2 and then of T_1. This composition of transformations is referred to as *multiplication*. The role of unity element for this multiplication is played by the identity transformation E. In terms of the functions f^i, the multiplication of transformations can be written as

$$f^i(x) = f^i{}_1(f^1{}_2(x), f^2{}_2(x), \dots, f^N{}_2(x)), \qquad i = 1, 2, \dots, N.$$

It follows from the above definition of multiplication as consecutive transformations that multiplication is associative,

$$T_1(T_2 T_3) = (T_1 T_2) T_3.$$

The definition also implies the inversion formula

$$(T_1 T_2)^{-1} = T_2^{-1} T_1^{-1} \tag{0.1}$$

for the product of transformations.

Now consider a family $\{T_a\}$ of transformations depending on a real parameter a ranging in an interval Δ. The family $\{T_a\}$ is said to be *locally closed with respect to multiplication* if there exists a subinterval $\delta \in \Delta$ such that the product $T_a T_b$ belongs to $\{T_a\}$ for any $a, b \in \delta$. In coordinate form,

$$
\begin{aligned}
T_a: & \qquad x^{i*} = f^i(x, a), & i &= 1, 2, \ldots, N, \\
T_a T_b: & \quad f^i(f(x, b), a) = f^i(x, \phi(a, b)), & i &= 1, 2, \ldots, N.
\end{aligned}
\tag{0.2}
$$

Thus, there is a function $\phi(a, b)$ determining the *multiplication law* for the transformations in the family $\{T_a\}$ by the formula $T_b T_a = T_c$, $c = \phi(a, b)$. We assume that this function is three times continuously differentiable.

A family $\{T_a\}$ of transformations is called a *local one-parameter continuous transformation group* (a local Lie transformation group) if

1. $\{T_a\}$ is locally closed with respect to multiplication.

2. There exists a unique parameter value $a_0 \in \delta$ determining the identity transformation T_{a_0}.

3. The equation $\phi(a, b) = a_0$ has a unique solution $b = a^{-1}$ for each $a \in \delta$. This means that every transformation T_a, $a \in \delta$, is invertible, $(T_a)^{-1} = T_{a^{-1}}$.

Note that the multiplication and inversion of transformations is defined only for $a \in \delta$ rather than on the entire admissible interval Δ. For δ we can take any smaller interval containing a_0; i.e., we are only interested in some small neighborhood of a_0. Accordingly, the object introduced above is not a group in general; it is called a *local group*. We denote a local one-parameter continuous transformation group by G_1.

The group parameter can be transformed with the use of a three times continuously differentiable function, $\bar{a} = \bar{a}(a)$. In particular, using the transformation $\bar{a} = a - a_0$, we can ensure that the identity transformation is associated with the zero parameter value. As the group parameter varies, each point $(x^1, x^2, \ldots, x^n)$ moves in R^n along a smooth one-parameter curve, which is called an *orbit of the group G_1*.

The definition of G_1 implies the following properties of the function $\phi(a, b)$ determining the multiplication law for the transformations:

$$\phi(0, 0) = 0, \qquad \phi(a, 0) = a, \qquad \phi(0, b) = b, \qquad \phi(a, a^{-1}) = \phi(a^{-1}, a) = 0.$$

The parameter a is said to be *canonical* if the multiplication law is just the addition $\phi(a, b) = a + b$. In this case, $a^{-1} = -a$, and formulas (0.1) can be rewritten as

$$f^i(f(x, a), b) = f^i(x, a + b), \qquad i = 1, 2, \ldots, N. \tag{0.3}$$

One can show that there exists a canonical parameter for any one-parameter group. The transformation yielding the canonical parameter is given by the formula

$$\bar{a}(a) = \int_0^a V(s)\,ds, \qquad \text{where} \quad V(b) = \left.\frac{\partial\phi(a, b)}{\partial b}\right|_{a=b^{-1}}.$$

It follows that every one-parameter transformation group is *Abelian* (commutative).

Some examples of one-parameter transformation groups include
1. The translations along a vector $(\gamma^1, \gamma^2, \ldots, \gamma^N)$ in R^N,

$$x^{i*} = x^i + \gamma^i a, \qquad i = 1, 2, \ldots, N.$$

(As a special case, this includes the translations $x^* = x + a$ on the real line.)

2. The dilations $x^* = e^a x$ on the real line and the inhomogeneous dilations $x^{i*} = e^{s^i a} x^i$ in R^N.

3. The Galilei translations on the plane,

$$x^* = x + ay, \quad y^* = y.$$

4. The rotations on the plane,

$$x^* = x \cos a + y \sin a, \qquad y^* = -x \sin a + y \cos a.$$

5. The Lorentz transformations on the plane,

$$x^* = x \cosh a + y \sinh a, \qquad y^* = x \sinh a + y \cosh a.$$

6. The projective transformations on the plane,

$$x^* = \frac{x}{1 - ax}, \qquad y^* = \frac{y}{1 - ax}.$$

Note that $\phi(a, b) = a + b$ in all these examples.

0.1.2. Infinitesimal operator of a group

With the group G_1 determined by the transformations (0.1), we associate the auxiliary functions

$$\xi^i(x) = \left.\frac{\partial f^i(x, a)}{\partial a}\right|_{a=0}, \qquad i = 1, 2, \ldots, N. \tag{0.4}$$

The following theorem establishes a one-to-one correspondence between vector fields (0.4) and one-parameter transformation groups G_1.

THEOREM. *The functions $f^i(x, a)$ determining the transformation group satisfy the system of differential equations (which are called the Lie equations)*

$$\frac{\partial f^i}{\partial a} = \xi^i(f), \qquad i = 1, 2, \ldots, N, \tag{0.5}$$

with the initial conditions

$$f^i\big|_{a=0} = x^i, \qquad i = 1, 2, \ldots, N. \tag{0.6}$$

Conversely, for any set of sufficiently smooth functions $\xi^i(x)$, system (0.5)–(0.6) has a solution $f^i(x, a)$, which determines the group G_1.

This theorem establishes the most important relationship between the transformation group and the tangent vector field (0.4).

Along with the tangent field (0.4), consider the linear differential operator

$$X = \xi^i(x)\frac{\partial}{\partial x_i}, \tag{0.7}$$

which is called the *infinitesimal operator* (or the *generator*) of the group G_1. (In formula (0.7) and in what follows, summation over repeated indices is assumed.) The functions $\xi^i(x)$ are called the *coordinates* of the operator X.

Let us write out the infinitesimal operators for the above examples.

1. The operator of the group of translations along the vector $(\gamma^1, \gamma^2, \ldots, \gamma^N)$ in R^N has the form

$$X = \gamma^i\frac{\partial}{\partial x_i}.$$

2. The inhomogeneous dilation transformation in R^N has the infinitesimal operator

$$X = s^i x^i\frac{\partial}{\partial x^i}.$$

3. The Galilei translations on the plane have the generator

$$X = y\frac{\partial}{\partial x}.$$

4. The rotations on the plane are generated by the operator

$$X = y\frac{\partial}{\partial x} - x\frac{\partial}{\partial y}.$$

5. The infinitesimal Lorentz transformation has the form

$$X = y\frac{\partial}{\partial x} + x\frac{\partial}{\partial y}.$$

6. The infinitesimal operator corresponding to projective transformations on the plane is given by

$$X = x^2\frac{\partial}{\partial x} + xy\frac{\partial}{\partial y}.$$

The action of a transformation T_a of the group G_1 on a scalar function $F(x)$ is defined as $T_a F(x) = F(T_a x)$. The infinitesimal operator gives the principal linear part of the increment of the function; indeed,

$$T_a F(x) = e^{aX} F(x) = F(x) + aXF(x) + \frac{a^2}{2!}X^2 F(x) + \cdots + \frac{a^n}{n!}X^n F(x) + \cdots.$$

If we make a change of variables $y^i = y^i(x)$ in R^N, then the coordinates of the infinitesimal operator are changed by the formulas

$$\bar{\xi}^i = X(y^i): \quad \overline{X} = \bar{\xi}^i(y)\frac{\partial}{\partial y^i}. \tag{0.8}$$

If one group can be obtained from another group by a smooth invertible point change of variables, then these groups are said to be *similar.*

THEOREM. *Each one-parameter transformation group is similar to the group of translations along one of the coordinates.*

Remark. The desired change of variables can be found from the linear system

$$X(y^i(x)) = 0, \quad i = 1, 2, \ldots, N-1, \qquad X(y^N(x)) = 1.$$

0.1.3. Group invariants and invariant manifolds

A locally analytic function $F(x) \neq 0$ is said to be *group invariant* if $F(x^*) = F(x)$ for any transformations of the group.

THEOREM. *For $F(x)$ to be group invariant, it is necessary and sufficient that*

$$XF(z) = 0, \tag{0.9}$$

where X is the operator of the group (0.3).

It is well known that the linear partial differential equation (0.9) has $N-1$ functionally independent solutions $I^1(x), I^2(x), \ldots, I^{N-1}(x)$ and that the general solution has the form

$$F(x) = \Phi(I^1(x), I^2(x), \ldots, I^{N-1}(x)),$$

where $\Phi(z^1, \ldots, z^{N-1})$ is an arbitrary differentiable function. Thus, the group G_1 has $N-1$ functionally independent invariants.

Consider some examples of solutions of Eq. (0.9) used to calculate group invariants.

1. The group of translations along a vector $(\gamma^1, \gamma^2, \ldots, \gamma^N)$ in R^N has $N - 1$ independent invariants. For example (assuming that all γ_i are nonzero), one can take

$$I_i = \gamma^{i+1} x^i - \gamma^i x^{i+1}, \qquad i = 1, 2, \ldots, N - 1.$$

2. The inhomogeneous dilation transformations in R^N have the invariants

$$I_i = \frac{(x^i)^{s^{i+1}}}{(x^{i+1})^{s^i}}, \qquad i = 1, 2, \ldots, N - 1.$$

3. The Galilei translations on the plane have one invariant

$$I_1 = y.$$

4. The rotations on the plane have the obvious invariant

$$I_1 = x^2 + y^2.$$

5. The Lorentz transformations on the plane have one invariant

$$I_1 = x^2 - y^2.$$

6. The projective transformations on the plane have the invariant

$$I_1 = \frac{x}{y}.$$

Remark. One can also find group invariants in a different way, by considering finite transformations of the group and by eliminating the group parameter. We illustrate this by the example of projective transformations on the plane,

$$x^* = \frac{x}{1 - ax}, \qquad y^* = \frac{y}{1 - ax}.$$

By eliminating the group parameter a, we obtain

$$\frac{x^*}{y^*} = \frac{x}{y}.$$

This relation means exactly that the expression x/y is an invariant of the one-parameter group considered.

This idea was generalized in [54, 109, 110] to multiparameter groups acting in a space with a larger number of variables.

The manifold defined in R^N by some functions $\phi^s(x)$ according to the formulas

$$\phi^s(x) = 0, \qquad s = 1, 2, \ldots, A, \tag{0.10}$$

is said to be invariant if the following relation holds for all group transformations:

$$\phi^s(x^*) = 0, \qquad s = 1, 2, \ldots, A.$$

In other words, the group transformations take the manifold (0.8) to itself.

DEFINITION. The manifold (0.10) is said to be *regularly defined* if the functions $\phi^s(x)$ are continuously differentiable and the matrix $\|\partial\phi^s/\partial x^i\|$ has rank A (equal to the number of equations in system (0.10)).

A criterion for a regular manifold to be invariant can be written in terms of the group operator.

THEOREM. *For a manifold regularly defined by Eqs. (0.8) to be invariant under a group G_1 with operator X, it is necessary and sufficient that*

$$X\phi^s(x)\big|_{\phi^s(x)=0} = 0. \tag{0.11}$$

The geometric meaning of condition (0.11) is that the vector field $\xi^i(x)$ is tangent to the surface (0.10).

0.1.4. Prolongation of the transformation group to derivatives

Now we divide the coordinates in R^N into two types, independent variables x^i, $i = 1, 2, \ldots, n$, and dependent (differential) variables u^k, $k = 1, 2, \ldots, m$, $N = m+n$. Accordingly, we divide the transformations in the group G_1 into two types,

$$x^{*i} = f^i(x, u, a), \qquad i = 1, 2, \ldots, n,$$
$$u^{*k} = g^k(x, u, a), \qquad k = 1, 2, \ldots, m.$$

We prolong the space R^N to the derivatives, i.e., supplement it with differential variables $u^k{}_i$,

$$u^k{}_i = \frac{\partial u^k}{\partial x^i}, \qquad i = 1, 2, \ldots, n, \quad k = 1, 2, \ldots, m.$$

The space prolonged to the first derivatives has dimension $\overline{N} = N + mn$. In the prolonged space, consider the one-parameter transformation group defined by the operator

$$X = \xi^i \frac{\partial}{\partial x_i} + \eta^k \frac{\partial}{\partial u^k} + \zeta^k{}_i \frac{\partial}{\partial u^k{}_i}, \tag{0.12}$$

where the $\zeta^k{}_i$ are some functions of x_i, u^k, and $u^k{}_i$.

To ensure that the group transformations preserve the definition and the geometric meaning of the derivatives, we require that the following relations be invariant under the transformations in G_1:

$$du^k = u^k{}_i \, dx^i, \qquad k = 1, 2, \ldots, m. \tag{0.13}$$

The invariance of Eqs. (0.13) under the group G_1 with operator (0.12) implies the following *prolongation formulas*:

$$\zeta^k{}_i = D_i(\eta^k) - u^k{}_j D_i(\xi^j), \qquad i = 1, 2, \ldots, n, \quad k = 1, 2, \ldots, m, \tag{0.14}$$

where

$$D_i = \frac{\partial}{\partial x^i} + u_i{}^k \frac{\partial}{\partial u^k}, \qquad i = 1, 2, \ldots, n,$$

is the operator of total differentiation with respect to the variable x^i.

Prolonging this process to the second derivatives, in a similar way we obtain expressions for the coordinates of the group operator which determine the transformation of the second derivatives:

$$\zeta^k{}_{ji} = D_i(\zeta^k_j) - u^k{}_{sj} D_i(\xi^s), \qquad i, j = 1, 2, \ldots, n, \quad k = 1, 2, \ldots, m. \tag{0.15}$$

In the same manner, one can obtain formulas of prolongation to third and higher derivatives. The operator of the group G_1 prolonged to the desired number of derivatives (which will be clear from the context) will be indicated by a tilde,

$$\widetilde{X} = \xi^i \frac{\partial}{\partial x_i} + \eta^k \frac{\partial}{\partial u^k} + \zeta^k{}_i \frac{\partial}{\partial u^k{}_i} + \zeta^k{}_{ij} \frac{\partial}{\partial u^k{}_{ij}} + \cdots.$$

The operation of prolongation is linear and homogeneous in the coordinates of the original operator, which can readily be seen from formulas (0.14) and (0.15).

In the prolonged space, there are more group invariants. The invariants of the prolonged group that are not invariants of the original group (i.e., of the group acting in the space of independent and dependent variables alone) are called *differential invariants* of the group G_1. In a similar way, a *differential invariant manifold* of the group is defined to be an invariant manifold that is not an invariant manifold of the original group (i.e., contains derivatives).

0.1.5. Transformation groups admitted by differential equations

Consider the system of differential equations

$$F_\alpha(x, u, u_1, u_2, \ldots, u_s) = 0, \qquad \alpha = 1, 2, \ldots, m, \tag{0.16}$$

where $x \in R^n$, $u \in R^m$, and u_s is the set of sth partial derivatives.

We treat Eqs. (0.16) as the equations of a manifold in the corresponding prolonged space.

DEFINITION. One says that system (0.16) *admits a group* G_1 if the corresponding manifold is a differential invariant manifold of G_1, i.e., if Eqs. (0.16) remain unchanged under the action of any group transformation appropriately prolonged to the derivatives.

Another equivalent definition can be stated as follows: system (0.16) admits the group G_1 if the group action takes every solution of the system to a solution of the same system (see [73, 107, 111]).

In connection with this definition, the following main problem of group analysis arises: for a given system of equations, find all transformation groups admitted by this system. We point out that, for this problem to be solved, it is insignificant whether the system has solutions. The only significant characteristic is the possibility to rewrite (0.16) in the form of a regularly defined manifold.

Since finding a symmetry group is equivalent to finding its infinitesimal operator, we continue this operator to the derivatives up to and including u_s,

$$\widetilde{X} = \xi^i \frac{\partial}{\partial x_i} + \eta^k \frac{\partial}{\partial u^k} + \zeta^k_{\,i} \frac{\partial}{\partial u^k_{\,i}} + \cdots + \zeta^k_{\,s} \frac{\partial}{\partial u^k_{\,s}}, \tag{0.17}$$

and rewrite the criterion for system (0.16) of differential equations to be invariant in the form

$$\widetilde{X} F_\alpha(x, u, u_1, u_2, \ldots, u_s)\big|_{(0.16)} = 0, \qquad \alpha = 1, 2, \ldots, m. \tag{0.18}$$

The invariance criterion (0.18) is an overdetermined system of *linear equations* for the coordinates of the operator (0.17). Therefore, the solutions of system (0.18) form a linear vector space L_r of some dimension r. Thus, the problem of finding a symmetry group (operator) is always linear, regardless of whether the system itself is linear or nonlinear. The efficiency of group analysis is a consequence of this fact.

System (0.18) is called the *system of determining equations*. In general, there is no relationship between the dimension r of the space of symmetry operators and the dimension of system (0.16).

We illustrate the process of solving the determining equations by an example of a nonlinear ordinary differential equation.

EXAMPLE. Consider the ordinary differential equation

$$u'' = \frac{1}{u^3}. \tag{0.19}$$

We seek the operator of the symmetry group in the form

$$X = \xi(x, u) \frac{\partial}{\partial x} + \eta(x, u) \frac{\partial}{\partial u},$$

which we have to prolong to the first and second derivatives:

$$\widetilde{X} = \xi(x, u) \frac{\partial}{\partial x} + \eta(x, u) \frac{\partial}{\partial u} + \zeta_1 \frac{\partial}{\partial u'} + \zeta_2 \frac{\partial}{\partial u''}, \tag{0.20}$$

where

$$\zeta_1 = D(\eta) - u'D(\xi), \qquad D = \frac{\partial}{\partial x} + u'\frac{\partial}{\partial u} + u''\frac{\partial}{\partial u'} + u'''\frac{\partial}{\partial u''} + \cdots,$$

$$\zeta_2 = D(\zeta_1) - u''D(\xi) = D^2(\eta) - 2u''D(\xi) - u'D^2(\xi).$$

We act by the prolonged operator (0.20) on Eq. (0.19):

$$D^2(\eta) - 2u''D(\xi) - u'D^2(\xi) = -\frac{3\eta}{u^4},$$

or, in expanded form,

$$\eta_{xx} + 2\eta_{xu}u' + \eta_{uu}(u')^2 + \eta_u u'' - 2u''(\xi_x + \xi_u u')$$
$$- u'(\xi_{xx} + 2\xi_{xu}u' + \xi_{uu}(u')^2 + \xi_u u'') + \frac{3}{u^4}\eta = 0. \quad (0.21)$$

Now it is necessary to "introduce the manifold," i.e., write out the action of the operator at the points of Eq. (0.19). To this end, we can, for example, express u'' from (0.19) and substitute it into Eq. (0.21):

$$\eta_{xx} + 2\eta_{xu}u' + \eta_{uu}u'^2 + \eta_u\frac{1}{u^3} - 2\frac{1}{u^3}(\xi_x + \xi_u u')$$
$$- u'(\xi_{xx} + 2\xi_{xu}u' + \xi_{uu}(u')^2 + \xi_u\frac{1}{u^3}) + \frac{3}{u^4}\eta = 0. \quad (0.22)$$

Thus, the determining equation has been obtained. Any of its nonzero solutions gives the coordinates of an operator generating a one-parameter group.

The desired coordinates of the operator depend only on x and u but are independent of u', and the determining equation (0.22) should be satisfied identically in the variables x, u, and u'. This permits splitting Eq. (0.22) into several simpler equations. By matching the coefficients of like powers of u', we readily obtain the following overdetermined system of equations:

$$\xi_{uu} = 0, \qquad \eta_{uu} - 2\xi_{xu} = 0,$$

$$2\eta_{xu} - 3\frac{\xi_u}{u^3} - \xi_{xx} = 0, \qquad\qquad (0.23)$$

$$\eta_{xx} + (\eta_u - 2\xi_x)\frac{1}{u^3} + \frac{3\eta}{u^4} = 0.$$

From the first two equations in (0.23), we obtain

$$\xi = \alpha(x)u + \beta(x), \qquad \eta = \alpha_x u^2 + \gamma(x)u + \delta(x).$$

By substituting these expressions into the remaining equations in system (0.23) and by matching the coefficients of like powers of u, we obtain the general solution in the form

$$\xi(x) = Ax^2 + 2Bx + C, \qquad \eta(x, u) = (Ax + B)u,$$

where A, B, and C are arbitrary constants.

Thus, we have obtained a three-dimensional space of operators. By setting any two of the three constants to zero, we obtain a basis of symmetry operators:

$$X_1 = \frac{\partial}{\partial x}, \qquad X_2 = 2x\frac{\partial}{\partial x} + u\frac{\partial}{\partial u}, \qquad X_3 = x^2\frac{\partial}{\partial x} + xu\frac{\partial}{\partial u}. \tag{0.24}$$

Thus, Eq. (0.19) admits three one-parameter transformation groups with operators (0.24). The finite transformations on the (x, u)-plane for each of these one-parameter groups can readily be obtained by solving the corresponding Lie equation (0.5). Thus, we obtain three families of transformations,

$$
\begin{aligned}
x^* &= x + a, & u^* &= u; \\
x^* &= e^{2a}x, & u^* &= e^{a}u; \\
x^* &= \frac{x}{1 - ax}, & u^* &= \frac{u}{1 - ax}.
\end{aligned}
\tag{0.25}
$$

Let us prolong the transformations (0.25) to the first and second derivatives:

$$
\begin{aligned}
x^* = x + a, \quad u^* = u, \quad (u')^* = u', \quad (u'')^* = u''; \\
x^* = e^{2a}x, \quad u^* = e^{a}u, \quad (u')^* = e^{-a}u', \quad (u'')^* = e^{-3a}u''; \\
x^* = \frac{x}{1 - ax}, \quad u^* = \frac{u}{1 - ax}, \\
(u')^* = au + (1 - ax)u', \quad (u'')^* = (1 - ax)^3 u''.
\end{aligned}
\tag{0.26}
$$

By substituting the transformations (0.26) into Eq. (0.19), we readily see that it is invariant.

We point out the following obvious fact, which we need in the subsequent analysis of invariance properties of finite-difference equations. Under the transformations (0.26), Eq. (0.19) becomes the same equation

$$(u'')^* = \frac{1}{(u^*)^3}$$

but at a different point of the same prolonged space (x, u, u', u''). Thus, the group action does not change the invariant equation but transforms the point at which it is written, $(x, u, u', u'') \to (x^*, u^*, (u')^*, (u'')^*)$.

0.1.6. Lie algebra of infinitesimal operators

Thus, the symmetry of given differential equations is described by a vector space of infinitesimal operators, which was confirmed by an example. Along with the operation of addition of operators and their multiplication by numbers, one more operation can be introduced in this space.

DEFINITION. The *commutator* of operators $X_1 = \xi_1^i \partial/\partial x_i$ and $X_2 = \xi_2^i \partial/\partial x_i$ is the operator

$$[X_1, X_2] = X_1 X_2 - X_2 X_1 = (X_1 \xi_2^i - X_2 \xi_1^i) \frac{\partial}{\partial x_i}.$$

This definition readily implies the following properties of the commutation operation:

1. The commutator is bilinear,

$$[aX_1 + bX_2, X_3] = a[X_1, X_3] + b[X_2, X_3], \qquad a, b = \text{const.}$$

2. The commutator is antisymmetric,

$$[X_1, X_2] = -[X_2, X_1].$$

3. The Jacobi identity

$$[[X_1, X_2], X_3] + [[X_2, X_3], X_1] + [[X_3, X_1], X_2] = 0$$

is satisfied.

A linear space L of operators containing all commutators of these operators is called a *Lie algebra* of operators.

In the above example, we obtained three linearly independent operators X_1, X_2, and X_3. Now let us calculate the following commutators of these operators:

$$[X_1, X_2] = 2X_1, \qquad [X_1, X_3] = X_2, \qquad [X_2, X_3] = 2X_3.$$

The other commutators can be obtained from the property that the commutator is antisymmetric; in particular, $[X_i, X_i] = 0$, $i = 1, 2, 3$.

Thus, we have shown that the commutator of any two operators admitted by our equation can be expressed via the basis operators and is also admitted by the equation. This assertion is also true in general [111].

THEOREM. *If a manifold is invariant under operators X_1 and X_2, then it is also invariant under their commutator $[X_1, X_2]$.*

This means that, for any system of differential equations, the set of infinitesimal operators admitted by it is a Lie algebra.

Note two more properties of the commutator, which are useful in further analysis:

1. The commutator is invariant under changes of the coordinate system.

2. The operation of prolongation to the derivatives commutes with the operation of commutation [111].

Table 0.1: Commutators of the group G_3

	X_1	X_2	X_3
X_1	0	$2X_1$	X_2
X_2	$-2X_1$	0	$2X_3$
X_3	$-X_2$	$-2X_3$	0

It is convenient to arrange the commutators of the operators under study in a table where the commutator $[X_i, X_j]$ is placed at the intersection of the ith row with the jth column. Table 0.1 shows the commutators of the operators X_1, X_2, X_3.

Lie algebras have been studied sufficiently well in the general theory. Here we need only the simplest of their properties. In particular, if a subspace of operators itself forms a Lie algebra, then it is called a subalgebra. In our example, the subspaces spanned by the basis operators X_1, X_2 and X_2, X_3 are subalgebras, while the operators X_1 and X_3 do not span a subalgebra, because their commutator cannot be expressed as a linear combination of X_1 and X_3.

0.1.7. Local Lie transformation groups

In the general theory of Lie transformation groups, one-parameter transformations groups are generalized to the multiparameter groups. In the Euclidean space R^N of points $x = (x^1, x^2, x^3, \ldots, x^N)$, one introduces transformations taking R^N to itself:

$$x^{i*} = f^i(x, a) = f^i(x^1, x^2, \ldots, x^N; a^1, a^2, \ldots, a^r), \qquad i = 1, 2, \ldots, N. \quad (0.27)$$

We assume that the transformations (0.27) satisfy the same axioms as for one-parameter transformation groups. The main novelty is that the parameter is a vector $(a^1, a^2, \ldots, a^r) \in R^r$ ranging in a small neighborhood of the point $(0, 0, \ldots, 0)$ corresponding to the identity transformation.

By $G_r{}^N$ we denote a set of transformations (0.27) satisfying the axioms of a local group with the usual law $\phi(a, b)$ of multiplication of transformations:

$$T_b T_a = T_{\phi(a,b)}: \qquad f^i(f(x, a), b) = f^i(x, \phi(a, b)), \quad i = 1, 2, \ldots, N.$$

The definition of the group $G_r{}^N$ implies the following properties of the function $\phi(a, b) = (\phi^1, \phi^2, \ldots, \phi^r)$ determining the multiplication law:

$$\phi(0, 0) = 0, \qquad \phi(a, 0) = a, \qquad \phi(0, b) = b.$$

With the group $G_r{}^N$ of transformations (0.27), we associate the auxiliary functions

$$\xi_\alpha{}^i(x) = \left. \frac{\partial f^i(x, a)}{\partial a^\alpha} \right|_{a=0}, \qquad i = 1, 2, \ldots, N, \qquad \alpha = 1, 2, \ldots, r.$$

These functions are used to define the linear operators

$$X_\alpha = \xi_\alpha{}^i(x)\frac{\partial}{\partial x_i}, \qquad i = 1, 2, \ldots, N, \quad \alpha = 1, 2, \ldots, r, \qquad (0.28)$$

which are called the *basis operators of the group* G_r^N.

To state the multidimensional analog of Lie equations, we also need the auxiliary functions

$$V_\beta{}^\alpha(b) = \left.\frac{\partial \phi^\alpha(a, b)}{\partial b^\beta}\right|_{a=b^{-1}}, \qquad V_\beta{}^\alpha(0) = \delta_\beta{}^\alpha, \qquad \alpha, \beta = 1, 2, \ldots, r.$$

Using these functions, we can write out the Lie equations

$$\frac{\partial f^i}{\partial a^\alpha} = \xi_\sigma{}^i(f)V_\beta{}^\sigma(a) \qquad (0.29)$$

with the initial conditions

$$f^i|_{a=0} = x^i, \qquad i = 1, 2, \ldots, N, \qquad (0.30)$$

and state the following theorem.

THEOREM. *The functions $f^i(x, a)$ determining a transformation group satisfy system (0.29) of differential Lie equations with the initial conditions (0.30). Conversely, if there are given auxiliary functions $V_\beta{}^\sigma(a)$ and linearly independent vectors $\xi_\alpha{}^i(x)$, then the solution of system (0.29), (0.30) determines a local Lie transformation group.*

In the group G_r^N, a one-parameter subgroup can be chosen as follows. In the space of the parameters $(a^1, a^2, \ldots, a^r)$, take a directing vector $e = (e^1, e^2, \ldots, e^r)$ and consider the straight line $a^\alpha = e^\alpha t$, where t is a parameter. Then the group transformations become the one-parameter transformations

$$x^{i*} = f^i(x, et), \qquad i = 1, 2, \ldots, N,$$

and the Lie equations acquire the form

$$\frac{\partial f^i}{\partial t} = e^\alpha \xi_\alpha{}^i(f), \quad f^i|_{t=0} = x^i, \qquad i = 1, 2, \ldots, N.$$

We denote the Lie algebra generated by the basis operators (0.28) of the group G_r^N by the symbol L_r. Note that the general theories of Lie groups and Lie algebras are completely parallel: there is a full correspondence between the structures of Lie algebras and Lie groups; in particular, to any subalgebra there corresponds a subgroup of the Lie transformation group up to the choice of a coordinate system. In fact, this reduces the problem of finding the transformation group admitted by a given system of equations to the problem of finding one-parameter subgroups admitted by the system.

An r-parameter Lie transformation group can be constructed from the given basis operators (0.28) by various methods. For example, one can construct one-parameter transformation groups corresponding to the basis operators and then use the multiplication of the corresponding transformations. Thus the so-called *canonical coordinates of the second kind* [111] are introduced:

$$T_a = T_{a^1} T_{a^2} \cdots T_{a^r}. \tag{0.31}$$

Note that the introduction of an r-parameter group in such a way is not unique; namely, a permutation of transformations in (0.31) generally leads to a different representation of the group $G_r{}^N$.

EXAMPLE. We return to our example of the three-dimensional Lie algebra admitted by Eq. (0.19). Each of the basis operators generates its own one-parameter subgroup:

$$
\begin{aligned}
X_1: && x^* &= x + a, & u^* &= u; \\
X_2: && x^* &= e^{2a}x, & u^* &= e^a u; \\
X_3: && x^* &= \frac{x}{1 - ax}, & u^* &= \frac{u}{1 - ax}.
\end{aligned}
$$

Two-parameter subgroups $G_2{}^2$ can be constructed from the subalgebras spanned by X_1, X_2 or X_2, X_3, respectively,

$$
\begin{aligned}
x^* &= e^{2a}x + b, & u^* &= e^a u; \\
x^* &= \frac{x e^{2b}}{1 - ax}, & u^* &= \frac{u e^b}{1 - ax}.
\end{aligned}
$$

The full three-parameter group $G_3{}^2$ can be represented as the superposition of all three one-parameter transformations,

$$x^* = \frac{x e^{2b}}{1 - ax} + c, \quad u^* = \frac{u e^b}{1 - ax},$$

where a, b, c are the parameters of the group $G_3{}^2$.

0.1.8. Group invariants

An invariant of the group $G_r{}^N$ is a function $I(x)$ that is not identically constant and satisfies $I(T_a x) = I(x)$ for any transformation $T_a \in G_r{}^N$.

Since a one-parameter subgroup can be drawn through any element of the group (sufficiently close to the identity element), it follows that a function $I(x)$ is an invariant of the group if and only if it is an invariant of any subgroup $G_1{}^N$. Thus, a necessary and sufficient condition for $I(x)$ to be invariant can be written as

$$X_\alpha I(x) = 0, \qquad \alpha = 1, 2, \ldots, r, \tag{0.32}$$

where the X_α are the basis operators of the group.

System (0.32) is a system of linear first-order partial differential equations. If $I^1(x), I^2(x), \ldots, I^s(x)$ are some solutions, then any function of them is also a solution of this system. Therefore, it is meaningful to speak only about functionally independent solutions, i.e., about solutions for which the relation

$$F(I^1(x), I^2(x), \ldots, I^s(x)) = 0$$

implies that $F(y^1, y^2, \ldots, y^s)$ is zero as a function of the independent variables $y^1, y^2, \ldots, y^s$. If there exists a function $F(y^1, y^2, \ldots, y^s)$ that is not identically zero but $F(I^1(x), I^2(x), \ldots, I^s(x)) = 0$, then such solutions are said to be functionally dependent. The solutions of systems of the form (0.32) have been well studied in the classical literature. In particular, the following assertion holds.

PROPOSITION. *For functions $I^1(x), I^2(x), \ldots, I^s(x)$ to be functionally independent, it is necessary and sufficient that the Jacobi matrix $J = \|\partial I^i/\partial x^J\|$ have general rank equal to s (the number of functions), $R(J) = s$. If $R(J) < s$, then there exist $s - R(J)$ independent functions $F_\alpha(y^1, y^2, \ldots, y^s)$ satisfying the condition $F_\alpha(I^1(x), I^2(x), \ldots, I^s(x)) = 0$, $\alpha = 1, 2, \ldots, s - R$.*

DEFINITION. Operators X_α, $\alpha = 1, 2, \ldots, r$, are said to be *linearly connected* if there exist functions $\Phi^\alpha(x)$ of which not all are identically zero such that $\Phi^\alpha X_\alpha = 0$. If such functions do not exist, then the operators X_α are said to be *linearly unconnected*.[2]

DEFINITION. Operators X_α, $\alpha = 1, 2, \ldots, r$, form a complete system if they are linearly unconnected and their commutators satisfy the representation

$$[X_\alpha, X_\beta] = \phi_{\alpha\beta}{}^\sigma X_\sigma$$

with some functions $\phi_{\alpha\beta}^\sigma(x)$.

The above-introduced definitions permit stating the following lemma.

LEMMA. *If the system of equations*

$$X_\alpha I(x) = 0, \qquad \alpha = 1, 2, \ldots, r,$$

is generated by a complete system of operators, then, for $s \leq N$, there exist $N - s$ functionally independent solutions such that any other solution is a function of them. (For $N = s$, the system does not have functionally independent solutions.)

---

[2]Note that linearly connected operators may or may not be linearly dependent over $\mathbb{C}$.

The maximum number of linearly unconnected basis operators of the group $G_r{}^N$ is determined by the general rank $R(M)$ of the function matrix

$$M = \|\xi_\alpha{}^i\|, \qquad \alpha = 1, 2, \ldots, r, \quad i = 1, 2, \ldots, N.$$

The number $R(M)$ is used to solve the problem on the number of functionally independent invariants of $G_r{}^N$.

THEOREM. *The group $G_r{}^N$ has invariants if and only if $R(M) < N$. In this case, there exist $t = N - R$ functionally independent invariants $I^1(x), I^2(x), \ldots, I^t(x)$ of the group such that any invariant of $G_r{}^N$ is a function of them.*

EXAMPLE. We return to our example of the three-dimensional Lie algebra admitted by Eq. (0.19). The three-parameter group $G_3{}^N$ can have invariants only for $N > 3$. Therefore, we need to consider the prolonged space x, u, u', u'' and the corresponding prolonged operators

$$X_1 = \frac{\partial}{\partial x}, \qquad X_2 = 2x\frac{\partial}{\partial x} + u\frac{\partial}{\partial u} - u'\frac{\partial}{\partial u'} - 3u''\frac{\partial}{\partial u''},$$

$$X_3 = x^2\frac{\partial}{\partial x} + xu\frac{\partial}{\partial u} + (u - xu')\frac{\partial}{\partial u'} - 3xu''\frac{\partial}{\partial u''}. \tag{0.33}$$

It is easily seen that the operators (0.33) are linearly unconnected, $R(M) = 3$, and hence the group $G_3{}^4$ has $4 - 3 = 1$ invariant. Since the invariant depends on the derivatives, it follows that this is a differential invariant. To find the invariant, one has to solve the system of linear partial differential equations

$$X_j\big(I(x, u, u', u'')\big) = 0, \qquad j = 1, 2, 3.$$

We solve it successively. It follows from the equations $X_1(I(x, u, u', u'')) = 0$ that the invariant is independent of x. Then the second equation $X_2(I(u, u', u'')) = 0$ has two solutions, $J_1 = uu'$ and $J_2 = u^3u''$. We prolong the action of X_3 to the new variables J_1, J_2,

$$\widetilde{X}_3 = x^2\frac{\partial}{\partial x} + xu\frac{\partial}{\partial u} + (u - xu')\frac{\partial}{\partial u'} - 3xu''\frac{\partial}{\partial u''} + u^2\frac{\partial}{\partial J_1} + 0\frac{\partial}{\partial J_2}.$$

It follows that the only common differential invariant is $I = J_2 = u^3u''$.

Note that each group has differential invariants. Indeed, in the case of successive prolongations of the group to derivatives provided that the rank R is bounded, the group begins to acquire differential invariants after a certain increase in the dimension of the space. In our example, this occurs in the prolongation to the second derivative.

0.1.9. Invariant manifolds of a group

DEFINITION. A manifold K is said to be invariant under the group $G_r{}^N$ if $T_a x \in K$ for any $x \in K$ and $T_a \in G_r{}^N$.

Just as in the case of invariants, the problem on the invariance of a manifold regularly defined by the equations

$$\phi^k(x) = 0, \qquad k = 1, 2, \ldots, s, \tag{0.34}$$

can be solved by using one-parameter subgroups.

THEOREM. *For the manifold regularly defined by Eqs.* (0.34) *to be an invariant manifold of the group* $G_r{}^N$, *it is necessary and sufficient that*

$$X_\alpha \phi^k(x)\big|_{(0.34)} = 0, \qquad k = 1, 2, \ldots, s, \quad \alpha = 1, 2, \ldots, r.$$

The following definition essentially distinguishes the invariance of a manifold with respect to multiparameter groups and the invariance with respect to the action of $G_1{}^N$.

DEFINITION. A manifold K is called a *nonsingular* manifold of the group $G_r{}^N$ if $R(M)\big|_K = R$. Otherwise, if the rank of the matrix M decreases at the points of the manifold compared with the general rank, then the manifold is said to be *singular*.

If we have a complete set of invariants $I^1(x), I^2(x), \ldots, I^t(x)$ of the group $G_r{}^N$, then, obviously, each system of equations of the form

$$\Phi^k(I^1(x), I^2(x), \ldots, I^t(x)) = 0, \qquad k = 1, 2, \ldots, s, \tag{0.35}$$

is an invariant manifold. It turns out that this method for constructing invariant manifolds is actually most general.

THEOREM (on the representation of nonsingular invariant manifolds of a group). *For a group* $G_r{}^N$ *to have nonsingular invariant manifolds, it is necessary and sufficient that* $R < N$. *In this case, a nonsingular invariant manifold can be defined by* (0.35), *where* $I^1(x), I^2(x), \ldots, I^t(x)$ *is a complete set of functionally independent invariants of the group* $G_r{}^N$.

This theorem shows that the problem of constructing invariant manifolds is rather simple and can be solved constructively. It suffices to find a complete set of group invariants. In the case of differential equations, the problem can be solved in a similar way, only the group transformations are understood as appropriately prolonged transformations and one needs a complete set of functionally independent differential invariants of the prolonged group $G_r{}^N$.

EXAMPLE. We return to our example of the three-dimensional algebra admitted by Eq. (0.19). The only differential invariant in the space (x, u, u', u'') is $I = u^3 u''$. Any sufficiently smooth function of this invariant determines an invariant of the second-order differential equation

$$F(u^3 u'') = 0.$$

A special case of this equation coincides with Eq. (0.19),

$$I = u^3 u'' = 1.$$

The equation thus obtained is an invariant representation of Eq. (0.19).

This example shows that invariant equations can very easily be constructed for a given group; it suffices to calculate the differential invariants of the desired order and use them to construct an equation satisfying the conditions of a given problem. It is this simple idea that we use to construct finite-difference equations and meshes preserving the symmetry of the original differential equations.

0.1.10. Group classification of differential equations

For the three-dimensional Lie algebra (0.24), we have constructed the most general differential equation admitting the corresponding three-parameter transformation group. This raises the natural question as to whether it is possible to list all equations of given order that are invariant under a set of groups. Lie classified all groups on the line and on the plane (x, y) with respect to their dimension and structure. Starting from the classification of algebras (and the corresponding groups), he classified invariant second-order differential equations [93]. In each class of similar subalgebras, by choosing an appropriate point change of variables, he found the simplest representatives of invariant equations containing the minimum possible number of arbitrary constants. The result of this classification is shown in Table 0.2, where one can readily see the equation considered in our example. The dimension of a Lie algebra of symmetries is equal to $1, 2, 3$, or 8. The maximal eight-dimensional algebra is admitted only by a linear equation and by any other equation related to the linear equation by a point change of variables $x, y \to x^*, y^*$. The same pertains to all other equations in the list of invariant second-order equations. Thus, invariant second-order differential equations have been classified up to arbitrary point changes of variables. Such transformations take the corresponding subgroup to a similar subgroup together with the corresponding equation.

In the case of partial differential equations, the situation is more complicated, because no classification of Lie algebras and Lie groups in the space (x, y, z) has been obtained yet.

But in this case, it is also possible to pose the problem of group classification for equations in a certain class. Apparently, this problem was first put forward

Table 0.2: Group classification of second-order ordinary differential equations

Group	Basis operators	Equation
G_1	$X_1 = \frac{\partial}{\partial x}$	$y'' = F(y, y')$
G_2 (a)	$X_1 = \frac{\partial}{\partial x}, \quad X_2 = \frac{\partial}{\partial y}$	$y'' = F(y')$
G_2 (b)	$X_1 = \frac{\partial}{\partial y}, \quad X_2 = x\frac{\partial}{\partial x} + y\frac{\partial}{\partial y}$	$y'' = \frac{1}{x}F(y')$
G_3 (a)	$X_1 = \frac{\partial}{\partial x} + \frac{\partial}{\partial y}, \quad X_2 = x\frac{\partial}{\partial x} + y\frac{\partial}{\partial y},$ $X_3 = x^2\frac{\partial}{\partial x} + y^2\frac{\partial}{\partial y}$	$y'' + 2\frac{y'+Cy'\sqrt{y'}+y'^2}{x-y} = 0$
G_3 (b)	$X_1 = \frac{\partial}{\partial x}, \quad X_2 = 2x\frac{\partial}{\partial x} + y\frac{\partial}{\partial y},$ $X_3 = x^2\frac{\partial}{\partial x} + xy\frac{\partial}{\partial y}$	$y'' = y^{-3}$
G_3 (c)	$X_1 = \frac{\partial}{\partial x}, \quad X_2 = \frac{\partial}{\partial y},$ $X_3 = x\frac{\partial}{\partial x} + (x+y)\frac{\partial}{\partial y}$	$y'' = C\exp(-y')$
G_3 (d)	$X_1 = \frac{\partial}{\partial x}, \quad X_2 = \frac{\partial}{\partial y},$ $X_3 = x\frac{\partial}{\partial x} + ky\frac{\partial}{\partial y}, \quad k \neq 0, \frac{1}{2}, 1, 2$	$y'' = Cy'^{\frac{k-2}{k-1}}$
G_8	$X_1 = \frac{\partial}{\partial x}, \quad X_2 = \frac{\partial}{\partial y}, \quad X_3 = x\frac{\partial}{\partial y},$ $X_4 = x\frac{\partial}{\partial x}, \quad X_5 = y\frac{\partial}{\partial x},$ $X_6 = y\frac{\partial}{\partial y}, \quad X_7 = x^2\frac{\partial}{\partial x} + xy\frac{\partial}{\partial y},$ $X_8 = xy\frac{\partial}{\partial x} + x^2\frac{\partial}{\partial y}$	$y'' = 0$

by Ovsyannikov [112], and its solution was demonstrated for the nonlinear heat equation

$$u_t = (k(u)u_x)_x, \qquad k \neq \text{const.} \tag{0.36}$$

In this equation, the unspecified coefficient $k = k(u)$ is called an *arbitrary element*, and the equations in this class are classified with respect to this element. The group classification problem is posed as follows: find the transformation group admitted by Eq. (0.36) for an arbitrary $k = k(u)$ (the so-called main group), and also find all special cases of $k = k(u)$ in which the symmetry group can be extended and find these larger transformation groups. Technically, the computation of the group by solving the determining system is complicated only by the presence of an arbitrary element. Hence, in the solution process, there arise additional equations for the arbitrary element. Just their solution gives all special cases in which the main group can be extended.

To concentrate the results of the group classification, Ovsyannikov [112] proposed to write out the corresponding groups up to some "external" transformations that transform only the arbitrary element but do not change the type and structure of the equation under study. The group of such transformations was called the equivalence group. For Eq. (0.36), the following group was chosen as the equivalence group:

$$\bar{t} = at + e, \qquad \bar{x} = bx + f, \qquad \bar{u} = cu + g; \quad \bar{k} = \frac{b^2}{a}k;$$

$$a, b, c, e, f, g - \text{const}, \qquad abc \neq 0.$$

The classification of Eq. (0.36) up to transformations of the equivalence group has led to the following result. For arbitrary $k = k(u)$, Eq. (0.36) admits the three-dimensional Lie algebra of operators

$$X_1 = \frac{\partial}{\partial t}, \qquad X_2 = \frac{\partial}{\partial x}, \qquad X_3 = 2t\frac{\partial}{\partial t} + x\frac{\partial}{\partial x}.$$

The algebra of symmetry operators extends to L_4 in the following cases:

$$k = e^u: \qquad X_4 = x\frac{\partial}{\partial x} + 2\frac{\partial}{\partial u},$$

$$k = u^\sigma, \sigma \neq 0, -4/3: \qquad X_4 = \frac{\sigma}{2}x\frac{\partial}{\partial x} + u\frac{\partial}{\partial u}.$$

The space of symmetry operators becomes five dimensional if

$$k = u^{-4/3}: \qquad X_4 = -\frac{2}{3}x\frac{\partial}{\partial x} + u\frac{\partial}{\partial u}, \qquad X_5 = -x^2\frac{\partial}{\partial x} + 3xu\frac{\partial}{\partial u}.$$

The above results completely solve the problem of group properties of an equation of the form (0.36). It should be noted that the publication of [112] caused a wave of studies on group properties of equations of mathematical physics and mechanics. The results of these studies are largely reflected in the reference book [74].

0.1.11. Integration of ordinary differential equations with the use of a symmetry group

Integration of nonlinear ordinary differential equations and systems is a rather difficult problem. Therefore, numerous examples of successful integration were collected in various reference books, which, as a rule, contain hundreds of equations and the corresponding solutions obtained by various specific methods. Part of these specific integration methods are also traditionally presented in manuals on differential equations. It is remarkable that an absolute majority of these integration methods can be considered from a unique viewpoint, namely, from the viewpoint of the transformation group admitted by a given equation.

Integrating factor

It is well known that any first-order ordinary differential equation

$$y' = f(x, y) \tag{0.37}$$

admits an infinite point group (e.g., see [107, 111]). Let us find the relationship between its symmetry and integrability.

We seek the operator of a symmetry group in the form

$$X = \xi(x, y)\frac{\partial}{\partial x} + \eta(x, y)\frac{\partial}{\partial y}.$$

Prolonging it to the derivative

$$\widetilde{X} = \xi(x, y)\frac{\partial}{\partial x} + \eta(x, y)\frac{\partial}{\partial y} + \big(D(\eta) - y'D(\xi)\big)\frac{\partial}{\partial y'},$$

we apply it to Eq. (0.37) and replace y' by $f(x, y)$,

$$\eta_x + (\eta_y - \xi_x)f - \xi_y f^2 = \xi f_x + \eta f_y. \tag{0.38}$$

This determining equation contains two unknown functions ξ and η; therefore, it is clear that the symmetry group is infinite-dimensional. We introduce the new function

$$\theta(x, y) = \eta - \xi f;$$

then the determining equation (0.38) acquires the form

$$\theta_x + f\theta_y = f_y\theta. \tag{0.39}$$

The zero solution $\theta = 0$ of this equation gives the symmetry operator

$$X = \xi\frac{\partial}{\partial x} + \xi f\frac{\partial}{\partial y}, \tag{0.40}$$

with an arbitrary function $\xi(x, y)$.

But if a nonzero solution $\theta(x, y) \neq 0$ of Eq. (0.39) is known, then the function

$$M = \frac{1}{\theta} = \frac{1}{\eta - \xi f} \tag{0.41}$$

is an integrating factor for the original equation. Indeed, we can rewrite (0.37) as

$$dy - f \, dx = 0$$

and multiply by M,

$$M \, dy - M f \, dx = 0. \tag{0.42}$$

The well-known integrating factor condition

$$\frac{\partial M}{\partial x} + \frac{\partial (M f)}{\partial y} = 0$$

is transformed into the determining equation (0.39) by the change (0.41).

We rewrite the total differential equation (0.42) as

$$A(x, y) \, dx + B(x, y) \, dy = 0, \tag{0.43}$$

where $A(x, y) = V_x(x, y) = -M f$, $B(x, y) = V_y(x, y) = M$, $A_y = B_x$, and $V(x, y)$ is a function implicitly determining the general solution of Eq. (0.37),

$$V(x, y) = c = \text{const.}$$

In particular, Eq. (0.43) admits the one-parameter group generated by the operator

$$X = B(x, y) \frac{\partial}{\partial x} - A(x, y) \frac{\partial}{\partial y},$$

which is obtained from (0.40) for $\xi = M(x, y) = B$.

It is easily seen that in the space (x, y) this operator has the unique invariant

$$J_1 = V(x, y)$$

determining the general solution of Eq. (0.37).

Thus, the knowledge of a nonzero solution of the determining equation is equivalent to the knowledge of an integrating factor, which permits integrating the first-order equation. Unfortunately, the problem of finding a nonzero solution of the determining equation is not at all simpler than the problem of integration of the original equation. But if some nontrivial symmetry of the original equation is known from some auxiliary considerations, then the change (0.41) gives an efficient integration method for the equation. This property underlies all elementary integration methods for first-order ordinary differential equations presented in numerous manuals. The idea of integrating factor for ordinary differential equations of higher order was developed in [13].

EXAMPLE. Consider the nonlinear equation

$$y' = \frac{2}{x^2} - y^2.$$

It admits the dilation operator

$$X = x\frac{\partial}{\partial x} - y\frac{\partial}{\partial y},$$

which has two invariants $x^2 y'$ and xy in the space (x, y, y'). In this case,

$$\theta = -y - \frac{2}{x} + xy^2, \qquad M = \frac{x}{x^2 y^2 - xy - 2},$$

which permits rewriting the equation as the total differential equation

$$\frac{x}{x^2 y^2 - xy - 2}\, dy + \frac{xy^2 - 2/x}{x^2 y^2 - xy - 2}\, dx = 0.$$

Integration gives the solution

$$V(x, y) = \ln|x| + \frac{1}{3}\ln\left|\frac{xy - 2}{xy + 1}\right| = \text{const.}$$

Integration of second-order ordinary differential equations with the use of a symmetry group

The above Table 0.2 displaying Lie's group classification of second-order ordinary differential equations contains an exhaustive list of equations with symmetries. As was already noted, the table contains the resulting examples of equations "up to arbitrary point transformations"; i.e., any other ordinary differential equation obtained from the given equation by a point transformation admits a similar algebra of operators. This is very inconvenient in practical applications, where a given ordinary differential equation may be related to one of the equations in the table by a rather unobvious transformation. By distinguishing between four types of L_2, Lie developed a unique approach to the integration of second-order ordinary differential equations with a symmetry that contains two-dimensional subalgebras (e.g., see [77]).

We assume that the symmetry of a given equation contains a two-dimensional subalgebra of operators

$$X_1 = \xi_1(x, y)\frac{\partial}{\partial x} + \eta_1(x, y)\frac{\partial}{\partial y}, \qquad X_2 = \xi_2(x, y)\frac{\partial}{\partial x} + \eta_2(x, y)\frac{\partial}{\partial y}.$$

To classify the algebras L_2, we need to calculate the commutator and the pseudoscalar (skew) product

$$X_1 \vee X_2 = \xi_1 \eta_2 - \xi_2 \eta_1$$

of the operators.

Then, determining the type of the symmetry subalgebra in Table 0.3, we can make the change of variables and pass to the *canonical variables* t, u, in which this equation takes one of the integrable forms listed in Table 0.3.

Table 0.3: Standard forms of L_2 and canonical forms of second-order ODE

Structure of L_2	Standard form of L_2	Canonical second-order ODE
$[X_1, X_2] = 0, \ \xi_1\eta_2 - \xi_2\eta_1 \neq 0$	$X_1 = \dfrac{\partial}{\partial t}, \ X_2 = \dfrac{\partial}{\partial u}$	$u'' = F(u')$
$[X_1, X_2] = 0, \ \xi_1\eta_2 - \xi_2\eta_1 = 0$	$X_1 = \dfrac{\partial}{\partial u}, \ X_2 = t\dfrac{\partial}{\partial u}$	$u'' = F(t)$
$[X_1, X_2] = X_1, \ \xi_1\eta_2 - \xi_2\eta_1 \neq 0$	$X_1 = \dfrac{\partial}{\partial u},$ $X_2 = t\dfrac{\partial}{\partial t} + u\dfrac{\partial}{\partial u}$	$u'' = \dfrac{1}{t}F(u')$
$[X_1, X_2] = X_1, \ \xi_1\eta_2 - \xi_2\eta_1 = 0$	$X_1 = \dfrac{\partial}{\partial u}, \ X_2 = u\dfrac{\partial}{\partial u}$	$u'' = u'F(t)$

Thus, to integrate a second-order ordinary differential equation, we should

- Find the transformation group admitted by the equation and single out a two-dimensional subalgebra L_2 if any.

- Determine the type of the obtained subalgebra L_2 according to Table 0.3.

- Find canonical variables by solving the corresponding equation in Table 0.4 and make the corresponding change of variable.

- Integrate the resulting equation and return to the original variables.

Let us illustrate Lie's method by an example.

EXAMPLE. Consider the nonlinear equation

$$y'' = \frac{y'^3}{y}.$$

Table 0.4: Equations to be solved for obtaining the canonical variables

Standard form of L_2	Equations to be solved
$X_1 = \dfrac{\partial}{\partial t},\ X_2 = \dfrac{\partial}{\partial u}$	$X_1(t) = 1,\ X_2(t) = 0,\ X_1(u) = 0,\ X_2(u) = 1$
$X_1 = \dfrac{\partial}{\partial u},\ X_2 = t\dfrac{\partial}{\partial u}$	$X_1(t) = 0,\ X_2(t) = 0,\ X_1(u) = 1,\ X_2(u) = t$
$X_1 = \dfrac{\partial}{\partial u},\ X_2 = t\dfrac{\partial}{\partial t} + u\dfrac{\partial}{\partial u}$	$X_1(t) = 0,\ X_2(t) = t,\ X_1(u) = 1,\ X_2(u) = u$
$X_1 = \dfrac{\partial}{\partial u},\ X_2 = u\dfrac{\partial}{\partial u}$	$X_1(t) = 0,\ X_2(t) = 0,\ X_1(u) = 1,\ X_2(u) = u$

By solving the determining equation, we find the symmetry algebra

$$X_1 = \frac{\partial}{\partial x}, \qquad X_2 = y\frac{\partial}{\partial x}.$$

Then we determine the type of the two-dimensional algebra:

$$[X_1, X_2] = 0, \qquad \xi_1\eta_2 - \xi_2\eta_1 = 0.$$

We find the canonical variables by solving the following equations in the second row of Table 0.4:

$$t = y(x), \qquad u(t) = x.$$

In the canonical variables, the operators acquire the form

$$X_1 = \frac{\partial}{\partial u}, \qquad X_2 = t\frac{\partial}{\partial u},$$

and the equation becomes

$$u'' = -\frac{1}{t}$$

and has the general solution

$$u = t - t\ln|t| + C_1 t + C_2.$$

Returning to the original variables, we obtain

$$y - y\ln|y| + C_1 y - x = C_2.$$

The Lie method is discussed in more detail in [77].

0.1.12. Symmetry reduction and invariant solutions of partial differential equations

The best-known application of transformation groups admitted by partial differential equations is the construction of invariant solutions. Let the following system of differential equations be given:

$$F_\alpha(x, u, u_1, u_2, \ldots, u_s) - 0, \qquad \alpha = 1, 2, \ldots, m, \qquad (0.44)$$

where $x \in R^n$, $u \in R^m$, and u_s is the set of sth partial derivatives.

We assume that Eqs. (0.44) admit a transformation group $G_r{}^N$, and H is a subgroup of $G_r{}^N$.

DEFINITION. A solution $u = \Phi(x)$ of system (0.44) is called an *invariant solution* if it is an invariant manifold of the subgroup H.

We restrict our consideration to solutions that form a nonsingular invariant manifold. Moreover, the manifolds given by Eqs. (0.44) are also assumed to be nonsingular.

The nonsingular manifold $u = \Phi(x)$ has some rank ρ, which is called the *rank of the invariant solution*. Let the subgroup H be generated by the subalgebra of operators

$$X_\alpha = \xi_\alpha^i \frac{\partial}{\partial x_i} + \eta_\alpha^k \frac{\partial}{\partial u^k}, \qquad \alpha = 1, 2, \ldots, r. \qquad (0.45)$$

Let $R = R(M)$ be the general rank of the matrix

$$M = \|\xi_\alpha{}^i, \ \eta_\alpha{}^k\|$$

of coordinates of the operators (0.45); then the following assertion holds.

PROPOSITION. *For an invariant solution to exist, it is necessary that the group H have invariants $I^\tau(x, u)$ (i.e., $R(M) < N$) and*

$$R\left(\left\|\frac{\partial I^\tau(x, u)}{\partial u^k}\right\|\right) = m. \qquad (0.46)$$

Under these conditions, there exists an invariant solution in principle. To construct such a solution, one has to calculate a complete set of independent invariants of the subgroup H and perform *the symmetry reduction of the original system* by means of this subgroup H, i.e., pass to a system of equations relating the invariants, which is ensured by condition (0.46). We illustrate this by an example.

EXAMPLE (symmetry reduction).

A. Consider a solution invariant under the operator $X_1 = \partial/\partial t$ for the heat equation with a power-law coefficient,

$$u_t = (u^\sigma u_x)_x.$$

In this case, the invariants are $J_1 = x$ and $J_2 = u$. One can readily verify that both necessary conditions for the existence of an invariant solution are satisfied:

$$R = 1 < 3, \qquad R\left(\left\|\frac{\partial I^\tau(x, u)}{\partial u}\right\|\right) = 1.$$

To pass to the space of invariants, we let the second invariant to be a function of the first: $u = V(x)$. Substituting this representation into the heat equation, we obtain the ordinary differential equation

$$(V^\sigma V_x)_x = 0$$

for the unknown function $V(x)$. By integrating this equation, we obtain the *stationary solution*

$$u = (C_1 x + C_2)^{1/(\sigma+1)}, \quad \sigma \neq -1; \qquad u = C_2 e^{C_1 x}, \quad \sigma = -1,$$

where C_1 and C_2 are arbitrary constants.

B. For the same heat equation, let us construct a *self-similar solution*, i.e., a solution that is invariant under the dilation operator

$$X_4 = \frac{\sigma}{2} x \frac{\partial}{\partial x} + u \frac{\partial}{\partial u}.$$

One can readily verify that both necessary conditions for the existence of an invariant solution are satisfied. In this case, the invariants are

$$J_1 = t, \qquad J_2 = u x^{-2/\sigma}.$$

We seek the solution as a relation between the two invariants, $ux^{-2/\sigma} = V(t)$. We reduce the original equation to an ordinary differential equation for the unknown function $V(t)$ by substituting the invariant representation $u = V(t)x^{2/\sigma}$ of the solution into the heat equation,

$$V' = \frac{2(\sigma + 2)}{\sigma^2} V^{\sigma+1}.$$

By integrating this equation, we obtain the self-similar solution

$$u = \left(\frac{x^2}{C - 2(\sigma + 2)t/\sigma}\right)^{1/\sigma}, \qquad C = \text{const}.$$

C. Remark about self-similar solutions. Self-similar solutions are often used to solve problems in mechanics and physics. Their popularity is based on the fact that the form of solutions can be guessed by analyzing the dimensions of the variables determining the solution of the problem. The central theorem in this approach, the so-called π-theorem, is a special case of the theorem on the invariant representation of the solution determined by the dilation group. This approach is not invariant under the change of the coordinate system; it suffices to pass to another coordinate system, and the group ceases to be the dilation group. In our last example, the change of variables

$$\bar{x} = \ln x, \qquad \bar{t} = \ln t, \qquad \bar{u} = \ln u$$

transforms the operator X_4 into the translation operator

$$X_4 = \frac{\sigma}{2} \frac{\partial}{\partial \bar{x}} + \frac{\partial}{\partial \bar{u}}.$$

For the translation operator, dimension analysis does not work any more, but the theorem on the invariant representation of the solution remains valid. A detailed discussion of this problem can be found in [111].

The preceding examples show that if the necessary conditions are satisfied, then invariant solutions can be constructed for any subgroup of the group admitted by the equation or system under study. In the general case, the subgroups can have different dimensions, and for the operators one can take any linear combinations of basis operators. Therefore, the following problem of classification of the set of invariant solutions arises.

The basic classification of invariant solutions is usually performed with respect to their rank $\rho = n - R$. Therefore, the rank of solutions may vary in the range $\rho = 0, 1, 2, \ldots, n - 1$. The set of invariant solutions of given rank ρ can also be arranged in a certain order. To this end, L. Ovsyannikov proposed to divide a subgroup $H \in G_r{}^N$ of given rank into classes of equivalent subgroups in the sense of the following definition.

Two subgroups H and H^* are said to be similar in the group $G_r{}^N$ if there exists a transformation $T \in G_r{}^N$ such that $H^* = THT^{-1}$. In this case, the invariant solutions corresponding to the subgroups H and H^* are obviously related to each other by the same transformation T. Thus, the problem is reduced to listing all nonsimilar subgroups of given dimension. Using the algorithm for obtaining all nonsimilar subgroups of given dimension (see [111]), upon which we do not dwell here, we present the result of such calculations for the nonlinear heat equation with an arbitrary coefficient $k(u)$ and for one-dimensional subalgebras. This set of subalgebras, which is called an *optimal system of subalgebras*, is generated by the following operators:

$$X_1 = \frac{\partial}{\partial t}, \quad X_2 = \frac{\partial}{\partial x}, \quad X_3 = 2t\frac{\partial}{\partial t} + x\frac{\partial}{\partial x}, \quad X_1 + X_2 = \frac{\partial}{\partial t} + \frac{\partial}{\partial x}. \quad (0.47)$$

All other invariant solutions of the first rank are similar in the above sense to the solutions invariant under the four subalgebras (0.47). In all cases where one has a larger symmetry group, an optimal system of subalgebras can also be constructed.

The knowledge of the transformation group admitted by a system of differential equations also permits constructing solutions that are noninvariant in the rigorous sense. The most famous extension of the set of solutions is given by the so-called *partially invariant solutions*: these are solutions lying in a certain invariant manifold but not coinciding with it. An algorithm for constructing partially invariant solutions can be found in [111]. Moreover, to reduce a given system of equations, one can use differential rather than finite invariants. In this case, *differential-invariant solutions* are constructed with the use of relations between differential invariants.

0.1.13. Contact symmetries of differential equations

Contact (tangent) transformations have been widely used in mechanics and the theory of differential equations for a long time. Sophus Lie used the group of contact transformations of the form

$$x^{*i} = f^i(x, u, u_1, a), \qquad i = 1, 2, \dots, n,$$
$$u^* = g(x, u, u_1, a), \qquad u_i{}^* = h_i(x, u, u_1, a), \tag{0.48}$$

where u_1 denotes the set of all partial derivatives u_i. It turns out that the transformations (0.48) can be different from point transformations prolonged to the first derivatives only in the case of a single dependent variable u (e.g., see [111]).

Consider the one-parameter group operator

$$X = \xi^i(x_i, u, u_i)\frac{\partial}{\partial x_i} + \eta(x_i, u, u_i)\frac{\partial}{\partial u} + \zeta_i(x_i, u, u_i)\frac{\partial}{\partial u_i}. \tag{0.49}$$

The requirement that the first derivatives are preserved, or the manifold is invariant,

$$du = u_i \, dx^i, \qquad i = 1, 2, \dots, n,$$

results in the prolongation formulas

$$\zeta_i = D_i(\eta) - u_j D_i(\xi^j), \qquad i = 1, 2, \dots, n,$$

where

$$D_i = \frac{\partial}{\partial x_i} + u_i\frac{\partial}{\partial u} + u_{ij}\frac{\partial}{\partial u_j}, \qquad i = 1, 2, \dots, n.$$

THEOREM. *For the operator* (0.49) *to generate a group of contact transformations, it is necessary and sufficient that*

$$\xi^i = -\frac{\partial W}{\partial u_i}, \quad \eta = W - u_i\frac{\partial W}{\partial u_i}, \quad \zeta^i = \frac{\partial W}{\partial x^i} + u_i\frac{\partial W}{\partial u}, \qquad i = 1, 2, \dots, n,$$

where $W = W(x^i, u, u_i)$ *is the so-called characteristic function of the group of contact transformations (see* [73, 107, 111]).

Note that the point transformation groups form a subset of contact groups.

The invariance criterion for a group of contact transformations can be written in the same form as in the case of point groups. Any second-order ordinary differential equation has infinitely many contact symmetries (just as a first-order ordinary differential equation admits infinitely many point transformations), but usually one does not succeed in finding them. For third- and higher-order ordinary differential equations, contact symmetries can be found (in this case, the determining equation splits), because the characteristic W is independent of the second derivatives. As an example of solving such a problem, we present the symmetry of a linear third-order equation.

EXAMPLE. The ordinary third-order equation $u''' = 0$ admits the symmetry

$$X_1 = \frac{\partial}{\partial u}, \quad X_2 = x\frac{\partial}{\partial u}, \quad X_3 = x^2\frac{\partial}{\partial u}, \quad X_4 = u\frac{\partial}{\partial u},$$

$$X_5 = \frac{\partial}{\partial x}, \quad X_6 = x\frac{\partial}{\partial x}, \quad X_7 = x^2\frac{\partial}{\partial x} + xu\frac{\partial}{\partial u}, \quad X_8 = 2u'\frac{\partial}{\partial x} + u'^2\frac{\partial}{\partial u},$$

$$X_9 = (u - xu')\frac{\partial}{\partial x} - \frac{xu'^2}{2}\frac{\partial}{\partial u} - \frac{u'^2}{2}\frac{\partial}{\partial u'},$$

$$X_{10} = \left(xu - \frac{x^2u'}{2}\right)\frac{\partial}{\partial x} + \left(u^2 - \frac{x^2u'^2}{4}\right)\frac{\partial}{\partial u} + \left(uu' - \frac{xu'^2}{2}\right)\frac{\partial}{\partial u'}.$$

We note that the operators $X_1, X_2, \ldots, X_7$ correspond to point transformation groups and the operators X_8, X_9, X_{10} correspond to contact groups.

0.1.14. Groups of formal power series and higher-order symmetries of differential equations

A nontrivial generalization of contact transformation groups are groups of higher-order symmetries, or Lie–Bäcklund groups [5, 73].

Consider the space Z of sequences $(x, u, u_1, u_2, \ldots)$, where x is an independent variable, u is a dependent variable, and $u_1, u_2, \ldots$ are differential variables; here u_s is the sth derivative. By z we denote some vector consisting of *finitely many* coordinates of the vector $(x, u, u_1, u_2, \ldots)$, and by z^i we denote its ith coordinate.

In the space Z, we introduce a mapping D (differentiation) by the rule

$$D(x) = 1, \quad D(u) = u_1, \quad \ldots, \quad D(u_s) = u_{s+1}, \quad s = 1, 2, \ldots.$$

Let A be the space of analytic functions $F(z)$ of finitely many variables z. (Different functions $F(z)$ contained in A can depend on different variables in the sequence $(x, u, u_1, u_2 \ldots)$, but the set of arguments is always finite.) Identifying D with the action of the first-order linear differential operator

$$D = \frac{\partial}{\partial x} + u_1\frac{\partial}{\partial u} + u_2\frac{\partial}{\partial u_1} + \cdots + u_{s+1}\frac{\partial}{\partial u_s} + \cdots,$$

we generalize the differentiation D to functions in A, and $D(F(z)) \in A$ in this case.

Consider sequences of formal power series

$$f^i(z, a) = \sum_{k=0}^{\infty} A_k^i(z) a^k, \qquad i = 1, 2, \ldots, \qquad (0.50)$$

in a single symbol (parameter) a, where $A_k^i(z) \in A$ and $A_0^i \equiv z^i$. Here z^i is the ith coordinate of a vector in Z.

We denote the space of sequences

$$(f^1(z, a), f^2(z, a), \ldots, f^s(z, a), \ldots)$$

of formal power series (0.50) by $\widetilde{Z}$. The sequences $(x, u, u_1, u_2, \ldots)$ are a special case of such sequences, $Z \subset \widetilde{Z}$.

In $\widetilde{Z}$, by definition, we introduce the operations of addition and multiplication by a number and the product of formal series as follows (these operations coincide with the operations for converging series):

$$\alpha \left(\sum_{k=0}^{\infty} A_k^i a^k \right) + \beta \left(\sum_{k=0}^{\infty} B_k^i a^k \right) = \sum_{k=0}^{\infty} \left(\alpha A_k^i + \beta B_k^i \right) a^k,$$

$$\left(\sum_{k_1=0}^{\infty} A_{k_1}^i a^{k_1} \right) \left(\sum_{k_2=0}^{\infty} B_{k_2}^i a^{k_2} \right) = \sum_{k=0}^{\infty} \left(\sum_{k_1+k_2=k} A_{k_1}^i B_{k_2}^i \right) a^k,$$

where $i = 1, 2, \ldots$ and $\alpha, \beta = \text{const}$. We also introduce the operations

$$D \left(\sum_{k=0}^{\infty} A_k^i(z) a^k \right) = \sum_{k=0}^{\infty} D(A_k^i(z)) a^k, \qquad \frac{\partial}{\partial a} \left(\sum_{k=0}^{\infty} A_k^i(z) a^k \right) = \sum_{k=1}^{\infty} k A_k^i(z) a^{k-1},$$

$$\frac{\partial}{\partial a} \left(\sum_{k=0}^{\infty} A_k^i(z) a^k \right) \bigg|_{a=0} = A_1^i(z), \qquad i = 1, 2, \ldots,$$

of differentiation of the series (0.50). An equality of formal power series is understood in the sense that the coefficients of like powers of a coincide. In particular, the series (0.50) is equal to 0 if $A_k^i = 0$, $k = 0, 1, 2, \ldots$.

In $\widetilde{Z}$, consider the transformation

$$z^{i*} = f^i(z, a), \qquad i = 1, 2, \ldots, \qquad (0.51)$$

determined by the series (0.50). This transformation takes a sequence z^i to the sequence z^{i*}.

The above-introduced operations on the series (0.50) permit considering powers of series of the form (0.50), monomials, polynomials, and even analytic functions (or formal power series) of finitely many variables z^i. For example, the composition of transformations of the form (0.51) is well defined:

$$z^{i**} = f^i(z^*, b) = \sum_{k=0}^{\infty} A_k^i(z^*)b^k = \sum_{k=0}^{\infty} A_k^i(f(z, a))b^k, \qquad i = 1, 2 \dots .$$

In general, such a composition takes a one-parameter series (0.50) out of $\widetilde{Z}$. We consider only such series (0.50) (and the corresponding transformations (0.51)) whose coefficient structure ensures that the transformations (0.51) are closed in $\widetilde{Z}$:

$$z^{i**} = f^i(z^*, b) = f^i(z, a + b) = \sum_{k=0}^{\infty} A_k^i(z)(a + b)^k, \qquad i = 1, 2, \dots . \quad (0.52)$$

Property (0.52) of the formal series (0.50) means that the transformations (0.51) form a *formal one-parameter group* in $\widetilde{Z}$.

Property (0.52) is equivalent to the following exponential representation of the power series (0.50) (see [107]):

$$z^{i*} = f^i(z, a) = e^{aX}(z^i) \equiv \sum_{s=0}^{\infty} \frac{a^s}{s!} X^{(s)}(z^i), \qquad i = 1, 2, \dots, \quad (0.53)$$

where X is the infinitesimal operator (generator) of the group:

$$X = \xi^i(z)\frac{\partial}{\partial x_i}, \qquad \xi^i(z) = \frac{\partial f^i(z, a)}{\partial a}\bigg|_{a=0} \in A, \qquad i = 1, 2, \dots . \quad (0.54)$$

Let us show that the representation (0.53) is equivalent to definition (0.54) of the formal group. To this end, we consider the superposition of transformations of the form (0.52):

$$z^{i**} = e^{bX}(e^{aX}(z^i)) = \sum_{l=0}^{\infty} \frac{b^l}{l!} X^{(l)} \left(\sum_{s=0}^{\infty} \frac{a^s}{s!} X^{(s)}(z^i) \right)$$

$$= \sum_{k=0}^{\infty} \left(\sum_{s+l=k} \frac{a^s b^l}{s!l!} X^{s+l}(z^i) \right) = \sum_{k=0}^{\infty} \frac{(a + b)^k}{k!} X^k(z^i) = e^{(a+b)X}(z^i).$$

Hence the representation (0.53) is equivalent to (0.54). Moreover, it is obvious that $f^i(z, a)\big|_{a=0} = z^i$ and $f^i(f^i(z, a), -a) = z^i$.

Thus, for formal one-parameter groups, the exponential representation in terms of the infinitesimal operator holds just as for the classical local Lie transformation groups.

Note that the exponential representation (0.53) implies the recursive chain

$$A_k^i(z) = \frac{1}{k}X(A_{k-1}^i(z)), \qquad k = 1, 2, \dots, \qquad i = 1, 2, \dots,$$

for the coefficients of the formal group (0.54) and the following formula convenient for calculating the coefficients of the formal series (0.50):

$$A_k^i(z) = \frac{1}{k!} X^{(k)}(z^i), \qquad k = 1, 2, \ldots, \qquad i = 1, 2, \ldots.$$

In the theory of formal groups, it is shown (see [72]) that the tangent vector field

$$\xi^i(z) = \left. \frac{\partial f^i(z, a)}{\partial a} \right|_{a=0}$$

is related to formal series (0.50) by the Lie equations

$$\frac{\partial f^i}{\partial a} = \xi^i(f), \qquad f^i(z, a)\big|_{a=0} = z^i, \qquad i = 1, 2, \ldots; \qquad (0.55)$$

i.e., the sequence of formal series $(f^1, f^2, \ldots)$, which form a group with tangent field $\xi^i(z)$, satisfy system (0.55), and conversely, for an arbitrary sequence $(\xi^1(z), \xi^2(z), \ldots)$, $\xi^i(z) \in A$ of functions, the solution of system (0.55) is a formal one-parameter group.

Thus, the same relationship between the group and the operator as in the case of classical local Lie groups holds for formal one-parameter groups. If the formal power series converge and give sufficiently smooth differentiable functions, then we deal with Lie point or tangent transformation groups. Thus, point and contact transformations are part of formal groups but do not exhaust them completely. The class additional to point and tangent transformation groups is formed by groups of "higher-order symmetries," or by Lie–Bäcklund groups [5, 73, 107].

For formal groups, just as for Lie point and tangent transformation groups, one can introduce the notion of invariants and invariant manifolds.

A locally analytic function $F(z)$ of finitely many variables is called an *invariant* of a formal group if $F(z^*) = F(z)$ for any transformations of the group (0.54).

For a function $F(z) \in A$ to be invariant, it is necessary and sufficient that

$$XF(z) = 0,$$

where X is the group operator (0.54).

A manifold defined in $\widetilde{Z}$ by a function $\phi(z) \in A$ according to the formula

$$\phi(z) = 0,$$

is called an invariant manifold if

$$\phi(z^*) = 0$$

for all solutions (0.50) and all transformations of the formal group. A criterion for a manifold to be invariant can also be written with the use of the group operator (see [73, 107]):

$$X\phi(z)\big|_{\phi(z)=0} = 0.$$

In the set of formal groups whose transformations are described by formal power series (0.50), a special place is occupied by point and contact groups and by higher-order symmetries. While the first two classes of transformations can be considered in the finite-dimensional part of $\widetilde{Z}$, any nontrivial higher-order symmetry can be realized only in the entire infinite-dimensional space Z.

In what follows, we present an example of a nontrivial Lie–Bäcklund group, i.e., of a group that cannot be reduced to a point or contact group.

EXAMPLE (of a Lie–Bäcklund group). Consider the heat equation

$$u_t = u_{xx}$$

and seek a symmetry operator in the form

$$X = \eta(t, x, u, u_1, u_2, \ldots u_k)\frac{\partial}{\partial u},$$

where the spatial derivatives are denoted by $u_k = \partial u_{k-1}/\partial x$. The invariance criterion acquires the form

$$(D_t - D^2)\eta\big|_{u_t=u_2} = 0, \tag{0.56}$$

where

$$D = \frac{\partial}{\partial x} + u_1\frac{\partial}{\partial u} + u_2\frac{\partial}{\partial u_1} + \cdots, \qquad D_t = \frac{\partial}{\partial t} + u_t\frac{\partial}{\partial u} + u_{t1}\frac{\partial}{\partial u_1} + \cdots.$$

One solution of the determining equation (0.56) is obvious, $\eta_0 = u$, which implies the *point* dilation group. One can readily see that the action of the operator D takes each solution of the determining equation to a solution, because the following commutation relation holds:

$$[(D_t - D^2), D]\big|_{u_t=u_2} = 0.$$

Thus, the solutions of the determining equation form the infinite series

$$\eta_k = D^k(u) = u_k, \qquad k = 1, 2, \ldots.$$

This series of solutions suggests the following approach to solving (0.56). One seeks an operator L commuting with the operators of the determining equation,

$$[(D_t - D^2), L]\big|_{u_t=u_2} = 0. \tag{0.57}$$

Then the action of such an operator L generates an infinite series of solutions of the determining equation (0.56). One can show (see [73, 107]) that (0.57) has the solution

$$L = \alpha D + \beta(2tD + x),$$

where α and β are arbitrary constants.

One can acquaint oneself with higher-order symmetry groups and a vast bibliography on the subject in [73, 106, 107].

0.1.15. Invariant variational problems and conservation laws

One of the great achievements in group analysis, which strongly impressed physicists and mechanicians, was the discovery of the relationship between symmetries of differential equations and the existence of conservation laws for their solutions. This relationship is established not directly but with the use of a variational functional. More precisely, E. Noether's well-known theorem [104] establishes a relationship between the invariance of the variational functional and the conservativity of the corresponding Euler differential equations (i.e., the fact that the conservation laws are satisfied on their solutions).

Suppose that a differential equation

$$F(x, u, u_1, \ldots, u_s) = 0$$

of some order s is given and its solutions are extremals of some functional

$$L(u) = \int_\Omega \mathcal{L}(x, u, u_1, \ldots, u_k)\, dx, \qquad (0.58)$$

where the integral is taken over a domain $\Omega \subset R^n$. We assume there is a given group $G_r{}^N$ of transformations of the N-dimensional space of n independent and m dependent variables, $n+m = N$, which is generated by the Lie algebra of operators

$$X_\alpha = \xi_\alpha{}^i \frac{\partial}{\partial x_i} + \eta_\alpha{}^k \frac{\partial}{\partial u^k}, \qquad \alpha = 1, 2, \ldots, r. \qquad (0.59)$$

The following theorem [104] presents a necessary and sufficient condition for the functional (0.58) to be invariant. Without loss of generality, we restrict our consideration to functionals independent of the higher derivatives,

$$L(u) = \int_\Omega \mathcal{L}(x, u, u_1)\, dx. \qquad (0.60)$$

THEOREM. *For the functional* (0.60) *to be invariant under the group $G_r{}^N$, it is necessary and sufficient that*

$$\widetilde{X}_\alpha \mathcal{L} + \mathcal{L} D_i(\xi_\alpha{}^i) = 0, \qquad \alpha = 1, 2, \ldots, r, \qquad (0.61)$$

where the operator $\widetilde{X}_\alpha$ is the operator X_α prolonged to the first derivatives and D_i is the operator of total differentiation with respect to the variable x^i.

Simple transformations reduce condition (0.61) to the equivalent condition

$$(\eta_\alpha{}^k - u_i{}^k \xi_\alpha{}^i)\frac{\delta \mathcal{L}}{\delta u^k} + D_i A_\alpha^i = 0, \qquad \alpha = 1, 2, \ldots, r, \qquad (0.62)$$

where

$$A_\alpha^i = (\eta_\alpha^k - u_i^k \xi_\alpha^i) \frac{\partial \mathcal{L}}{\partial u_i^k} + \mathcal{L} \xi_\alpha^i.$$

This condition explicitly contains the Euler equation

$$\frac{\delta \mathcal{L}}{\delta u^k} = \frac{\partial \mathcal{L}}{\partial u^k} - D_i \left(\frac{\partial \mathcal{L}}{\partial u_i^k} \right) = 0, \qquad k = 1, 2, \ldots, m, \tag{0.63}$$

which provides an extremum of the functional, and the vector A_α^i, which determines the conservation law. These relations allow us to state the following theorem due to E. Noether.

THEOREM (E. Noether). *Let the variational functional (0.60) be invariant under the group G_r^N with the operators (0.59). Then the r linearly independent vectors A_α^i give r conservation laws*

$$D_i A_\alpha^i \big|_{(0.63)} = 0, \qquad \alpha = 1, 2, \ldots, r, \tag{0.64}$$

on the solutions of the Euler equation (0.63).

Thus, Noether's theorem gives sufficient conditions for the existence of conservation laws. We point out that the invariance of the functional rather than of the Euler equations is required. The Euler equations may admit a group, but this does not ensure that the functional is invariant. The invariance of the Euler equations is a *necessary* condition for the variational functional to be invariant.

THEOREM. *The group G_r^N takes each extremal of an invariant functional to an extremal of the same functional.*

Thus, the invariance of a variational functional implies the invariance of the corresponding Euler equations.

In the proof of Noether's theorem, one needs to show that formulas (0.61) and (0.62) are equivalent. Ibragimov [72, 73] noticed that the relationship between these formulas is of identical character regardless of the function $\mathcal{L}$ on which the operators act. This operator identity was called the Noether identity in [73]:

$$\xi_\alpha^i \frac{\partial \mathcal{L}}{\partial x_i} + \eta_\alpha^k \frac{\partial \mathcal{L}}{\partial u^k} + [D_i(\eta_\alpha^k) - u_j^k D_i(\xi_\alpha^j)] \frac{\partial \mathcal{L}}{\partial u_i^k} + \mathcal{L} D_i(\xi_\alpha^i)$$

$$\equiv (\eta_\alpha^k - u_i^k \xi_\alpha^i) \frac{\delta \mathcal{L}}{\delta u^k} + D_i \left((\eta_\alpha^k - u_i^k \xi_\alpha^i) \frac{\partial \mathcal{L}}{\partial u_i^k} + \mathcal{L} \xi_\alpha^i \right),$$

$\alpha = 1, 2, \ldots, r$. This identity permits stating a necessary and sufficient condition for the existence of conservation laws.

THEOREM ([73]). *The invariance of the variational functional* (0.60) *with respect to the group* G_r^N *on the solutions of the Euler equations* (0.63) *is a necessary and sufficient condition for the existence of the conservation laws* (0.64).

The Noether identity gives a relationship between three types of differentiation: the total differentiation D_i, the variational differentiation $\frac{\delta}{\delta u^k}$, and the Lie derivative $\widetilde{X}_\alpha$. This permits one to formalize the theory in general without relating the problem of constructing conservation laws to the statement of the variational problem.

The Noether identity readily allows one to generalize it to the case in which both of its parts are equal not to zero but to the divergence of a vector $B^i(x, u, u_i^k)$ (see [11]). Introducing this vector on the right-hand side in (0.62),

$$(\eta_\alpha^k - u_i^k \xi_\alpha^i) \frac{\delta \mathcal{L}}{\delta u^k} + D_i \left((\eta_\alpha^k - u_i^k \xi_\alpha^i) \frac{\partial \mathcal{L}}{\partial u_i^k} + \mathcal{L} \xi_\alpha^i - B^i(x, u, u_i^k) \right) = 0,$$

$\alpha = 1, 2, \ldots, r$, we see that, for such (divergence-invariant) Lagrangians, the conservation law takes the form

$$D_i (A_\alpha^i - B^i)\big|_{(0.63)} = 0.$$

The fact that the functional is divergence-invariant means that there exists another Lagrangian function for which the exact invariance is satisfied (see [107]).

EXAMPLE. The second-order ordinary differential equation

$$u'' = \frac{1}{u^3} \tag{0.65}$$

admits the Lie algebra L_3 with the basis operators

$$X_1 = \frac{\partial}{\partial x}, \qquad X_2 = 2x \frac{\partial}{\partial x} + u \frac{\partial}{\partial u}, \qquad X_3 = x^2 \frac{\partial}{\partial x} + xu \frac{\partial}{\partial u}.$$

One can readily verify that the Lagrangian function

$$\mathcal{L} = u'^2 - \frac{1}{u^2}$$

satisfies the condition that the corresponding functional is invariant with respect to X_1 and X_2. By Noether's theorem, this implies the following first integrals:

$$J_1 = u'^2 + \frac{1}{u^2} = A^0, \qquad J_2 = \frac{x}{u^2} - (u - u'x)u' = B^0. \tag{0.66}$$

The action of the third operator X_3 gives the total derivative

$$X_3 \mathcal{L} + 2x\mathcal{L} = 2uu' = D_x(u^2).$$

The divergence-invariant Lagrangian generates the first integral

$$J_3 = \frac{x^2}{u^2} + (u - xu')^2 = C^0. \tag{0.67}$$

The same integral can be obtained starting from the other Lagrangian

$$\mathcal{L} = \left(\frac{u}{x} - u'\right)^2 - \frac{1}{u^2},$$

which is invariant under X_3 without any additional divergence.

Note that Noether's theorem guarantees the independence of the first integrals only under the condition that they were obtained from the same Lagrangian invariant under all symmetries under study. The violation of this condition in our case implies that the three integrals are dependent and satisfy the relation

$$J_1 J_3 - J_2{}^2 = 1.$$

But for any two integrals in (0.66) and (0.67) it suffices, eliminating u', to construct the general solution of Eq. (0.65):

$$A_0 u^2 = (A_0 x + B_0)^2 + 1.$$

Invariance of Euler–Lagrange equations. There exists a relation between the invariance of the Lagrangian functional (0.60) and invariance of the corresponding Euler–Lagrange equations (0.63).

THEOREM ([73, 107]). *If the Lagrangian L is invariant with respect to the operator (0.59), i.e., if condition (0.61) is satisfied, then the Euler–Lagrange equations (0.63) are also invariant.*

Remark. If the Lagrangian L is divergence invariant, i.e., satisfies condition (0.62), then the Euler–Lagrange equations (0.63) are also invariant. This follows from the fact that full divergences belong to the kernel of variational operators.

Thus, if X is a variational or divergence symmetry of the functional $\mathbb{L}(\mathbf{u})$, it is also a symmetry of the corresponding Euler–Lagrange equations (0.63). The symmetry group of the Euler–Lagrange equations can, of course, be larger than the group generated by the variational and divergence symmetries of the Lagrangian.

It is of interest to establish a *necessary* and sufficient condition for the invariance of the Euler–Lagrange equations. We will need the following lemma [45]:

LEMMA. *For any smooth function $L(t, \mathbf{u}, \mathbf{u_1})$, it is true that*

$$\frac{\delta}{\delta u^j}\left(X(L) + LD(\xi)\right) \equiv X\left(\frac{\delta L}{\delta u^j}\right) + \left(\frac{\partial \eta^i}{\partial u^j} + \delta_{ij} D(\xi) - \frac{\partial \xi}{\partial u^j} u_1^i\right)\frac{\delta L}{\delta u^i},$$

$j = 1, \ldots, n$, *where δ_{ij} is the Kronecker delta.*

Proof. The result can be established by a straightforward computation. □

The above theorem and remark follow from this lemma. The lemma also provides a *necessary and sufficient* condition for the invariance of the Euler–Lagrange equations.

THEOREM ([45]). *The Euler–Lagrange equations* (0.63) *are invariant with respect to a symmetry* (0.59) *if and only if the following conditions are true (on the solutions of Eqs.* (0.63)):

$$\left[\frac{\delta}{\delta u^j}\left(X(L) + LD(\xi)\right)\right]_{\frac{\delta L}{\delta u^1}=\cdots=\frac{\delta L}{\delta u^n}=0} = 0, \qquad j = 1,\ldots,n. \tag{0.68}$$

Proof. The assertion follows from the identities in the preceding lemma. □

EXAMPLE. The equation

$$u'' = \frac{1}{u^2} \tag{0.69}$$

is the Euler–Lagrange equation for the Lagrangian function

$$L(t, u, u') = \frac{u'^2}{2} - \frac{1}{u}.$$

The equation admits the symmetries

$$X_1 = \frac{\partial}{\partial t}, \qquad X_2 = 3t\frac{\partial}{\partial t} + 2u\frac{\partial}{\partial u}.$$

The operator X_1 is a symmetry of the Lagrangian L and hence a symmetry of Eq. (0.69). The symmetry X_2 is not a symmetry of the Lagrangian,

$$X_2(L) + LD(\xi_2) = L.$$

However, it is a symmetry of the equation, as follows from the theorem:

$$\left[\frac{\delta}{\delta u}\left(X_2(L) + LD(\xi_2)\right)\right]_{\frac{\delta L}{\delta u}=0} = \left[\frac{\delta L}{\delta u}\right]_{\frac{\delta L}{\delta u}=0} = 0.$$

0.1.16. Symmetries and first integrals of canonical Hamiltonian equations

In this section, we consider the relationship between the symmetries of canonical Hamiltonian equations and the first integrals. We do not follow the traditional way accepted in the literature but develop a method based on an operator identity [44,45] that is a Hamiltonian analog of the Noether identity for Lagrangian structures. This allows us to draw a complete analogy between the Hamiltonian and the Lagrangian formalism for obtaining the first integrals of ordinary differential equations with symmetries. This approach is illustrated by several examples.

In the present section, we are interested in the canonical Hamiltonian equations

$$\dot{q}^i = \frac{\partial H}{\partial p_i}, \quad \dot{p}_i = -\frac{\partial H}{\partial q^i}, \qquad i = 1, \ldots, n. \tag{0.70}$$

These equations can be obtained by the variational principle from the action functional

$$\delta \int_{t_1}^{t_2} \left(p_i \dot{q}^i - H(t, \mathbf{q}, \mathbf{p})\right) dt = 0 \tag{0.71}$$

in the phase space $(\mathbf{q}, \mathbf{p})$, where $\mathbf{q} = (q^1, q^2, \ldots, q^n)$, $\mathbf{p} = (p_1, p_2, \ldots, p_n)$ (e.g., see [60,98]). The variations δq^i and δp_i are arbitrary and satisfy $\delta q^i(t_1) = \delta q^i(t_2) = 0$, $i = 1, \ldots, n$. Then we have

$$\delta \int_{t_1}^{t_2} \left(p_i \dot{q}^i - H(t, \mathbf{q}, \mathbf{p})\right) dt = \int_{t_1}^{t_2} \left(\delta p_i \dot{q}^i + p_i \delta \dot{q}^i - \frac{\partial H}{\partial q^i} \delta q^i - \frac{\partial H}{\partial p_i} \delta p_i\right) dt$$

$$= \int_{t_1}^{t_2} \left[\left(\dot{q}^i - \frac{\partial H}{\partial p_i}\right) \delta p_i - \left(\dot{p}_i + \frac{\partial H}{\partial q^i}\right) \delta q^i\right] dt + \left[p_i \delta q^i\right]_{t_1}^{t_2}.$$

The last term vanishes, because $\delta q^i = 0$ at the endpoints. Since the variations δq^i and δp_i are arbitrary, it follows that the variational principle (0.71) is equivalent to the canonical Hamiltonian equations (0.70).

Note that the canonical Hamiltonian equations (0.70) can be obtained by applying the variational operators

$$\frac{\delta}{\delta p_i} = \frac{\partial}{\partial p_i} - D \frac{\partial}{\partial \dot{p}_i}, \qquad i = 1, \ldots, n, \tag{0.72}$$

and

$$\frac{\delta}{\delta q^i} = \frac{\partial}{\partial q^i} - D \frac{\partial}{\partial \dot{q}^i}, \qquad i = 1, \ldots, n, \tag{0.73}$$

where

$$D = \frac{\partial}{\partial t} + \dot{q}^i \frac{\partial}{\partial q^i} + \dot{p}_i \frac{\partial}{\partial p_i} + \cdots \tag{0.74}$$

is the operator of total differentiation with respect to time, to the function

$$p_i \dot{q}^i - H(t, \mathbf{q}, \mathbf{p}).$$

The Legendre transformation relates the Hamiltonian and Lagrange functions,

$$L(t, \mathbf{q}, \dot{\mathbf{q}}) = p_i \dot{q}^i - H(t, \mathbf{q}, \mathbf{p}),$$

where

$$\mathbf{p} = \frac{\partial L}{\partial \dot{\mathbf{q}}}, \qquad \dot{\mathbf{q}} = \frac{\partial H}{\partial \mathbf{p}}.$$

This permits one to establish the equivalence of the Euler–Lagrange and Hamiltonian equations [6]. Indeed, from the Euler–Lagrange equations in the form ($m = 1$)

$$\frac{\delta L}{\delta q^k} = \frac{\partial L}{\partial q^k} - D\left(\frac{\partial L}{\partial \dot{q}^k}\right) = 0, \qquad k = 1, \dots, n,$$

we can obtain the canonical Hamiltonian equations (0.70) by using the Legendre transformation. It should be noted that the Legendre transformation is not a point transformation. Hence, there is no conservation of Lie group properties of the corresponding Euler–Lagrange equations and Hamiltonian equations within the class of point transformations.

The Lie group symmetries in the space $(t, \mathbf{q}, \mathbf{p})$ are generated by operators of the form

$$X = \xi(t, \mathbf{q}, \mathbf{p})\frac{\partial}{\partial t} + \eta^i(t, \mathbf{q}, \mathbf{p})\frac{\partial}{\partial q^i} + \zeta_i(t, \mathbf{q}, \mathbf{p})\frac{\partial}{\partial p_i}. \tag{0.75}$$

The standard approach to the symmetry properties of Hamiltonian equations is to consider so-called *Hamiltonian symmetries* [107]. In the case of canonical Hamiltonian equations, these are the evolution ($\xi = 0$) symmetries (0.75)

$$X = \eta^i(t, \mathbf{q}, \mathbf{p})\frac{\partial}{\partial q^i} + \zeta_i(t, \mathbf{q}, \mathbf{p})\frac{\partial}{\partial p_i} \tag{0.76}$$

with

$$\eta^i = \frac{\partial I}{\partial p_i}, \quad \zeta_i = -\frac{\partial I}{\partial q^i}, \qquad i = 1, \dots, n \tag{0.77}$$

for some function $I(t, \mathbf{q}, \mathbf{p})$, namely, symmetries of the form

$$X_I = \frac{\partial I}{\partial p_i}\frac{\partial}{\partial q^i} - \frac{\partial I}{\partial q^i}\frac{\partial}{\partial p_i}. \tag{0.78}$$

These symmetries are restricted to the phase space $(\mathbf{q}, \mathbf{p})$ and are generated by the function $I = I(t, \mathbf{q}, \mathbf{p})$. For the symmetry (0.78), the independent variable t is invariant and plays the role of a parameter.

Noether's theorem [107, Theorem 6.33] relates Hamiltonian symmetries of the Hamiltonian equations to their first integrals. Being restricted to the case of the canonical Hamiltonian equations, it can be stated as follows.

PROPOSITION. *An evolution vector field X of the form (0.76) generates a Hamiltonian symmetry group of the canonical Hamiltonian system (0.70) if and only if there exists a first integral $I(t, \mathbf{q}, \mathbf{p})$ such that $X = X_I$ is the corresponding Hamiltonian vector field. Another function $\widetilde{I}(t, \mathbf{q}, \mathbf{p})$ determines the same Hamiltonian symmetry if and only if $\widetilde{I} = I + F(t)$ for some time-dependent function $F(t)$.*

Indeed, the invariance of the canonical Hamiltonian equations (0.70) with respect to the symmetry (0.78) implies that

$$
\begin{aligned}
\frac{\partial^2 I}{\partial t \partial p_i} + \frac{\partial H}{\partial p_j}\frac{\partial^2 I}{\partial q^j \partial p_i} - \frac{\partial H}{\partial q^j}\frac{\partial^2 I}{\partial p_j \partial p_i} &= \frac{\partial I}{\partial p_j}\frac{\partial^2 H}{\partial q^j \partial p_i} - \frac{\partial I}{\partial q^j}\frac{\partial^2 H}{\partial p_j \partial p_i}, \\
-\frac{\partial^2 I}{\partial t \partial q^i} - \frac{\partial H}{\partial p_j}\frac{\partial^2 I}{\partial q^j \partial q^i} + \frac{\partial H}{\partial q^j}\frac{\partial^2 I}{\partial p_j \partial q^i} &= \frac{\partial I}{\partial q^j}\frac{\partial^2 H}{\partial p_j \partial q^i} - \frac{\partial I}{\partial p_j}\frac{\partial^2 H}{\partial q^j \partial q^i},
\end{aligned}
\qquad i = 1, \dots, n.
$$

These equations can be rewritten as

$$
\begin{aligned}
\frac{\partial}{\partial p_i}\left(\frac{\partial I}{\partial t} + \frac{\partial H}{\partial p_j}\frac{\partial I}{\partial q^j} - \frac{\partial H}{\partial q^j}\frac{\partial I}{\partial p_j} \right) &= 0, \\
\frac{\partial}{\partial q^i}\left(\frac{\partial I}{\partial t} + \frac{\partial H}{\partial p_j}\frac{\partial I}{\partial q^j} - \frac{\partial H}{\partial q^j}\frac{\partial I}{\partial p_j} \right) &= 0,
\end{aligned}
\qquad i = 1, \dots, n.
$$

Therefore, we obtain

$$
\frac{\partial I}{\partial t} + \frac{\partial H}{\partial p_j}\frac{\partial I}{\partial q^j} - \frac{\partial H}{\partial q^j}\frac{\partial I}{\partial p_j} = f(t).
$$

The left-hand side is the total time derivative of I on the solutions of the canonical Hamiltonian equations,

$$
\frac{\partial I}{\partial t} + \frac{\partial H}{\partial p_j}\frac{\partial I}{\partial q^j} - \frac{\partial H}{\partial q^j}\frac{\partial I}{\partial p_j} = D(I)\big|_{(0.70)}.
$$

Thus, we conclude that a Hamiltonian symmetry determines a first integral of the canonical Hamiltonian equations up to some time-dependent function that can be found with the help of these equations. As a disadvantage of such approach, one can note the loss of the geometrical meaning of transformations in the evolution form (0.78) and the necessity of integration to find first integrals with the help of (0.77). In this approach, it is also not clear why some point symmetries of Hamiltonian equations yield first integrals while others do not.

In the present section, we consider symmetries of the general form (0.75) which are not restricted to the phase space and can also transform the time variable t. In contrast to Hamiltonian symmetries in the form (0.78), the underlying symmetries have a clear geometric meaning in finite space and do not require integration for the calculation of first integrals. We shall provide a Hamiltonian version of Noether's theorem (in the strong statement) based on a newly established Hamiltonian identity, which is an analog of the Noether identity in the Lagrangian approach. The Hamiltonian identity directly links an invariant Hamiltonian function to the first integrals of the canonical Hamiltonian equations. This approach provides a simple and clear way to construct first integrals by means of purely algebraic manipulations with symmetries of the action functional. The approach will be illustrated in a number of examples, including the equations of three-dimensional Kepler motion.

Invariance of elementary Hamiltonian action

As an analog of the elementary Lagrangian action [73], consider the *elementary Hamiltonian action*

$$p_i \, dq^i - H \, dt, \tag{0.79}$$

which can be invariant or not with respect to the group generated by an operator of the form (0.75).

DEFINITION. We say that a Hamiltonian function is invariant with respect to a symmetry (0.75) if the elementary action (0.79) is an invariant of the group generated by the operator (0.75).

THEOREM 0.1. *A Hamiltonian is invariant with respect to a group with operator* (0.75) *if and only if*

$$\zeta_i \dot{q}^i + p_i D(\eta^i) - X(H) - H D(\xi) = 0. \tag{0.80}$$

Proof. The invariance condition follows directly from the action of X prolonged to the differentials dt and dq^i, $i = 1, \ldots, n$:

$$X = \xi(t, \mathbf{q}, \mathbf{p}) \frac{\partial}{\partial t} + \eta^i(t, \mathbf{q}, \mathbf{p}) \frac{\partial}{\partial q^i} + \zeta_i(t, \mathbf{q}, \mathbf{p}) \frac{\partial}{\partial p_i} + D(\xi) dt \frac{\partial}{\partial (dt)} + D(\eta^i) dt \frac{\partial}{\partial (dq^i)}.$$

An application of this equation gives

$$X(p_i \, dq^i - H \, dt) = \left(\zeta_i \dot{q}^i + p_i D(\eta^i) - X(H) - H D(\xi) \right) dt = 0.$$

□

COROLLARY. *It follows from the relation*

$$L(t, \mathbf{q}, \dot{\mathbf{q}}) \, dt = p_i \, dq^i - H(t, \mathbf{q}, \mathbf{p}) \, dt \tag{0.81}$$

that if the Lagrangian is invariant, then the Hamiltonian is also invariant with respect to the same group. Conversely, invariant Hamiltonians provide invariant Lagrangians by means of (0.81).

Proof. This follows from the action of the operator (0.75) on relation (0.81).   □

Remark 0.2. The total differentiation operator (0.74) applied to the Hamiltonian H on the solutions of Hamiltonian equations (0.70) coincides with the partial differentiation with respect to time,

$$D(H)\big|_{(0.70)} = \left[\frac{\partial H}{\partial t} + \dot{q}^i\frac{\partial H}{\partial q^i} + \dot{p}_i\frac{\partial H}{\partial p_i}\right]_{(0.70)} = \frac{\partial H}{\partial t}.$$

Remark 0.3. Condition (0.80) means that $p_i\,dq^i - H\,dt$ is an invariant in the space $(\mathbf{p}, \mathbf{q}, d\mathbf{q}, dt)$. Meanwhile, this condition can be obtained as an invariance condition for the *manifold*

$$h = p_i\dot{q}^i - H \tag{0.82}$$

under the action of the operator (0.75), which is specially prolonged to the new variable h in the following way:

$$X = \xi(t, \mathbf{q}, \mathbf{p})\frac{\partial}{\partial t} + \eta^i(t, \mathbf{q}, \mathbf{p})\frac{\partial}{\partial q^i} + \zeta_i(t, \mathbf{q}, \mathbf{p})\frac{\partial}{\partial p_i} - hD(\xi)\frac{\partial}{\partial h}. \tag{0.83}$$

Indeed, an application of the operator (0.83) to (0.82) yields

$$-hD(\xi) = \zeta_i\dot{q}^i + p_i\big(D(\eta^i) - \dot{q}_iD(\xi)\big) - X(H). \tag{0.84}$$

Then the substitution of h from (0.82) gives condition (0.80).

Hamiltonian identity and Noether-type theorem

Now we can relate the conservation properties of the canonical Hamiltonian equations to the invariance of the Hamiltonian function.

LEMMA 0.4. *The identity*

$$\zeta_i\dot{q}^i + p_iD(\eta^i) - X(H) - HD(\xi) = \xi\left(D(H) - \frac{\partial H}{\partial t}\right) - \eta^i\left(\dot{p}_i + \frac{\partial H}{\partial q^i}\right)$$
$$+ \zeta_i\left(\dot{q}^i - \frac{\partial H}{\partial p_i}\right) + D\left[p_i\eta^i - \xi H\right] \tag{0.85}$$

is true for any smooth function $H = H(t, \mathbf{q}, \mathbf{p})$.

Proof. The identity can be established by a straightforward computation.   □

We call this identity the *Hamiltonian identity*. This identity permits one to prove the following result.

THEOREM 0.5. *The canonical Hamiltonian equations* (0.70) *possess a first integral of the form*

$$I = p_i \eta^i - \xi H$$

if and only if the Hamiltonian function is invariant with respect to the operator (0.75) *on the solutions of the canonical equations* (0.70).

Proof. The result follows from identity (0.85) in view of Remark 0.2. □

Theorem 0.5 corresponds to the strong version of Noether's theorem (i.e., a necessary and sufficient condition) for invariant Lagrangians and Euler–Lagrange equations [73].

COROLLARY. *Theorem 0.5 can be generalized to the case of divergence invariance of the Hamiltonian action,*

$$\zeta_i \dot{q}^i + p_i D(\eta^i) - X(H) - H D(\xi) = D(V), \tag{0.86}$$

where $V = V(t, \mathbf{q}, \mathbf{p})$. *If this condition holds on the solutions of the canonical Hamiltonian equations* (0.70), *then one has the first integral*

$$I = p_i \eta^i - \xi H - V.$$

Invariance of the canonical Hamiltonian equations

In the Lagrangian framework, the variational principle gives the Euler–Lagrange equations. It is known that the invariance of the Euler–Lagrange equations follows from the invariance of the action integral. The following Lemma 0.6 and Theorem 0.7 establish a sufficient condition for the canonical Hamiltonian equations to be invariant.

LEMMA 0.6. *The application of the variational operators* (0.72) *to* (0.85) *gives*

$$\frac{\delta}{\delta p_j} \left(\zeta_i \dot{q}^i + p_i D(\eta^i) - X(H) - H D(\xi) \right)$$

$$= D(\eta^j) - \dot{q}^j D(\xi) - X\left(\frac{\partial H}{\partial p_j} \right) + \xi_{p_j} \left(D(H) - \frac{\partial H}{\partial t} \right)$$

$$- \eta^i_{p_j} \left(\dot{p}_i + \frac{\partial H}{\partial q^i} \right) + \left((\zeta_i)_{p_j} + \delta_{ij} D(\xi) \right) \left(\dot{q}^i - \frac{\partial H}{\partial p_i} \right),$$

$$j = 1, \ldots, n. \tag{0.87}$$

Likewise, the application of the variational operators (0.73) to identity (0.85) gives

$$\frac{\delta}{\delta q^j}\left(\zeta_i \dot{q}^i + p_i D(\eta^i) - X(H) - H D(\xi)\right)$$

$$= -D(\zeta_j) + \dot{p}_j D(\xi) - X\left(\frac{\partial H}{\partial q^j}\right) + \xi_{q^j}\left(D(H) - \frac{\partial H}{\partial t}\right)$$

$$- (\eta^i_{q^j} + \delta_{ij} D(\xi))\left(\dot{p}_i + \frac{\partial H}{\partial q^i}\right) + (\zeta_i)_{q^j}\left(\dot{q}^i - \frac{\partial H}{\partial p_i}\right),$$

$$j = 1, \ldots, n. \qquad (0.88)$$

Here δ_{ij} is the Kronecker delta.

The above identities are true for any smooth function $H = H(t, \mathbf{q}, \mathbf{p})$.

THEOREM 0.7. *If the Hamiltonian is invariant with respect to the symmetry (0.75), then Eqs. (0.70) are also invariant.*

Proof. We start from the invariance of the canonical Hamiltonian equations (0.70). An application of the symmetry operator to these equations yields

$$D(\eta^j) - \dot{q}^j D(\xi) - X\left(\frac{\partial H}{\partial p_j}\right) = 0, \qquad j = 1, \ldots, n, \qquad (0.89)$$

$$D(\zeta_j) - \dot{p}_j D(\xi) + X\left(\frac{\partial H}{\partial q^j}\right) = 0, \qquad j = 1, \ldots, n. \qquad (0.90)$$

Both conditions obtained should be true on the solutions of Eqs. (0.70).

Let the Hamiltonian be invariant; then all left-hand sides of identities (0.87) and (0.88) are zero on the solutions of Eqs. (0.70). All right-hand sides of (0.87) and (0.88) are also zero. By substituting (0.70) into the right-hand sides of (0.87) and (0.88), we obtain the invariance conditions (0.89) and (0.90). □

Remark 0.8. Theorem 0.7 remains valid if we consider divergence symmetries of the Hamiltonian, i.e., condition (0.86), because the term $D(V)$ on the right-hand side belongs to the kernel of the variational operators (0.72) and (0.73).

The invariance of the Hamiltonian on Eqs. (0.70) is a *sufficient condition* for the canonical Hamiltonian equations to be invariant. The symmetry group of the canonical Hamiltonian equations can of course be larger than that of the Hamiltonian. The following Theorem 0.9 establishes a *necessary and sufficient* condition for the canonical Hamiltonian equations to be invariant.

THEOREM 0.9. *The canonical Hamiltonian equations (0.70) are invariant with respect to a symmetry (0.75) if and only if the following conditions are true (on the solutions of the canonical Hamiltonian equations*):

$$\left[\frac{\delta}{\delta p_j}\left(\zeta_i \dot{q}^i + p_i D(\eta^i) - X(H) - HD(\xi)\right)\right]_{(0.70)} = 0,$$

$$\left[\frac{\delta}{\delta q^j}\left(\zeta_i \dot{q}^i + p_i D(\eta^i) - X(H) - HD(\xi)\right)\right]_{(0.70)} = 0. \qquad i,j = 1,\ldots,n. \quad (0.91)$$

Proof. The proof follows from identities (0.87) and (0.88). □

We point out that conditions (0.91) are true for all symmetries of canonical Hamiltonian equations. But not all of these symmetries yield the "variational integral" of these conditions, i.e.,

$$\left(\zeta_i \dot{q}^i + p_i D(\eta^i) - X(H) - HD(\xi)\right)\big|_{(0.70)} = 0,$$

which gives first integrals by Theorem 0.5. That is why not all symmetries of the canonical Hamiltonian equations provide first integrals.

Symplecticity of the canonical Hamiltonian equations

The solutions of the canonical Hamiltonian equations

$$\dot{q}^i = \frac{\partial H}{\partial p_i}, \quad \dot{p}_i = -\frac{\partial H}{\partial q^i}, \qquad i = 1,\ldots,n. \quad (0.92)$$

possess the important property called *symplecticity*.

For the solution $(\mathbf{q}(t), \mathbf{p}(t))$ of system (0.92) with the initial data $\mathbf{q}(t_0) = \mathbf{q}_0$, $\mathbf{p}(t_0) = \mathbf{p}_0$, we introduce the solution operator $\psi(t, t_0)$ by means of the transformation

$$(\mathbf{p}, \mathbf{q}) = \psi(t, t_0)(\mathbf{p}_0, \mathbf{q}_0). \quad (0.93)$$

For *autonomous* Hamiltonians, the mapping (0.93) depends on the difference $t - t_0$ and possesses the group property

$$\psi(t_2 - t_0) = \psi(t_2 - t_1)\psi(t_1 - t_0).$$

We denote the Jacobi matrix of the transformation ψ on the phase space R^{2n} by

$$\psi' = \frac{\partial(p, q)}{\partial(p_0, q_0)}.$$

Such a transformation ψ is said to be *symplectic* if

$$\psi'^T J \psi' = J, \quad (0.94)$$

where

$$J = \begin{pmatrix} 0 & I \\ -I & 0 \end{pmatrix}$$

and I is the $n \times n$ identity matrix.

Consider the special case in which $n = 1$. The transformation

$$(p^*, q^*) = \psi(t, t_0)(p, q),$$

defined in some domain Ω and possessing property (0.94) gives

$$\frac{\partial p^*}{\partial p} \frac{\partial q^*}{\partial q} - \frac{\partial p^*}{\partial q} \frac{\partial q^*}{\partial p} = 1. \tag{0.95}$$

This means that the Jacobian is identically 1; i.e., $\psi(t, t_0)$ is volume preserving.

Alternatively, the symplectic property of the Hamiltonian equations can be expressed as a conservation of a certain differential form. The differentials dp^* and dq^* of the components of the transformation ψ give the so-called one-forms

$$dp^* = \frac{\partial p^*}{\partial p} dp + \frac{\partial p^*}{\partial q} dq, \qquad dq^* = \frac{\partial q^*}{\partial p} dp + \frac{\partial q^*}{\partial q} dq.$$

Two one-forms by means of the *exterior*, or *wedge, product* $\wedge$ give rise to the so-called differential two-form

$$dp^* \wedge dq^* = \frac{\partial p^*}{\partial p} \frac{\partial q^*}{\partial p} dp \wedge dp + \frac{\partial p^*}{\partial p} \frac{\partial q^*}{\partial q} dp \wedge dq + \frac{\partial p^*}{\partial q} \frac{\partial q^*}{\partial p} dq \wedge dp + \frac{\partial p^*}{\partial q} \frac{\partial q^*}{\partial q} dq \wedge dq.$$

We can simplify this expression using the fact that the exterior product is skew-symmetric,

$$dp \wedge dq = -dq \wedge dp, \quad dp \wedge dp = dq \wedge dq = 0.$$

Consequently, the two-form becomes

$$dp^* \wedge dq^* = \left(\frac{\partial p^*}{\partial p} \frac{\partial q^*}{\partial q} - \frac{\partial p^*}{\partial q} \frac{\partial q^*}{\partial p} \right) dp \wedge dq. \tag{0.96}$$

From (0.95) and (0.96), we conclude that

$$dp^* \wedge dq^* = dp \wedge dq.$$

The geometric interpretation of symplecticity of the transformation $\psi(t, t_0)$ is that the oriented area of the projection of a parallelogram onto the (p, q)-plane is preserved by $\psi(t, t_0)$.

In the multidimensional case, we have the conservation of the two-form

$$\omega^2 = \sum_{i=1}^{n} (dp)_i^* \wedge (dq^i)^* = \sum_{i=1}^{n} dp_i \wedge dq^i;$$

i.e., the two-form ω^2 is invariant.

The geometric interpretation of the symplecticity of the mapping $\psi(t, t_0)$ in the multidimensional case is that the sum of oriented areas of its projections onto the (p_i, q^i)-planes is conserved throughout the evolution (see [6, 98, 128, 132] for details). This property is significant for long-time behavior of solutions of Hamiltonian systems. Moreover, all properties of Hamiltonian equations can be derived from the area preservation property.

The key theorem for Hamiltonian equations is the following.

THEOREM (Poincaré, 1899). *The flow $\psi(t, t_0)$ generated by Eqs. (0.92) with the Hamiltonian function H is symplectic.*

EXAMPLE (harmonic oscillator). We rewrite the ordinary differential equation

$$u'' + u = 0$$

as the Hamiltonian system

$$\dot{q} = p, \qquad \dot{p} = -q$$

with Hamiltonian

$$H(t, q, p) = \frac{1}{2}(p^2 + q^2).$$

The corresponding Hamiltonian equations have the general solution

$$p = A \sin t + B \cos t, \qquad q = B \sin t - A \cos t.$$

From the initial data

$$p_0 = A \sin t_0 + B \cos t_0, \qquad q_0 = B \sin t_0 - A \cos t_0$$

at $t = t_0$, we obtain the representation of the solution operator as

$$p = -q_0 \sin(t - t_0) + p_0 \cos(t - t_0), \qquad q = p_0 \sin(t - t_0) + q_0 \cos(t - t_0),$$

which demonstrates the local group property.

The corresponding one-forms are as follows:

$$dp = \cos(t - t_0)\, dp_0 - \sin(t - t_0)\, dq, \qquad dq = \sin(t - t_0)\, dp + \cos(t - t_0)dq.$$

The two-form demonstrates its invariance

$$dp \wedge dq = \left(\sin^2(t - t_0) + \cos^2(t - t_0)\right) dp_0 \wedge dq_0$$

along the solution of the harmonic oscillator.

Examples

Here we provide examples showing how to find first integrals by using symmetries.

EXAMPLE 0.10 (a scalar ODE). As the first example, we consider the following second-order ordinary differential equation

$$\ddot{u} = \frac{1}{u^3},$$

which admits the Lie algebra L_3 with basis operators

$$X_1 = \frac{\partial}{\partial t}, \quad X_2 = 2t\frac{\partial}{\partial t} + u\frac{\partial}{\partial u}, \quad X_3 = t^2\frac{\partial}{\partial t} + tu\frac{\partial}{\partial u}.$$

Hamiltonian framework. Let us transfer this example to the Hamiltonian framework. We change the variables by setting

$$q = u, \qquad p = \frac{\partial L}{\partial \dot{u}} = \dot{u}.$$

The corresponding Hamiltonian is

$$H(t, q, p) = \dot{u}\frac{\partial L}{\partial \dot{u}} - L = \frac{1}{2}\left(p^2 + \frac{1}{q^2}\right).$$

The Hamiltonian equations

$$\dot{q} - p, \qquad \dot{p} = \frac{1}{q^3} \tag{0.97}$$

admit the symmetries

$$X_1 = \frac{\partial}{\partial t}, \quad X_2 = 2t\frac{\partial}{\partial t} + q\frac{\partial}{\partial q} - p\frac{\partial}{\partial p}, \quad X_3 = t^2\frac{\partial}{\partial t} + tq\frac{\partial}{\partial q} + (q - tp)\frac{\partial}{\partial p}. \tag{0.98}$$

We check the invariance of H in accordance with Theorem 0.1 and find that condition (0.80) is satisfied for the operators X_1 and X_2. Then, using Theorem 0.5, we calculate the corresponding first integrals

$$I_1 = -H = -\frac{1}{2}\left(p^2 + \frac{1}{q^2}\right), \qquad I_2 = pq - t\left(p^2 + \frac{1}{q^2}\right). \tag{0.99}$$

For the third symmetry operator, the Hamiltonian is divergence invariant with $V_3 = q^2/2$. In accordance with the Corollary of Theorem 0.5, this gives the conserved variable

$$I_3 = -\frac{1}{2}\left(\frac{t^2}{q^2} + (q - tp)^2\right). \tag{0.100}$$

Note that no integration is needed. As we indicated before, only two first integrals of a second-order ordinary differential equation can be functionally independent. Putting $I_1 = A/2$ and $I_2 = B$, we find the solution as

$$Aq^2 + (At - B)^2 + 1 = 0, \qquad p = \frac{B - At}{q}.$$

Evolution vector field approach. Consider the same example for evolution vector fields in the Hamiltonian form (0.78). We rewrite the operators (0.98) in the evolution form

$$\overline{X}_1 = -\dot{q}\frac{\partial}{\partial q} - \dot{p}\frac{\partial}{\partial p}, \qquad \overline{X}_2 = (q - 2t\dot{q})\frac{\partial}{\partial q} - (p + 2t\dot{p})\frac{\partial}{\partial p},$$

$$\overline{X}_3 = (tq - t^2\dot{q})\frac{\partial}{\partial q} + (q - tp - t^2\dot{p})\frac{\partial}{\partial p}. \tag{0.101}$$

The transformations corresponding to the symmetries (0.101) are not point transformations. Therefore, the Hamiltonian equations (0.97) are invariant with respect to (0.101) if they are considered together with their differential consequences. On the solutions of the canonical equations (0.97), these operators are equivalent to the set

$$\widetilde{X}_1 = -p\frac{\partial}{\partial q} - \frac{1}{q^3}\frac{\partial}{\partial p}, \qquad \widetilde{X}_2 = (q - 2tp)\frac{\partial}{\partial q} - \left(p + \frac{2t}{q^3}\right)\frac{\partial}{\partial p},$$

$$\widetilde{X}_3 = \left(tq - t^2p\right)\frac{\partial}{\partial q} + \left(q - tp - \frac{t^2}{q^3}\right)\frac{\partial}{\partial p}.$$

One should integrate Eqs. (0.77) for each operator to find the first integrals. Integration provides the three first integrals given in (0.99) and (0.100).

EXAMPLE 0.11 (Repulsive one-dimensional motion). As another example of an ordinary differential equation, we consider one-dimensional motion in the Coulomb field (the case of a repulsive force):

$$\ddot{u} = \frac{1}{u^2}, \tag{0.102}$$

which admits the Lie algebra L_2 with basis operators

$$X_1 = \frac{\partial}{\partial t}, \qquad X_2 = 3t\frac{\partial}{\partial t} + 2u\frac{\partial}{\partial u}.$$

We change the variables by the formula

$$q = u, \qquad p = \frac{\partial L}{\partial \dot{u}} = \dot{u}$$

and find the Hamiltonian function

$$H(t, q, p) = \dot{u}\frac{\partial L}{\partial \dot{u}} - L = \frac{p^2}{2} + \frac{1}{q}.$$

The Hamiltonian equations have the form

$$\dot{q} = p, \qquad \dot{p} = \frac{1}{q^2}. \tag{0.103}$$

We rewrite the symmetries in the canonical variables as the following algebra L_2:

$$X_1 = \frac{\partial}{\partial t}, \qquad X_2 = 3t\frac{\partial}{\partial t} + 2q\frac{\partial}{\partial q} - p\frac{\partial}{\partial p}.$$

The invariance of the Hamiltonian condition (0.80) holds only for the operator X_1. By applying Theorem 0.5, we calculate the corresponding first integral

$$I_1 = -\left(\frac{p^2}{2} + \frac{1}{q}\right).$$

An application of the operator X_2 to the Hamiltonian action gives

$$\zeta\dot{q} + pD(\eta) - X(H) - HD(\xi) = p\dot{q} - \left(\frac{p^2}{2} + \frac{1}{q}\right) \neq 0.$$

Meanwhile, by Theorem 0.9 we have

$$\left[\frac{\delta}{\delta p}(\zeta\dot{q} + pD(\eta) - X(H) - HD(\xi))\right]_{(0.103)} = \left[\frac{\delta}{\delta p}\left(p\dot{q} - \left(\frac{p^2}{2} + \frac{1}{q}\right)\right)\right]_{(0.103)} = 0.$$

We shall show that there exists a second integral of nonlocal character.

It was shown in [40] that Eq. (0.102) can be linearized by a contact transformation. For Eqs. (0.103), this transformation is as follows:

$$p(t) = P(s), \qquad Q^2(s) = \frac{2}{q(t)}, \qquad dt = -\frac{4}{Q^3}ds. \qquad (0.104)$$

The new Hamiltonian

$$H(s, Q, P) = \frac{1}{2}(P^2 + Q^2)$$

corresponds to the linear equations

$$\frac{dQ}{ds} = P, \qquad \frac{dP}{ds} = -Q,$$

which describe the one-dimensional harmonic oscillator. These equations have two first integrals

$$\tilde{I}_1 = \frac{P^2 + Q^2}{2}, \qquad \tilde{I}_2 = \arctan\frac{P}{Q} + s,$$

which let us write down the general solution

$$Q = A\sin s + B\cos s, \qquad P = A\cos s - B\sin s, \qquad A, B = \text{const.}$$

By applying the transformation (0.104) to the integral I_2, one obtains the nonlocal integral

$$I_2^* = \arctan\frac{p\sqrt{q}}{\sqrt{2}} - \frac{1}{\sqrt{2}}\int_{t_0}^t \frac{dt}{q^{3/2}}$$

of Eqs. (0.103).

EXAMPLE 0.12 (Three-dimensional Kepler motion). The three-dimensional Kepler motion of a body in Newton's gravitational field is described by the equations

$$\frac{d\mathbf{q}}{dt} = \mathbf{p}, \quad \frac{d\mathbf{p}}{dt} = -\frac{K^2}{r^3}\mathbf{q}, \quad r = |\mathbf{q}|, \tag{0.105}$$

where $\mathbf{q} = (q_1, q_2, q_3)$, $\mathbf{p} = (p_1, p_2, p_3)$, and $K = \text{const}$, with the initial data

$$\mathbf{q}(0) = \mathbf{q}_0, \quad \mathbf{p}(0) = \mathbf{p}_0.$$

These equations are Hamiltonian. They are defined by the Hamiltonian function

$$H(\mathbf{q}, \mathbf{p}) = \frac{1}{2}|\mathbf{p}|^2 - \frac{K^2}{r}. \tag{0.106}$$

Among symmetries admitted by the equations (0.105) there are

$$X_0 = \frac{\partial}{\partial t}, \quad X_1 = 3t\frac{\partial}{\partial t} + 2q_i\frac{\partial}{\partial q_i} - p_i\frac{\partial}{\partial p_i},$$

$$X_{ij} = -q_j\frac{\partial}{\partial q_i} + q_i\frac{\partial}{\partial q_j} - p_j\frac{\partial}{\partial p_i} + p_i\frac{\partial}{\partial p_j}, \quad i \neq j,$$

$$Y_l = (2q_lp_k - q_kp_l - (\mathbf{q}, \mathbf{p})\delta_{lk})\frac{\partial}{\partial q_k}$$

$$+ \left(p_lp_k - (\mathbf{p}, \mathbf{p})\delta_{lk} - \frac{K^2}{r^3}(q_lq_k - (\mathbf{q}, \mathbf{q})\delta_{lk})\right)\frac{\partial}{\partial p_k}, \quad l = 1, 2, 3,$$

where $(\mathbf{f}, \mathbf{g}) = \mathbf{f}^T\mathbf{g}$ is the inner product of vectors.

The Hamiltonian function (0.106) is invariant with respect to the symmetries X_0 and X_{ij}. Theorem 0.5 permits one to find the first integral for the symmetry X_0,

$$I_1 = -H,$$

which represents the conservation of energy in Kepler motion. For the symmetries X_{ij}, we obtain the first integrals

$$I_{ij} = q_ip_j - q_jp_i, \quad i \neq j,$$

which are components of the angular momentum

$$\mathbf{L}(\mathbf{q}, \mathbf{p}) = \mathbf{q} \times \mathbf{p}.$$

Conservation of the angular momentum shows that the orbit of motion of a body lies in a fixed plane perpendicular to the constant vector $\mathbf{L}$. It also follows that in this plane the position vector $\mathbf{q}$ sweeps out equal areas in equal time intervals, so that the sectorial velocity is constant [6]. Therefore, Kepler's second law can be considered as a geometric restatement of the conservation of angular momentum.

The scaling symmetry X_1 is not a Noether symmetry (neither variational, nor divergence symmetry) and does not lead to a conserved variable.

For each of the symmetries Y_l, the Hamiltonian is divergence invariant with the functions

$$V_l = q_l \left((\mathbf{p}, \mathbf{p}) + \frac{K^2}{r} \right) - p_l(\mathbf{q}, \mathbf{p}), \qquad l = 1, 2, 3.$$

Hence the operators Y_l yield the first integrals

$$I_l = q_l \left((\mathbf{p}, \mathbf{p}) - \frac{K^2}{r} \right) - p_l(\mathbf{q}, \mathbf{p}), \qquad l = 1, 2, 3,$$

which are the components of the Runge–Lenz vector

$$\mathbf{A}(\mathbf{q}, \mathbf{p}) = \mathbf{p} \times \mathbf{L} - \frac{K^2}{r}\mathbf{q} = \mathbf{q} \left(H(\mathbf{q}, \mathbf{p}) + \frac{1}{2}|\mathbf{p}|^2 \right) - \mathbf{p}(\mathbf{q}, \mathbf{p}). \qquad (0.107)$$

Physically, the vector $\mathbf{A}$ points along the major axis of the conic section determined by the body orbit. Its magnitude determines the eccentricity [134].

Note that not all first integrals are independent. There are two relations between them given by the equations

$$\mathbf{A}^2 - 2HL^2 = K^4, \qquad (\mathbf{A}, \mathbf{L}) = 0.$$

The two-dimensional Kepler motion can be treated in a similar way. Note that the symmetries and first integrals of the two-dimensional Kepler motion can be obtained by restricting the symmetries and first integrals of the three-dimensional Kepler motion to the space (t, q_1, q_2, p_1, p_2). As the conserved quantities of the two-dimensional Kepler motion, one obtains the energy

$$H(\mathbf{q}, \mathbf{p}) = \frac{1}{2}|\mathbf{p}|^2 - \frac{K^2}{r}, \qquad r = |\mathbf{q}|, \qquad \mathbf{q} = (q_1, q_2), \quad \mathbf{p} = (p_1, p_2),$$

one component

$$L_3 = q_1 p_2 - q_2 p_1$$

of the angular momentum, and two components

$$A_1 = q_1 \left(H(\mathbf{q}, \mathbf{p}) + \frac{1}{2}|\mathbf{p}|^2 \right) - p_1(\mathbf{q}, \mathbf{p}).$$

$$A_2 = q_2 \left(H(\mathbf{q}, \mathbf{p}) + \frac{1}{2}|\mathbf{p}|^2 \right) - p_2(\mathbf{q}, \mathbf{p})$$

of the Runge–Lenz vector. There is one relation between these conserved variables, namely,

$$A_1^2 + A_2^2 - 2HL_3^2 = K^4.$$

Further restriction to the one-dimensional Kepler motion leaves only one first integral, which is the Hamiltonian function.

0.2. Preliminaries: Heuristic Approach in Examples

Let us at once try to study the invariance of some difference equations using only basic knowledge about Lie groups of point transformations acting on a space of continuous variables.

EXAMPLE 0.13. As a first example, consider the ordinary differential equation

$$u'' = \frac{1}{u^3}. \tag{0.108}$$

Let us try to preserve its symmetries

$$X_1 = \frac{\partial}{\partial x}, \qquad X_2 = 2x\frac{\partial}{\partial x} + u\frac{\partial}{\partial u}, \qquad X_3 = x^2\frac{\partial}{\partial x} + xu\frac{\partial}{\partial u}$$

in a difference model of this equation.

We introduce the one-dimensional difference mesh

$$x^{i+1} = x^i + h^i, \qquad i = 0, \pm 1, \pm 2, \ldots, \tag{0.109}$$

where the mesh steps h^i are assumed to be given constants. To approximate a second-order equation at some point x, we need at least three mesh points. For the simplest finite-difference approximation to Eq. (0.108) we take the difference equation

$$\left(\frac{u_{i+1} - u_i}{h^{i+1}} - \frac{u_i - u_{i-1}}{h^i}\right)\frac{2}{h^i + h^{i+1}} = \frac{1}{u_i^3} \tag{0.110}$$

on the mesh (0.109). If all steps are equal (a uniform mesh), then the difference scheme has the extremely simple form

$$\frac{u_{i+1} - 2u_i + u_{i-1}}{h^2} = \frac{1}{u_i^3}.$$

The integer index i in Eq. (0.110) ranges over a finite or infinite interval; however, we shall consider this equation locally, in a neighborhood of some point (x, u), assuming that Eq. (0.110) has precisely the same form at all other points. (Such schemes are said to be homogeneous [122].) Accordingly, we rewrite Eq. (0.110) in the index-free form

$$\left(\frac{u^+ - u}{h^+} - \frac{u - u^-}{h^-}\right)\frac{2}{h^+ + h^-} = \frac{1}{u^3}, \tag{0.111}$$

where $u = u(x)$, $u^+ = u(x+h^+)$, $u^- = u(x-h^-)$, $h^+ = x^+ - x$, and $h^- = x - x^-$.

Now consider how group transformations act on the scheme (0.111), (0.109). The translations corresponding to X_1 act on each mesh point (because there is an orbit of the group through each point (x, u)) by the formulas

$$x^* = x + a, \qquad (x^+)^* = x^+ + a, \qquad (x^-)^* = x^- + a, \ldots$$

without affecting the mesh steps and Eq. (0.111). Thus, the scheme (0.111), (0.109) admits translations regardless of whether the mesh is uniform or not.

The dilations

$$x^* = xe^{2a}, \qquad (x^+)^* = x^+ e^{2a}, \qquad (x^-)^* = x^- e^{2a},$$
$$u^* = ue^a, \qquad (u^+)^* = u^+ e^a, \qquad (u^-)^* = u^- e^a$$

corresponding to X_2 change the mesh according to the formulas

$$(h^+)^* = h^+ e^{2a}, \qquad (h^-)^* = h^- e^{2a},$$

and so a mesh with fixed steps is not dilation invariant.

Now let us specify the mesh differently; we no longer try to fix the mesh steps (this is usually not important in computations) but preserve the mesh geometric structure. For example, let us force the mesh to be uniform by imposing the equation

$$h^+ = h^-. \tag{0.112}$$

It is easily seen that such a mesh is necessarily dilation invariant, for Eq. (0.112) remains the same, $(h^+)^* = (h^-)^*$, after the transformation.

For the mesh equation we could also take the difference equation

$$h^+ = kh^-, \qquad k = \text{const}, \tag{0.113}$$

specifying a nonuniform mesh whose steps correspond to a geometric progression. Let us verify the invariance of this mesh using the operator X_2 prolonged to all variables of the difference stencil. We obtain additional coordinates for h^+ and h^- by applying the operator to the relations $h^+ = x^+ - x$ and $h^- = x - x^-$:

$$\widetilde{X}_2 = 2x\frac{\partial}{\partial x} + u\frac{\partial}{\partial u} + u^+\frac{\partial}{\partial u^+} + u^-\frac{\partial}{\partial u^-} + 2h^+\frac{\partial}{\partial h^+} + 2h^-\frac{\partial}{\partial h^-}. \tag{0.114}$$

The operator (0.114) leaves invariant the mesh (0.112) (or (0.113). One can readily verify that Eq. (0.111) is invariant under the operator (0.114) as well. Thus, dilations are admitted by the difference model (0.111), (0.113).

Now consider how the transformations corresponding to X_3 act on the scheme (0.111), (0.113). We verify its invariance using the operator X_3 prolonged to the variables of the difference stencil,

$$\widetilde{X}_3 = x^2\frac{\partial}{\partial x} + (x^+)^2\frac{\partial}{\partial x^+} + (x^-)^2\frac{\partial}{\partial x^-} + xu\frac{\partial}{\partial u} + x^+u^+\frac{\partial}{\partial u^+}$$
$$+ x^-u^-\frac{\partial}{\partial u^-} + (2x + h^+)h^+\frac{\partial}{\partial h^+} + (2x - h^-)h^-\frac{\partial}{\partial h^-}.$$

An application of the operator $\widetilde{X}_3$ to Eq. (0.113) gives the relation $h^+(2x + h^+) = kh^-(2x - h^-)$, and the subsequent substitution of (0.113) into this relation yields

$h^+ + h^- = 0$. The application of finite transformations corresponding to $\widetilde{X}_3$ confirms the noninvariance of the mesh (0.113) for any k as well. The action of $\widetilde{X}_3$ on Eq. (0.111) yields the relation

$$\frac{x^+ u^+ - xu}{h^+} - \frac{xu - x^- u^-}{h^-} - \frac{u^+ - u}{h^+}(2x + h^+) + \frac{u - u^-}{h^-}(2x - h^-)$$
$$= \frac{h^+(x + h^+) + h^-(x - h^-)}{2u^2},$$

which cannot hold on the mesh (0.113) for any k.

Thus, only two out of the three symmetries present in the original differential model can be preserved on the uniform mesh and on its generalization (0.113). The conclusion suggests itself that, to preserve the symmetry corresponding to the operator $\widetilde{X}_3$, one needs some special mesh, but it is not clear how one can guess it. Later on in the book, we construct a completely invariant difference model of Eq. (0.108) by using a regular algorithm.

EXAMPLE 0.14. As another example, we consider a partial differential equation, namely, the linear heat equation

$$u_t = u_{xx}. \tag{0.115}$$

This equation is known to admit a six-parameter point transformation group and, in addition, an infinite-parameter group equivalent to the linearity of Eq. (0.115) (e.g., see [111]). This transformation group is completely described by the following Lie algebra of infinitesimal operators:

$$X_1 = \frac{\partial}{\partial t}, \qquad X_2 = \frac{\partial}{\partial x}, \qquad X_3 = 2t\frac{\partial}{\partial t} + x\frac{\partial}{\partial x}, \qquad X_4 = u\frac{\partial}{\partial u},$$

$$X_5 = 2t\frac{\partial}{\partial x} - xu\frac{\partial}{\partial u}, \qquad X_6 = 4t^2\frac{\partial}{\partial t} + 4tx\frac{\partial}{\partial x} - (x^2 + 2t)\frac{\partial}{\partial u}, \tag{0.116}$$

$$X_* = \alpha(x,t)\frac{\partial}{\partial u},$$

where α is an arbitrary solution of (0.115).

Let us consider some simplest difference model approximating Eq. (0.115) and see how the transformations determined by the operators (0.116) act on this model. For this simplest difference equation we take the explicit difference scheme (at the point $x = x^i$, $t = t^j$)

$$\frac{u_i^{j+1} - u_i^j}{\tau} = \frac{u_{i+1}^j - 2u_i^j + u_{i-1}^j}{h^2} \tag{0.117}$$

on an orthogonal difference mesh uniform in the t- and x-directions with mesh steps τ and h, respectively.

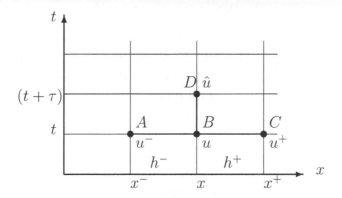

Figure 0.1: The difference stencil for Eq. (0.118)

Note an important difference between the discrete version (0.117) and the continuous version (0.115). It is meaningless to specify any function $u_i^j = \phi_i^j(t^j, x^i)$ without describing the discrete set $\{(t^j, x^i)\}$ on which u_i^j is defined. The same is true of the difference equation (0.117). Thus, we should supplement Eq. (0.117) by something completely characterizing the difference mesh. Take an orthogonal mesh (we simply write down the words

Orthogonal Difference Mesh

for now) uniform in x and t,

$$h^+ = h^-, \quad \tau^+ = \tau^-.$$

The integer indices i and j in Eq. (0.117) range over finite or infinite intervals, but we shall consider this equation in a neighborhood of some point (t, x, u) and rewrite Eq. (0.117) in the index-free form

$$\frac{\hat{u} - u}{\tau} = \frac{u^+ - 2u + u^-}{h^2}, \tag{0.118}$$

where the symbol $\hat{u}$ stands for passage to the "upper" layer with respect to t, while u^+ and u^- stand for the shifts to the right and left, respectively, with respect to x.

The difference stencil, i.e., the set of points of the (t, x)-plane used to describe Eq. (0.118), is shown in Fig. 0.1.

The set $(x, x+h, x-h, t, t+\tau)$ of points of the difference stencil is a "point" of the difference space on the (t, x)-plane where we consider Eq. (0.118). In contrast to the continuous case, a "point" of the difference space has quite a determined geometric structure. To describe this structure, let us denote the points of the difference stencil by A, B, C, and D. Then the mesh structure can be represented as follows:

$$(\overrightarrow{AB})(\overrightarrow{BD}) = (\overrightarrow{BC})(\overrightarrow{BD}) = 0,$$
$$h^+ = h^- = h, \qquad \tau^+ = \tau^-. \tag{0.119}$$

Now we can consider the action of the group G with infinitesimal operators (0.116) on Eq. (0.118) and relations (0.119). Note that the transformations (0.116) are continuous; i.e., the coordinates of all points of the (t, x, u)-space (in particular, of the points corresponding to the difference stencil in Fig. 0.1) are varied in a *continuous* manner (rather than discretely, where, say, some mesh points would be taken to other mesh points). By analogy with the case of differential equations, we say that Eqs. (0.119), (0.118) are invariant with respect to the corresponding transformations if these equations have the same form in the transformed variables. Consider the action of $X_1, \ldots, X_6$ on system (0.119), (0.118).

The translations $X_1 = \partial/\partial t$ and $X_2 = \partial/\partial x$ mean the corresponding shifts of the independent variables,

(i) $\quad t^* = t + a, \quad x^* = x, \qquad u^* = u;$

(ii) $\quad t^* = t, \qquad x^* = x + a, \quad u^* = u,$

where a is the group parameter.

It is fully obvious that the structure of the mesh (0.119), as well as Eq. (0.118), remains unchanged; i.e., Eqs. (0.119), (0.118) are invariant with respect to X_1 and X_2.

The dilation group with infinitesimal generator $X_3 = 2t\partial/\partial t + x\partial/\partial x$ is determined by the transformations

$$t^* = e^{2a}t, \qquad x^* = e^a x, \qquad u^* = u.$$

To find how the mesh steps are transformed, consider the orbits passing through the points of the difference stencil. Obviously, the operator X_3 can be prolonged to the coordinates of the points A, C, and D,

$$X_3 = 2t\frac{\partial}{\partial t} + x\frac{\partial}{\partial x} + 2\hat{t}\frac{\partial}{\partial \hat{t}} + 2\check{t}\frac{\partial}{\partial \check{t}} + x^+\frac{\partial}{\partial x^+} + x^-\frac{\partial}{\partial x^-},$$

where $\hat{t} = t + \tau^+$ and $\check{t} = t - \tau^-$. This means that the "additional" coordinates $\hat{t}$, $\check{t}$, x^+, and x^- are transformed in the same way as the "main" coordinates t and x,

$$\hat{t}^* = e^{2a}\hat{t}, \quad \check{t}^* = e^{2a}\check{t}, \qquad (x^+)^* = e^a x^+, \quad (x^-)^* = e^a x^-.$$

The transformation law for the mesh steps follows,

$$(\tau^+)^* = \hat{t}^* - t^* = e^{2a}(\hat{t} - t) = e^{2a}\tau^+, \qquad (\tau^-)^* = e^{2a}\tau^-,$$
$$(h^+)^* = (x^+)^* - x = e^a(x^+ - x) = e^a h^+, \qquad (h^-)^* = e^a h^-,$$

and the operation $\frac{\partial}{\partial a}\big|_{a=0}$ gives the additional coordinates of the operator X_3 extended to h^+, h^-, τ^+, and τ^-,

$$X_3 = 2t\frac{\partial}{\partial t} + x\frac{\partial}{\partial x} + \cdots + 2\tau^+\frac{\partial}{\partial \tau^+} + 2\tau^-\frac{\partial}{\partial \tau^-} + h^+\frac{\partial}{\partial h^+} + h^-\frac{\partial}{\partial h^-}.$$

From this, we see that the operator X_3, which dilates the (t, x)-plane, obviously dilates the difference mesh. The relations $h^+ = h^-$ and $\tau^+ = \tau^-$ are clearly invariant with respect to the prolonged operator X_3,

$$\left[X_3(h^+ - h^-)\right]\big|_{h^+=h^-} = 0, \qquad \left[X_3(\tau^+ - \tau^-)\right]\big|_{\tau^+=\tau^-} = 0.$$

The mesh orthogonality conditions are also preserved,

$$(\overrightarrow{A^*B^*})(\overrightarrow{B^*D^*}) = (\overrightarrow{B^*C^*})(\overrightarrow{B^*D^*}) = 0.$$

Thus, *the dilation group changes the mesh steps but does not affect the mesh orthogonality and uniformity.*

It is easily seen that the action of X_3 on the difference equation (0.118) does not change this equation as well,

$$\left[X_3\left(u_t - u_{x\bar{x}}\atop \tau \quad h\right)\right]_{u_t = u_{x\bar{x}} \atop \tau \quad h} = 0,$$

where we have introduced the shorthand notation

$$u_t \atop \tau = \frac{\hat{u} - u}{\tau}, \qquad u_{x\bar{x}} \atop h = \frac{u^+ - 2u + u^-}{h^2}$$

for the difference derivatives. Thus, Eqs. (0.119), (0.118) admit the dilations determined by the operator X_3.

The operator $X_4 = u\partial/\partial u$, which does not affect the independent variable t and x and dilates u (namely, $u^* = e^a u$), leaves unchanged relations (0.119) and is obviously admitted by Eq. (0.118),

$$\left[X_4\left(u_t - u_{x\bar{x}}\atop \tau \quad h\right)\right]_{u_t = u_{x\bar{x}} \atop \tau \quad h} = 0.$$

Now consider the subgroup generated by the operator X_5. The corresponding finite transformations have the form

$$t^* = t, \qquad x^* = x + 2ta, \qquad u^* = ue^{-xa-t^2a^2}.$$

Let us see what happens to the difference mesh under these transformations. They do not change t and hence do not affect the step τ. For each value of the parameter a, the mesh is shifted along x on each time layer t, the shift being proportional to t. The mesh structure after the transformations is shown in Fig. 0.2.

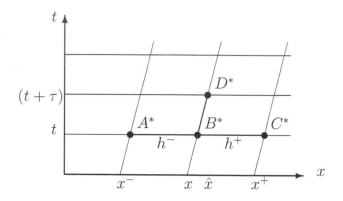

Figure 0.2: The mesh structure after the transformations generated by X_5

Clearly, the mesh uniformity in the x-direction is also preserved, $(h^+)^* = (x^+)^* - x^* = h^+$. However, the mesh is no longer orthogonal. Let us study the consequences. Equation (0.118) is changed by the transformations into the equation

$$\frac{\hat{u}e^{-xa-(t+\tau)^2 a^2} - ue^{-xa-t^2 a^2}}{\tau} = \frac{u^+ e^{-(x+h)a} - 2ue^{-xa} + u^- e^{-(x-h)a}}{h^2} e^{-t^2 a},$$

or

$$\frac{\hat{u}e^{-\tau(2t+\tau)a^2} - u}{\tau} = \frac{u^+ e^{-ha} - 2u + u^- e^{+ha}}{h^2}. \qquad (0.120)$$

Thus, the transformations generated by X_5 violate the mesh orthogonality and change Eq. (0.118). What does Eq. (0.120) on the transformed mesh approximate? First of all, note that the operator X_5, which violates the mesh orthogonality, does not change the meaning of the spatial difference derivatives (in particular, $u_{x\bar{x}}$) but distorts the geometric meaning of the time derivative $u_t = \frac{\hat{u}-u}{\tau}$, because the transformed expression $u_t^* = \frac{\hat{u}^*-u^*}{\tau}$ for the time derivative has a component not only in t but also in x. Now assume that Eq. (0.120) has a sufficiently smooth solution $u = u(x,t)$. By substituting the expansions $u^+ = u + hu_x + \frac{h^2}{2}u_{xx} + O(h^2)$, $u^- = u - hu_x + \frac{h^2}{2}u_{xx} + O(h^2)$, and $\hat{u} = u + u_t\tau + u_x 2\tau a + O(\tau)$ (here we have taken into account the spatial component) into Eq. (0.120), we obtain

$$u_t = u_{xx} - 4au_x + 2ta^2 u$$

modulo $O(\tau + h^2)$.

Thus, the difference equation (0.117) approximating Eq. (0.115) on the orthogonal mesh with accuracy $O(\tau + h^2)$ is transformed into the difference equation (0.120) on a "skew" mesh; Eq. (0.120) approximates a completely different differential equation, which, in addition, explicitly depends on the group parameter.

Thus, the operator X_5 is not admitted by the difference model (0.119), (0.118). A similar argument shows that the action of X_6 destroys the mesh orthogonality as well and also violates the mesh uniformity in the t-direction.

The operator $X_* = \alpha \partial/\partial u$ does not affect the variables (t, x) and adds an arbitrary solution of Eq. (0.115) to u,

$$u^* = u + \alpha(x, t)a.$$

This symmetry is equivalent to the fact that Eq. (0.115) is linear.

The difference equation (0.118) is linear as well, and so the substitution of the last transformation gives

$$\frac{\alpha(x, t + \tau) - \alpha(x, t)}{\tau} = \frac{\alpha(x + h, t) - 2\alpha(x, t) + \alpha(x - h, t)}{h^2},$$

where α is an arbitrary solution of the equation $\alpha_t = \alpha_{xx}$.

So far, it is not clear whether a solution of the differential equation (0.115) is an exact solution of the difference equation (0.118). However, we have clearly preserved linearity when passing to the difference model (0.119), (0.118), which is equivalent to the symmetry with the operator

$$\underset{h}{X}_* = A(t, x)\frac{\partial}{\partial u},$$

where $A(x, t)$ is an arbitrary solution of Eq. (0.118) on the mesh (0.119).

Thus, the difference model that we have considered — Eq. (0.118) on the uniform orthogonal mesh (0.119) —has preserved only four out of the six symmetries of the original equation (0.115) and also the linearity property. Note that the preserved symmetries form a Lie algebra that is a subalgebra of the original algebra.

We have chosen the simplest difference model, namely, an explicit scheme on an orthogonal mesh. However, it is clear that using an implicit scheme or combining it with an explicit one would not rectify the situation, for the transformations determined by X_5 and X_6 violate the mesh orthogonality, and this also irreparably destroys the structure of the difference equation. In the next chapter, we construct a difference model of Eq. (0.115) preserving all symmetries but defined on a nonorthogonal mesh depending on the solution.

Thus, *the geometric structure of the mesh* (and, naturally, the structure of the difference stencil, which is part of the mesh) *exerts substantial influence on the invariance of the difference model.* In the next chapter, we obtain necessary and sufficient conditions for the preservation of geometric characteristics of the difference mesh (stencil). Using these conditions, one can readily establish mesh classes suitable for a given symmetry without carrying out detailed constructions like those given above.

The preceding examples show that it is convenient to introduce new variables and notation. For example, the difference derivatives

$$\underset{h}{u_t} = \frac{\hat{u} - u}{\tau^+}, \qquad \underset{h}{u_x} = \frac{u^+ - u}{h^+}$$

(right time and space derivatives),

$$\underset{h}{u_{\bar{x}}} = \frac{u - u^-}{h^-}$$

(a left space derivative),

$$\underset{h}{u_{x\bar{x}}} = \frac{\underset{h}{u_x} - \underset{h}{u_{\bar{x}}}}{h} = \frac{u^+ - 2u + u^-}{h^2}$$

(the second space derivative), etc.

It is convenient to introduce the right shift operators $\underset{+\tau}{S}$ and $\underset{+h}{S}$ and the left shift operators $\underset{-\tau}{S}$ and $\underset{-h}{S}$. These operators act on (t, x) and translate these points to the right or left neighboring mesh points. The action of these shift operators on any functions of (t, x) is determined by their action on the arguments,

$$\underset{+\tau}{S} F(z) = F(\underset{+\tau}{S}(z)), \qquad \underset{-\tau}{S} F(z) = F(\underset{-\tau}{S}(z)),$$

$$\underset{+h}{S} F(z) = F(\underset{+h}{S}(z)), \qquad \underset{-h}{S} F(z) = F(\underset{-h}{S}(z)).$$

The operators

$$\underset{+h}{D} = \frac{\underset{+h}{S} - 1}{h}, \qquad \underset{-h}{D} = \frac{1 - \underset{-h}{S}}{h}, \qquad \underset{+\tau}{D} = \frac{\underset{+\tau}{S} - 1}{\tau}, \qquad \underset{-\tau}{D} = \frac{1 - \underset{-\tau}{S}}{\tau},$$

define right and left difference derivation, respectively; by using these operators, one can readily introduce the difference derivatives

$$\underset{h}{u_x} = \underset{+h}{D}(u), \qquad \underset{h}{u_{\bar{x}}} = \underset{-h}{D}(u),$$

$$\underset{h}{u_{x\bar{x}}} = \underset{h}{u_{\bar{x}x}} = \underset{-h+h}{D\,D}(u) = \underset{+h-h}{D\,D}(u),$$

etc. Note that the operators $\underset{+h}{D}$ and $\underset{-h}{D}$ on the uniform mesh commute.

In subsequent chapters, we show how one can naturally introduce difference derivatives starting from the "continuous" space and continuous differentiation.

Chapter 1

Finite Differences and Transformation Groups in Space of Discrete Variables

In this chapter, we find out how a local Lie transformation group acts on nonlocal objects such as discrete variables, finite-difference derivatives, lattice spacings, etc.

In contrast to differential operators, finite-difference operators are specified on a finite subset (a difference stencil) of the countable set of lattice points where the solution of the problem in question is to be sought. This *nonlocality* of operators (from the physical viewpoint, the presence of typical dimensional scales in the problem) results in specific properties of finite-difference operators, properties which are absent in the local differential model. In particular, we can mention right and left differentiations with the corresponding shifts, uniform and nonuniform lattices, and specific features of the difference Leibniz rule. As a result, there arises a specific *calculus of infinitesimal transformations of difference variables*.

The nonlocality of difference operators has the consequence that the transformation group can distort the proportions, orthogonality, and other geometric characteristics of the difference stencil. In turn, the violation of the lattice structural properties under transformations can distort the difference equations (say, by affecting the approximation order and the commutativity of difference derivatives); the lattice orthogonality distortion may result in a loss of geometric meaning of difference derivatives, etc. Hence a criterion for the invariance of difference models should necessarily include the invariance of the lattice (or the difference stencil viewed as an element of the lattice) on which the difference equations are written out. This gives rise to a peculiar *lattice geometry related to transformation groups*.

The next step is to find requirements that should be imposed on the transformation groups so as to preserve the *meaning of finite-difference derivatives* (i.e., ensure that the definition and geometric meaning of the finite-difference derivatives in the transformed variables will be the same). It turns out that this condition is equivalent to the preservation of infinite-order tangency. In the "continuous" space, such groups satisfy the prolongation formulas for point transformations, tangent transformations, and higher symmetries (or Lie–Bäcklund groups).

In this chapter, we also consider some structural properties of transformation groups in lattice spaces as well as difference integration and change-of-variables formulas needed for studying the group properties of finite-difference equations.

1.1. The Taylor Group and Introduction of Finite-Difference Derivatives

When trying to describe finite-difference objects (equations and lattices) in terms of differential operators, the main, almost insurmountable difficulty is the nonlocality of these objects. The nonlocality of objects needed in the description of difference equations, lattices, and functionals necessitates the use of infinite-dimensional spaces or spaces of sequences of functions and their derivatives.

1. In the space $\widetilde{Z}$ of formal series, consider the formal transformation group whose infinitesimal operator is the total derivative operator

$$D = \frac{\partial}{\partial x} + u_1 \frac{\partial}{\partial u} + u_2 \frac{\partial}{\partial u_1} + \cdots + u_{s+1} \frac{\partial}{\partial u_s} + \cdots . \tag{1.1}$$

For simplicity, we restrict ourselves to the case of one independent variable x and one dependent variable u for now.

According to the exponential representation, the transformations in this group are determined by the action of the operator $T_a \equiv e^{aD}$,

$$z^{i*} = e^{aD}(z^i) \equiv \sum_{s=0}^{\infty} \frac{a^s}{s!} D^{(s)}(z^i). \tag{1.2}$$

The point $z^* \in \widetilde{Z}$ has the coordinates

$$x^* = T_a(x) = x + a,$$

$$u^* = T_a(u) = \sum_{s=0}^{\infty} \frac{a^s}{s!} u_s,$$

$$u_1^* = T_a(u_1) = \sum_{s=0}^{\infty} \frac{a^s}{s!} u_{s+1}, \tag{1.3}$$

$$\dots\dots\dots\dots\dots$$

$$u_k^* = T_a(u_k) = \sum_{s=0}^{\infty} \frac{a^s}{s!} u_{k+s},$$

$$\dots\dots\dots\dots\dots$$

The transformations (1.3) considered on a smooth curve $u = u(x)$ are none other than the formal Taylor series expansions of the function $u = u(x)$ and its derivatives at the point $x + a$, and that is why the transformation group (1.3) with infinitesimal operator (1.1) was named the Taylor shift group, or simply the *Taylor group* [29].

In the theory of higher symmetries, or Lie–Bäcklund groups [73], of all the transformations comprising the group one distinguishes those which preserve the definition and geometric meaning of the derivatives $(u_1, u_2, \dots)$ in $\widetilde{Z}$, i.e., leave invariant the following infinite sequence (we still restrict ourselves to the one-dimensional case):

$$du = u_1\, dx,$$
$$du_1 = u_2\, dx,$$
$$\cdots\cdots\cdots$$
$$du_s = u_{s+1}\, dx,$$
$$\cdots\cdots\cdots$$

(1.4)

One can readily verify that the Taylor group leaves system (1.4) invariant and is a higher symmetry group.

Moreover, *the Taylor group is a nontrivial higher symmetry group*; i.e., it is not the prolongation to $\widetilde{Z}$ of a point or tangent transformation group. This follows from the fact that the sequence

$$\frac{dx^*}{da} = 1, \qquad x^*\big|_{a=0} = x,$$
$$\frac{du^*}{da} = u_1^*, \qquad u^*\big|_{a=0} = u,$$
$$\cdots\cdots\cdots\cdots\cdots$$
$$\frac{du_s^*}{da} = u_{s+1}^*, \qquad u_s^*\big|_{a=0} = u_s,$$
$$\cdots\cdots\cdots\cdots\cdots$$

of Lie equations determining the finite transformations in the Taylor group for a given group operator D has no solutions in the finite-dimensional part of $\widetilde{Z}$. In other words, the formal Taylor series expansions (1.3) form a one-parameter group only in the infinite-dimensional space $\widetilde{Z}$.

The Taylor group is a convenient tool in the study of properties of finite-difference objects.

2. Let us fix an arbitrary parameter value $a = h > 0$ and use the tangent field (1.1) of the Taylor group to form a pair of operators, which will be called the *right* and *left discrete shift operators*, respectively:

$$\underset{+h}{S} = e^{hD} \equiv \sum_{s=0}^{\infty} \frac{h^s}{s!} D^s, \qquad \underset{-h}{S} = e^{-hD} \equiv \sum_{s=0}^{\infty} \frac{(-h)^s}{s!} D^s,$$

(1.5)

where D is a derivation in $\widetilde{Z}$. The operators $\underset{+h}{S}$ and $\underset{-h}{S}$ commute with each other and with the operator $T_a = e^{aD}$, and moreover, $\underset{+h}{S}\underset{-h}{S} = \underset{-h}{S}\underset{+h}{S} = 1$. Furthermore,

$$\left(\underset{\pm h}{S}\right)^n = T_a\big|_{a=\pm nh}, \qquad n = 0, 1, 2, \dots.$$

(1.6)

Using $\underset{+h}{S}$ and $\underset{-h}{S}$, we form a pair of right and left discrete (finite-difference) differentiation operators by setting

$$
\begin{aligned}
\underset{+h}{D} &= \frac{1}{h}(\underset{+h}{S} - 1) \equiv \sum_{s=1}^{\infty} \frac{h^{s-1}}{s!} D^s, \\
\underset{-h}{D} &= \frac{1}{h}(1 - \underset{-h}{S}) \equiv \sum_{s=1}^{\infty} \frac{(-h)^{s-1}}{s!} D^s.
\end{aligned}
\tag{1.7}
$$

The operators $\underset{+h}{S}$, $\underset{-h}{S}$, $\underset{+h}{D}$, $\underset{-h}{D}$, and T_a pairwise commute, and

$$
\underset{+h}{D} = \underset{-h+h}{D\,S}, \qquad \underset{-h}{D} = \underset{+h-h}{D\,S}.
$$

Thus, the shift operators $\underset{\pm h}{S}$ and the discrete differentiation operators $\underset{\pm h}{D}$, which were introduced phenomenologically in the introduction, can be defined in $\widetilde{Z}$ as the power series (1.5) and (1.7) of the Taylor group operator (1.1).

The shift operators $\underset{\pm h}{S}$ and the differentiation operators $\underset{\pm h}{D}$ permit one to "discretize" the space $\widetilde{Z}$, i.e., introduce new variables (the difference derivatives and the lattice).

We denote the countable set $\{x_\alpha = \underset{+h}{S^\alpha}(x)\}$, $\alpha = 0, \pm 1, \pm 2, \ldots$, of values of the independent variable x by $\underset{h}{\omega}$ and call it a *uniform difference lattice* (or *mesh*).

Let us introduce formal power series of the following special form:

$$
\begin{aligned}
\underset{h}{u_1} &= \underset{+h}{D}(u), \\
\underset{h}{u_2} &= \underset{-h+h}{D\,D}(u), \\
\underset{h}{u_3} &= \underset{+h-h+h}{D\,D\,D}(u), \\
&\qquad \cdots\cdots\cdots\cdots
\end{aligned}
\tag{1.8}
$$

The variable $\underset{h}{u_s}$ will be called the *finite-difference* (or *discrete*, or *lattice*) *derivative of order s*. In the odd case, the formal series $\underset{h}{u_{2k+1}}$ will be called the *right difference derivative*. Using $\underset{h}{u_{2k}}$ and $\underset{h}{u_{2k+1}}$, one can introduce the *odd-order left difference derivatives*

$$
\underset{h}{u_{\overline{2k+1}}} = \underset{h}{u_{2k+1}} - h\underset{h}{u_{2k+2}}
$$

or the difference derivatives

$$
\underset{h}{u_{2k+1}^{(\sigma)}} = \sigma \underset{h}{u_{2k+1}} + (1-\sigma)\underset{h}{u_{\overline{2k+1}}} = \underset{h}{u_{2k+1}} + h(1-\sigma)\underset{h}{u_{2k+2}}, \qquad \sigma = \text{const},
$$

with weight σ, which are often used in the theory of finite-difference schemes.

We denote sequences $(u_1, u_2, u_3, \ldots)$ of formal series by $\underset{h}{Z}$ and the product of the spaces $\underset{h}{Z}$ and $\underset{h}{\widetilde{Z}}$ by $\underset{h}{\widetilde{Z}}$,

$$\underset{h}{\widetilde{Z}} = (x, u, u_1, u_2, \ldots ; \underset{h}{u_1}, \underset{h}{u_2}, \ldots).$$

If the series $\underset{h}{u_s}$ converges (in some norm), then it will be called a *continuous representation of the difference derivative* $\underset{h}{u_s}$.

Note that the formal series $\underset{h}{u_s}$ cannot be represented in exponential form, and so they do not form a group with parameter h and cannot be described in terms of a tangent field.

By definition, we extend the action of the discrete shift $\underset{\pm h}{S}$ to functions in A as follows:

$$\underset{\pm h}{S}(F(z)) = F(\underset{\pm h}{S}(z)).$$

This permits finding the difference derivatives of $F \in A$,

$$\underset{\pm h}{D}(F(z)) = \pm \frac{F(\underset{\pm h}{S}(z)) - F(z)}{h}.$$

3. Consider how the Taylor group is prolonged into $\underset{h}{Z}$. First, note that the Taylor group does not change the spacing h of the lattice $\underset{h}{\omega}$. Indeed, $h^* = x^*_{\alpha+1} - x^*_\alpha = h$, because $x^*_\alpha = x_\alpha + u$.

Having in mind the preservation of the lattice spacing, we define a transformation of the variables $\underset{h}{u_s}$ in $\underset{h}{\widetilde{Z}}$ by setting

$$\underset{h}{u^*_1} = \underset{+h}{D}(u^*) = u^*_1 + \frac{h}{2!}u^*_2 + \frac{h^2}{3!}u^*_3 + \cdots,$$

$$\underset{h}{u^*_2} = \underset{-h+h}{D\,D}(u^*) = u^*_2 + \frac{h^2}{12}u^*_4 + \cdots,$$

$$\cdots \cdots \cdots \cdots \cdots,$$

where the u^*_s are formal series of the form (1.3) whose transformations are determined by the tangent field of the Taylor group according to the formula

$$\zeta^s = \frac{\partial u^*_s}{\partial a}\bigg|_{a=0} = u_{s+1}, \qquad s = 1, 2, \ldots .$$

Starting from this, one can readily compute the additional coordinates of the operator for the variables $\underset{h}{u_s}$,

$$\underset{h}{\zeta^1} = \frac{\partial \underset{h}{u^*_1}}{\partial a}\bigg|_{a=0} = \zeta^1 + \frac{h}{2!}\zeta^2 + \cdots = u_2 + \frac{h}{2!}u_3 + \cdots = \underset{h}{D}(u_1),$$

$$\underset{h}{\zeta^2} = \frac{\partial \underset{h}{u^*_2}}{\partial a}\bigg|_{a=0} = \underset{h}{D}(u_2), \ldots, \qquad \underset{h}{\zeta^s} = \underset{h}{D}(u_s), \ldots .$$

(1.9)

Thus, the tangent field of the Taylor group prolonged into $\widetilde{Z}_h$ can be identified with the operator

$$\mathbf{D} = \sum_{i=1}^{\infty} D(z^i)\frac{\partial}{\partial z^i}, \tag{1.10}$$

where the z^i are the coordinates of the vector $(x, u, u_1, u_2, \dots, \underset{h}{u_1}, \underset{h}{u_2}, \dots)$. Note that the coordinates (1.9) are formal power series in h, and hence the series $\underset{h}{u_s}^* = \underset{h}{u_s}^*(h, a)$ are formal power series in *two symbols*, "group" series in a and "non-group" series in h.

4. Now consider the result of the action of the discrete shift operator $\underset{\pm h}{S}$ in the lattice space $\underset{h}{Z} = (x, u, \underset{h}{u_1}, \underset{h}{u_2}, \dots)$,

$$\underset{\pm h}{S}(x) = x \pm h,$$

$$\underset{+h}{S}(u) = u + h\sum_{s \geq 1}\frac{h^{s-1}}{s!}D^s(u) \equiv u + h\underset{h}{u_1}.$$

In a similar way, we obtain

$$\underset{-h}{S}(u) = u - h\underset{-h}{S}(\underset{h}{u_1}) = u - h\underset{h}{u_1} + h^2\underset{h}{u_2}.$$

By just the same procedure, we single out the formal series $\underset{h}{u_s}$ in the result of the action of $\underset{\pm h}{S}$ on $\underset{h}{u_k}$, $k = 1, 2, \dots$. As a result, we obtain a table showing how the discrete shift operator acts in the lattice space $\underset{h}{Z}$ (see Table 1.1).

From Table 1.1, we readily obtain a table showing how the discrete differentiation operator $\underset{\pm h}{D} = \pm h^{-1}(\underset{\pm h}{S} - 1)$ acts on the point $(x, u, \underset{h}{u_1}, \underset{h}{u_2}, \dots)$ (Table 1.2).

Remark 1.1. Under the action of the Taylor group in $\widetilde{Z}_h$, as a increases, the point $z = (x, u, u_1, u_2, \dots, \underset{h}{u_1}, \underset{h}{u_2}, \dots)$ draws a one-parameter curve (namely, the orbit of the point z). Since $(\underset{\pm h}{S})^n = T_a\big|_{a=\pm nh}$, we see that the orbit of the Taylor group is a "continuous shift" drawn through the "discrete shift" $(\underset{\pm h}{S})^n$. If the formal series in question converge, then one can speak of the geometric meaning of the lattice variables $\underset{h}{u_s}$. In particular, $\underset{h}{u_1}$ is the slope of the chord joining the points u and $\underset{+h}{S}(u)$ in the (x, u)-plane, onto which the orbit of the point z of the Taylor group is projected. Note that the action of the operators $\underset{\pm h}{S}$ does not form a group with parameter h in the space $\underset{h}{Z}$ of *difference* variables.

Indeed, consider, say, the following transformation of the variable u:

$$\underset{+h}{S}(u) = u + h\underset{h}{u_1}.$$

Table 1.1: Action of the discrete shift operator on the point $z = (x, u, \underset{h}{u_1}, \underset{h}{u_2}, \dots)$

$\underset{-h}{S}$ is the finite-difference left shift operator	$\underset{+h}{S}$ is the finite-difference right shift operator
$\underset{-h}{S}(x) = x - h$	$\underset{+h}{S}(x) = x + h$
$\underset{-h}{S}(u) = u - h\underset{h}{u_1} + h^2\underset{h}{u_2}$	$\underset{+h}{S}(u) = u + h\underset{h}{u_1}$
$\underset{-h}{S}(\underset{h}{u_1}) = \underset{h}{u_1} - h\underset{h}{u_2}$	$\underset{+h}{S}(\underset{h}{u_1}) = \underset{h}{u_1} + h\underset{h}{u_2} + h^2\underset{h}{u_3}$
$\underset{-h}{S}(\underset{h}{u_2}) = \underset{h}{u_2} - h\underset{h}{u_3} + h^2\underset{h}{u_4}$	$\underset{+h}{S}(\underset{h}{u_2}) = \underset{h}{u_2} + h\underset{h}{u_3}$
$\dots\dots\dots\dots\dots\dots\dots\dots\dots$	$\dots\dots\dots\dots\dots\dots\dots\dots\dots$
$\underset{-h}{S}(\underset{h}{u_{2k+1}}) = \underset{h}{u_{2k+1}} - h\underset{h}{u_{2k+2}}$	$\underset{+h}{S}(\underset{h}{u_{2k+1}}) = \underset{h}{u_{2k+1}} + h\underset{h}{u_{2k+2}} + h^2\underset{h}{u_{2k+3}}$
$\underset{-h}{S}(\underset{h}{u_{2k+2}}) = \underset{h}{u_{2k+2}} - h\underset{h}{u_{2k+3}} + h^2\underset{h}{u_{2k+4}}$	$\underset{+h}{S}(\underset{h}{u_{2k+2}}) = \underset{h}{u_{2k+2}} + h\underset{h}{u_{2k+3}}$
$\dots\dots\dots\dots\dots\dots\dots\dots\dots$	$\dots\dots\dots\dots\dots\dots\dots\dots\dots$

Table 1.2: Action of the discrete differentiation operator on the coordinates of the point $\underset{h}{z} = (x, u, \underset{h}{u_1}, \underset{h}{u_2}, \underset{h}{u_3}, \dots)$

$\underset{-h}{D}$ is the finite-difference left differentiation operator	$\underset{+h}{D}$ is the finite-difference right differentiation operator
$\underset{-h}{D}(x) = 1$	$\underset{+h}{D}(x) = 1$
$\underset{-h}{D}(u) = \underset{h}{u_1} - h\underset{h}{u_2}$	$\underset{+h}{D}(u) = \underset{h}{u_1}$
$\underset{-h}{D}(\underset{h}{u_1}) = \underset{h}{u_2}$	$\underset{+h}{D}(\underset{h}{u_1}) = \underset{h}{u_2} + h\underset{h}{u_3}$
$\underset{-h}{D}(\underset{h}{u_2}) = \underset{h}{u_3} - h\underset{h}{u_4}$	$\underset{+h}{D}(\underset{h}{u_2}) = \underset{h}{u_3}$
$\dots\dots\dots\dots\dots\dots\dots\dots$	$\dots\dots\dots\dots\dots\dots$
$\underset{-h}{D}(\underset{h}{u_{2k+1}}) = \underset{h}{u_{2k+2}}$	$\underset{+h}{D}(\underset{h}{u_{2k+1}}) = \underset{h}{u_{2k+2}}$
$\underset{-h}{D}(\underset{h}{u_{2k+2}}) = \underset{h}{u_{2k+3}} - h\underset{h}{u_{2k+4}}$	$\underset{+h}{D}(\underset{h}{u_{2k+2}}) = \underset{h}{u_{2k+3}}$
$\dots\dots\dots\dots\dots\dots\dots\dots$	$\dots\dots\dots\dots\dots$

The composition of such transformations has the form

$$\underset{+h}{S^2}(u) = \underset{+h}{S}(u + h\underset{h}{u_1}) = u + 2h\underset{h}{u_1} + h\underset{h}{u_2} + h^2\underset{h}{u_3} \neq u + 2h\underset{h}{u_1}.$$

However, the point $\underset{+h}{S^2}(u)$ can be reached with the help of the Taylor group with value $a = 2h$ of the group parameter,

$$\underset{+h}{S^2}(u) = T_a(u)\big|_{a=2h} = e^{2hD}(u).$$

5. *Remarks on the multidimensional case.* Let Z be the space of sequences $(x, u, u_1, u_2, \dots)$, where $x = \{x^i; i = 1, 2, \dots, n\}$, $u = \{u^k; k = 1, 2, \dots, m\}$, $u_1 = \{u_i^k\}$ is the set of mn first partial derivatives, $u_2 = \{u_{ij}^k\}$ is the set of second partial derivatives, etc.

The prolongation formulas obtained earlier can readily be generalized to the case of several dependent variables u^k; to this end, it suffices to treat the latter as the components of a vector u. However, one encounters essential changes when passing to the case of several variables x^i.

To avoid awkward formulas, we restrict ourselves to the case $n = 2$; i.e., $x = (x^1, x^2)$. The superscript k on u^k will be omitted.

We consider two types of differentiations,

$$\begin{aligned}
D_1 &= \frac{\partial}{\partial x^1} + u_1 \frac{\partial}{\partial u} + u_{11} \frac{\partial}{\partial u_1} + u_{21} \frac{\partial}{\partial u_2} + \cdots, \\
D_2 &= \frac{\partial}{\partial x^2} + u_2 \frac{\partial}{\partial u} + u_{12} \frac{\partial}{\partial u_1} + u_{22} \frac{\partial}{\partial u_2} + \cdots,
\end{aligned} \qquad (1.11)$$

where

$$u_1 = \frac{\partial u}{\partial x^1}, \qquad u_{11} = \frac{\partial^2 u}{\partial (x^1)^2}, \qquad u_{21} = \frac{\partial^2 u}{\partial x^2 \partial x^1}, \qquad \cdots$$

and summation over the omitted superscript k in (1.11) is assumed.

The operators D_1 and D_2 generate two commuting Taylor groups, whose finite transformations are determined by the action of the operators $T_a^1 = e^{aD_1}$ and $T_a^2 = e^{aD_2}$. Let us fix two arbitrary parameter values $h_1, h_2 > 0$ and form two kinds of discrete shift operators,

$$\underset{\pm h}{S}_1 = e^{\pm h_1 D_1} \equiv \sum_{s \geq 0} \frac{(\pm h_1)^s}{s!} D_1^s, \qquad \underset{\pm h}{S}_2 = e^{\pm h_2 D_2} \equiv \sum_{s \geq 0} \frac{(\pm h_2)^s}{s!} D_2^s. \qquad (1.12)$$

Accordingly, we have two pairs of discrete differentiation operators,

$$\underset{h}{D}_i = \pm \frac{1}{h_i} (\underset{\pm h}{S}_i - 1), \qquad i = 1, 2.$$

The set

$$\left\{ \underset{\pm h_1}{S}{}^\alpha (x^1), \underset{\pm h_2}{S}{}^\beta (x^2) \right\}, \qquad \alpha, \beta = 0, 1, 2, \dots,$$

of points in the (x^1, x^2)-plane will be called a uniform orthogonal difference mesh and denoted by $\underset{h}{\omega}$. The operators $\underset{h}{S}_i$ and $\underset{\pm h}{D}_i$ in any combination commute on the uniform mesh $\underset{h}{\omega}$.

By analogy with the one-dimensional case, we introduce the difference derivatives $\underset{h}{u}_1 = \underset{+h}{D}_1(u)$, $\underset{h}{u}_{1\bar{1}} = \underset{-h}{D}_1 \underset{+h}{D}_1(u)$, $\underset{h}{u}_{1\bar{2}} = \underset{-h}{D}_2 \underset{+h}{D}_1(u)$, $\underset{h}{u}_2 = \underset{+h}{D}_2(u)$, $\underset{h}{u}_{12} =$

$\underset{+h}{D_2}\underset{+h}{D_1}(u)$, etc. The analogs of Tables 1.1 and 1.2 of discrete shifts and differentiations can be formed accordingly.

Thus, the Taylor group permits naturally introducing the discrete shift operators $\underset{\pm h}{S}$ and the discrete differentiation operators $\underset{\pm h}{D}$. In turn, these operators allow one to consider new variables—difference derivatives, introduced as formal power series of a special form. Needless to say, there are various ways to introduce difference derivatives. For example, independently from the continuous variables, one can introduce difference derivatives $\underset{h}{u_s}$ and define the action of $\underset{\pm h}{S}$ and $\underset{\pm h}{D}$ on these derivatives by using Tables 1.1 and 1.2.

Another approach is to introduce the difference derivatives on the manifold $u = \phi(x)$. In this case, Tables 1.1 and 1.2 and the difference Leibniz rule can be obtained by a method usual in the theory of difference schemes (e.g., see [122]). Note that the method that we used in this section to introduce the variables $\underset{h}{u_s}$ is independent of a specific manifold.

The Taylor group considered above is unique in a sense. It is the simplest higher symmetry group, and the Taylor group is completely sufficient to extend the action of the point transformation group to the case of finite-difference variables. The algebra of operators corresponding to the Taylor group forms an ideal in the higher symmetry algebra.

A more general remark is also possible. As is known, the idea that the Lie transformation group is local consists in that the transformation superposition (and inversion) is possible only for elements sufficiently close to the unit element, i.e., for sufficiently small values of the group parameter a. Actually, the same idea also underlies the theory of functions when constructing analytic continuations. These typical ideas of two different fields in mathematics have a quite specific intersection: the Taylor group. The Taylor group considered on a manifold is the Taylor series used to construct the analytic continuation.

In what follows, we show that the Taylor group can be represented directly in the space of discrete variables; in this case, the transformation group isomorphic to it operates with the so-called Newton series.

1.2. Difference Analog of the Leibniz Rule and Discrete Differentiation Formulas

In the preceding section, the action of the discrete shift operator $\underset{\pm h}{S}$ was extended (by definition) to functions in A as follows: $\underset{\pm h}{S}(F(z)) = F(\underset{\pm h}{S}(z))$. This permits finding the difference derivatives of $F \in A$ in the form

$$\underset{\pm h}{D}(F(z)) = \pm \frac{F(\underset{\pm h}{S}(z)) - F(z)}{h}. \tag{1.13}$$

Starting from this definition, it is easy to introduce the *discrete* (difference) *Leibniz rule* for the operators of right and left discrete differentiation:

$$\boxed{\begin{aligned}
\underset{+h}{D}(FG) &= \underset{+h}{D}(F)G + F\underset{+h}{D}(G) + h\underset{+h}{D}(F)\underset{+h}{D}(G), \\
\underset{-h}{D}(FG) &= \underset{-h}{D}(F)G + F\underset{-h}{D}(G) - h\underset{-h}{D}(F)\underset{-h}{D}(G),
\end{aligned}}$$ (1.14)

where $F, G \in A$. Indeed, let us prove the first formula:

$$\underset{+h}{D}(FG) = \frac{\underset{+h}{S}(FG) - FG}{h} = \frac{\underset{+h}{S}(F)G - FG}{h} + \frac{F\underset{+h}{S}(G) - FG}{h}$$

$$+ h\frac{\left(\underset{+h}{S}(F) - F\right)\left(\underset{+h}{S}(G) - G\right)}{h^2} = \underset{+h}{D}(F)G + F\underset{+h}{D}(G) + h\underset{+h}{D}(F)\underset{+h}{D}(G).$$

The second equation is also obvious.

The Leibniz rule for the discrete differentiation can be written in the different equivalent form

$$\boxed{\underset{\pm h}{D}(FG) = \underset{\pm h}{D}(F)G + \underset{\pm h}{S}(F)\underset{\pm h}{D}(G) = \underset{\pm h}{D}(F)\underset{\pm h}{S}(G) + F\underset{\pm h}{D}(G),}$$ (1.15)

which readily follows from the above definitions.

Let us present several useful formulas of discrete differentiation, which can be proved by straightforward computations:

$$\underset{+h}{D}(uv^-) = u_x v + uv_{\bar{x}}, \qquad \underset{-h}{D}(uv^+) = u_{\bar{x}} v + uv_x,$$

where $z^+ = \underset{+h}{S}(z)$ and $z^- = \underset{-h}{S}(z)$;

$$\underset{+h}{D}(u^n) = \sum_{m=1}^{n}\binom{n}{m}u^{n-m}u_x h^{m-1}, \qquad \underset{-h}{D}(u^n) = \sum_{m=1}^{n}\binom{n}{m}u^{n-m}u_{\bar{x}}(-h)^{m-1},$$

$$\underset{-h}{D}\left(\frac{1}{x}\right) = -\frac{1}{x(x-h)}, \qquad \underset{+h}{D}\left(\frac{1}{x}\right) = -\frac{1}{x(x+h)},$$

$$\underset{-h}{D}\left(\frac{1}{u}\right) = -\frac{u_{\bar{x}}}{u^-u}, \qquad \underset{+h}{D}\left(\frac{1}{u}\right) = -\frac{u_x}{u^+u},$$

$$\underset{-h}{D}(a^x) = a^x\frac{1-a^{-h}}{h}, \qquad \underset{+h}{D}(a^x) = a^x\frac{a^{+h}-1}{h},$$

$$\underset{-h}{D}(\ln u) = \frac{1}{h}\ln\left(1 - h\frac{u_{\bar{x}}}{u}\right), \qquad \underset{+h}{D}(\ln u) = \frac{1}{h}\ln\left(1 + h\frac{u_x}{u}\right),$$

$$\underset{-h}{D}(\sin\alpha x) = \frac{2}{h}\sin\left(\frac{\alpha h}{2}\right)\cos\left(\alpha\left(x - \frac{h}{2}\right)\right),$$

$$\underset{+h}{D}(\sin\alpha x) = \frac{2}{h}\sin\left(\frac{\alpha h}{2}\right)\cos\left(\alpha\left(x + \frac{h}{2}\right)\right),$$

$$\underset{-h}{D}(\arctan x) = \arctan \frac{1}{1 + x(x - h)}, \qquad \underset{+h}{D}(\arctan x) = \arctan \frac{1}{1 + x(x + h)}.$$

The powers of the operators $\underset{+h}{D}$ and $\underset{+h}{S}$ are related as

$$\underset{+h}{D^n} = h^{-n} \sum_{m=0}^{n} (-1)^m \binom{n}{m} \underset{+h}{S^{n-m}}, \qquad \underset{+h}{S^n} = \sum_{m=0}^{n} \binom{n}{m} h^m \underset{+h}{D^m}.$$

The difference Leibniz rule significantly affects all constructions of the group analysis of difference equations. The distinction of the space of discrete variables from the space of continuous variables can be described in various languages. From the physical viewpoint, the space of discrete variables has a new scale, the mesh spacing, which is absent in the continuous model. From the functional-analytic viewpoint, the main distinction is that the objects are nonlocal (i.e., the space $\underset{h}{\widetilde{Z}}$ is in principle infinite-dimensional). From the algebraic viewpoint, this is a new Leibniz rule, which means that the action of the discrete differentiation on analytic functions (and on formal power series) is quite different from the "usual" Leibniz rule.

1.3. Invariant Difference Meshes

In this section, we study the relations between one-parameter groups and difference meshes preserving their geometric structure under group transformations. Considering several typical examples of difference meshes, we obtain criteria for their invariance.

1.3.1. Invariant uniform meshes and invariance criterion

The uniform difference mesh $\underset{h}{\omega}$ is the most widely used discretization method for spaces of independent variables. The formal one-parameter group transformations that change the independent variable x can distort the mesh by violating its uniformness, which affects finite-difference equations written on $\underset{h}{\omega}$. In particular, the operators $\underset{+h}{D}$ and $\underset{-h}{D}$ do not commute any more, and the approximation order can also be changed. Therefore, we need to single out the class of admissible transformations preserving the mesh uniformness. First, consider the case of a single independent variable.

We assume that there is a formal transformation group G_1 given in $\underset{h}{\widetilde{Z}}$,

$$
\begin{aligned}
x^* &= f(z, a), & \underset{h}{u_1}^* &= \psi_1(z, a), \\
u^* &= \varphi(z, a), & \underset{h}{u_2}^* &= \psi_2(z, a), \\
u_1^* &= \varphi_1(z, a), & &\dots\dots\dots\dots, \\
u_2^* &= \varphi_2(z, a), & &\dots\dots\dots\dots,
\end{aligned}
\tag{1.16}
$$

which is associated with an infinitesimal operator

$$X = \xi\frac{\partial}{\partial x} + \eta\frac{\partial}{\partial u} + \sum_{s\geq 1}\zeta^s\frac{\partial}{\partial u_s} + \sum_{m\geq 1}\underset{h}{\zeta^m}\frac{\partial}{\partial \underset{h}{u_m}}. \tag{1.17}$$

It is clear that if the independent variable is an invariant (i.e., $f(z, a) \equiv x$ and $\xi \equiv 0$), then such a class of transformations does not change the difference mesh; the same is true for uniform meshes. But this condition is not necessary.

Let us supplement the space $\underset{h}{Z}$ with new variables, the right spacing h_+ and the left spacing h_- at a point z^i: $(x, u, \underset{h}{u_1}, \underset{h}{u_2}, \ldots, h_+, h_-)$.

It is natural to define the mesh spacing transformations as follows:

$$h^{+*} = \underset{+h}{S}(x^*) - x^* = f(\underset{+h}{S}(z), a) - f(z, a),$$
$$h^{-*} = x^* - \underset{-h}{S}(x^*) = f(z, a) - f(\underset{-h}{S}(z), a);$$

then $h_+^*\big|_{a=0} = h_-^*\big|_{a=0} = h$. The additional coordinates of the operator (1.17) are

$$\frac{\partial h_+^*}{\partial a}\bigg|_{a=0} = (\underset{+h}{S} - 1)\xi(z) = \xi(\underset{+h}{S}(z)) - \xi(z) = h\underset{+h}{D}(\xi),$$
$$\frac{\partial h_-^*}{\partial a}\bigg|_{a=0} = (1 - \underset{-h}{S})\xi(z) = \xi(z) - \xi(\underset{-h}{S}(z)) = h\underset{-h}{D}(\xi).$$

If the tangent field for h_+ and h_- is known, then we can readily obtain the invariance criterion for the equality $h_+ = h_-$ viewed as a manifold in the extended space $\underset{h}{\widetilde{Z}} = (x, u, u_1, \ldots; \underset{h}{u_1}, \ldots, h_+, h_-)$. Indeed, by applying the infinitesimal criterion and by using the operator (1.17), we obtain the following second-order difference equation for the coordinate $\xi(z)$ of the operator (1.17):

$$(\underset{+h}{S} - 1)\xi(z) = (1 - \underset{-h}{S})\xi(z),$$

or $\underset{+h-h}{D D}(\xi(z)) = 0$. Thus, we arrive at the following assertion.

PROPOSITION 1.2. *For the mesh $\underset{h}{\omega}$ to remain uniform ($h_+^* = h_-^*$) under the action of the transformation group G_1, it is necessary and sufficient that the following condition be satisfied at each point $z \in \underset{h}{\widetilde{Z}}$:*

$$\underset{+h-h}{D D}(\xi(z)) = 0. \tag{1.18}$$

The meshes satisfying criterion (1.18) are said to be *invariantly uniform*.

Remark. Condition (1.18) means that an arbitrary uniform mesh preserves its uniformness in the entire space $\widetilde{Z}_h$. For a specific difference equation $F(z) = 0$ considered on a uniform mesh ω_h, condition (1.18) can be weakened to the condition

$$\underset{+h}{D}\underset{-h}{D}(\xi(z))\,\Big|_{F(z)=0} = 0. \tag{1.19}$$

In what follows, we consider an example of this weakening.

Consider examples of groups satisfying condition (1.18).

1. In particular, condition (1.18) is satisfied by the group G_1 with $\xi = \text{const}$, i.e., the group of shifts along the coordinate x. The simplest example is the translation along the independent variable, $X = \partial/\partial x$. But such a condition may be satisfied not only by a point transformation group. For example, for the Taylor group we have $\xi = 1$, $x^* = x + a$, and $h^* = h = \text{const}$.

2. In particular, condition (1.18) is also satisfied for the relations $\xi = Ax$, $A = \text{const}$, i.e., for the transformations under which the x-axis is extended. In this case, $h^{+*} = h^{-*} = e^{aA}(h)$, where a is the group parameter.

3. Condition (1.18) is satisfied by the group G_1 for which $\xi(x + h) = \xi(x)$ is a periodic function with period h.

4. In the more general case, $\xi(\underset{\pm h}{S}(z)) = \xi(z)$; i.e., $\xi(z)$ is invariant under the discrete shift operator $\underset{\pm h}{S}$.

5. Equation (1.18) is satisfied by the function $\xi(z) = A(z)x + B(z)$, where $A(z)$ and $B(z)$ are arbitrary invariants of the discrete shift operator $\underset{\pm h}{S}$.

It is of interest to note that criterion (1.18) is satisfied by the groups most widely considered in mathematical models of physics. We mean translations along independent variables and dilations, i.e., "self-similar" transformations; in the multidimensional case, as we shall see, they are supplemented with rotations and other well-known groups.

But, as the simple example given in the Introduction shows, it is impossible to *restrict oneself only to the case of uniform meshes.*

An example of a group (projective transformations) that does not satisfy condition (1.18) is as follows: $X = x^2 \partial/\partial x + \cdots$; in this case, criterion (1.18) is not satisfied: $\underset{+h}{D}\underset{-h}{D}(x^2) = 2$.

Proposition 1.2 can readily be generalized to the multidimensional case. Indeed, let a group G_1 be determined by the operator

$$X = \xi^i \frac{\partial}{\partial x^i} + \eta \frac{\partial}{\partial u} + \cdots, \qquad i = 1, 2, \ldots, n; \tag{1.20}$$

the number of dependent variables is of no importance here. In $\widetilde{Z}_h$, the operator (1.20) determines the finite transformations

$$x^{i*} = f^i(z,a), \qquad u^* = \phi(z,a), \qquad \dots\dots\dots,$$

which transform the spacings h_i^+ and h_i^- of the orthogonal mesh ω_h as follows:

$$h_i^{+*} = S_{+h}_i(x^{i*}) - x^{i*} = f^i(S_{+h}_i(z),a) - f^i(z,a),$$
$$h_i^{-*} = x^{i*} - S_{-h}_i(x^{i*}) = f^i(z,a) - f^i(S_{-h}_i(z),a).$$

Just as in the one-dimensional case, we say that the mesh ω_h preserves its uniformness in the direction x^i if the relation $h_i^+ = h_i^-$ is invariant under the action of G_1.

The use of an infinitesimal invariance criterion results in the following necessary and sufficient conditions for the invariance of the mesh uniformness in the direction x^i:

$$S_{+h}_i(\xi^i) - 2\xi^i + S_{-h}_i(\xi^i) = 0,$$

where i is fixed, or

$$\boxed{D_{+h}_i D_{-h}_i(\xi^i) = 0.} \tag{1.21}$$

Note that in the one-dimensional case criterion (1.18) completely solves the problem of invariance of the mesh geometric structure, while in the multidimensional case condition (1.21) solves this problem only partly. In particular, condition (1.21) does not guarantee that the mesh remains orthogonal under the action of the transformation group G_1.

1.3.2. Preservation of mesh orthogonality

Assume that there is given an orthogonal mesh ω_h, uniform or nonuniform. For simplicity, consider the case of two independent variables x^1 and x^2. Let us see how an arbitrary cell of the mesh ω_h is transformed under the action of G_1 with the operator (1.20).

Points A, B, C, and D of an arbitrary mesh cell (see Fig. 1.1) have the following coordinates:

$$A: (x^1, x^2), \quad B: (x^1, x^2 + h_2^+), \quad C: (x^1 + h_1^+, x^2 + h_2^+), \quad D: (x^1 + h_1^+, x^2).$$

The transformation group G_1 takes points A, B, C, and D into points A^*, B^*, C^*,

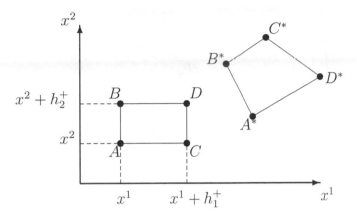

Figure 1.1

and D^* with the following coordinates:

$$A^*: \left(f^1(z, a), f^2(z, a)\right), \qquad\qquad B^*: \left(f^1(\underset{+h}{S}_2(z), a), f^2(\underset{+h}{S}_2(z), a)\right),$$

$$C^*: \left(f^1(\underset{+h}{S}_1 \underset{+h}{S}_2(z), a), f^2(\underset{+h}{S}_1 \underset{+h}{S}_2(z), a)\right), \quad D^*: \left(f^1(\underset{+h}{S}_1(z), a), f^2(\underset{+h}{S}_1(z), a)\right).$$

Now we write out the orthogonality condition for the angle $B^* A^* D^*$:

$$[f^1(\underset{+h}{S}_2(z), a) - f^1(z, a)][f^1(\underset{+h}{S}_1(z), a) - f^1(z, a)]$$
$$+ [f^2(\underset{+h}{S}_2(z), a) - f^2(z, a)][f^2(\underset{+h}{S}_1(z), a) - f^2(z, a)] = 0.$$

To obtain an infinitesimal characteristic of the last condition, we apply the operation $\partial/\partial a \big|_{a=0}$ to it:

$$\underset{+h}{D}_1(\xi^2) = -\underset{+h}{D}_2(\xi^1). \tag{1.22}$$

For the mesh orthogonality condition to be satisfied for all angles at this point, condition (1.22) must obviously hold for any combinations of the differentiation operators $\underset{-h}{D}_1$, $\underset{+h}{D}_1$, $\underset{-h}{D}_2$, and $\underset{+h}{D}_2$:

$$\underset{\pm h}{D}_1(\xi^2) = -\underset{\pm h}{D}_2(\xi^1). \tag{1.23}$$

Now we obtain the following obvious statement.

PROPOSITION 1.3. *For an orthogonal mesh $\underset{h}{\omega}$ to preserve its orthogonality in the plane (x^i, x^j) under any transformation of the group G_1 with the operator (1.20), it is necessary and sufficient that the following condition be satisfied at each mesh point:*

$$\boxed{\underset{\pm h}{D}_i(\xi^j) = -\underset{\pm h}{D}_j(\xi^i),} \qquad i \neq j. \tag{1.24}$$

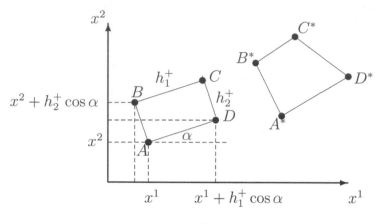

Figure 1.2

Condition (1.24) is a partial difference equation for the coordinate ξ^i of the generator of the group G_1. The meshes for which condition (1.24) is satisfied are said to be *invariantly orthogonal* under the action of G_1.

One can readily verify that condition (1.24) is satisfied for the translation, dilation, rotation and many other groups. But there exist transformations for which conditions (1.24) are not satisfied (i.e., orthogonal meshes do not preserve their structure). For example, for the Lorentz transformations

$$X = x^2 \frac{\partial}{\partial x^1} + x^1 \frac{\partial}{\partial x^2}$$

we have

$$\underset{+h}{D_2}(x^2) = 1, \qquad \underset{+h}{D_1}(x^1) = 1, \qquad \underset{+h}{D_2}(\xi^1) \neq -\underset{+h}{D_1}(\xi^2).$$

We note that we have obtained criterion (1.24) for an orthogonal mesh originally oriented along the coordinate axes. Therefore, if condition (1.24) for a given group G_1 is not satisfied, then this does not mean that there does not exist an orthogonal differently oriented mesh that preserves its orthogonality under the action of G_1.

Consider this situation.

We assume that there is given an orthogonal mesh $\underset{h}{\omega}$, uniform or nonuniform, titled with respect to the coordinate x^1- and x^2-axes (see Fig. 1.2).

In the plane (x^1, x^2), there is a pair of shift operators $\underset{\pm h}{S_1}$ and $\underset{\pm h}{S_2}$ acting so that the coordinates z^i of a vector in $\underset{h}{\widetilde{Z}}$ (or a finite set z of such vectors) move *along the mesh edges* $\underset{h}{\omega}$ but not along the coordinate axes.

Under the action of the transformation group G_1 with the operator (1.20), the nodes of an arbitrary cell of the mesh $\underset{h}{\omega}$ are taken to points A^*, B^*, C^*, and D^*

with the following coordinates:

$$A^*: (f^1(z,a), f^2(z,a)), \qquad\qquad B^*: (f^1(\underset{+h}{S}_2(z),a), f^2(\underset{+h}{S}_2(z),a)),$$

$$C^*: (f^1(\underset{+h}{S}_1\underset{+h}{S}_2(z),a), f^2(\underset{+h}{S}_1\underset{+h}{S}_2(z),a)), \quad D^*: (f^1(\underset{+h}{S}_1(z),a), f^2(\underset{+h}{S}_1(z),a)).$$

Let us write out the orthogonality condition for the angle $B^*A^*D^*$:

$$[f^1(\underset{+h}{S}_2(z),a) - f^1(z,a)][f^1(\underset{+h}{S}_1(z),a) - f^1(z,a)]$$

$$+ [f^2(\underset{+h}{S}_2(z),a) - f^2(z,a)][f^2(\underset{+h}{S}_1(z),a) - f^2(z,a)] = 0.$$

By applying the operation $\partial/\partial a \,\big|_{a=0}$ to the last equation, we obtain

$$(\underset{+h}{S}_2(\xi^1) - \xi^1)h_1^+ \cos\alpha - (\underset{+h}{S}_1(\xi^1) - \xi^1)h_2^+ \sin\alpha$$

$$+ (\underset{+h}{S}_2(\xi^2) - \xi^2)h_1^+ \sin\alpha + (\underset{+h}{S}_1(\xi^2) - \xi^2)h_2^+ \cos\alpha = 0,$$

or, by dividing by $h_1^+ h_2^+$,

$$\boxed{\underset{+h}{D}_2(\xi^1)\cos\alpha - \underset{+h}{D}_1(\xi^1)\sin\alpha + \underset{+h}{D}_2(\xi^2)\sin\alpha + \underset{+h}{D}_1(\xi^2)\cos\alpha = 0,} \qquad (1.25)$$

where the $\underset{+h}{D}_i$ are the differentiations along the mesh edges.

Thus, we obtain the following assertion.

PROPOSITION 1.4. *For an orthogonal mesh $\underset{h}{\omega}$ oriented at an angle α with the co-ordinate axes (according to Fig. 1.2) to preserve its orthogonality under any transformation of the group G_1 with the operator (1.20), it is necessary and sufficient that conditions (1.25) be satisfied.*

Note that conditions (1.25) for $\alpha = 0$ imply conditions (1.24).

EXAMPLE 1.5. We return to our example of an orthogonal mesh subjected to the Lorentz transformations:

$$X = x^2 \frac{\partial}{\partial x^1} + x^1 \frac{\partial}{\partial x^2}.$$

In this case, it follows from conditions (1.25) that

$$\underset{+h}{D}_2(x^2)\cos\alpha - \underset{+h}{D}_1(x^2)\sin\alpha + \underset{+h}{D}_2(x^1)\sin\alpha + \underset{+h}{D}_1(x^1)\cos\alpha = 0,$$

which, with the "inclined" differentiation $\underset{+h}{D}_i$ (see Fig. 1.2) taken into account, implies that $\sin^2\alpha = \cos^2\alpha$ and hence $\alpha = \pi/4 + k\pi/2$; i.e., the mesh must be oriented at an angle of 45° with respect to the coordinate axes: it is only in this case that it preserves its orthogonality under the Lorentz transformations.

Consider the one-parameter dilation group with the operator

$$X = Ax^1 \frac{\partial}{\partial x^1} + x^2 \frac{\partial}{\partial x^2}, \tag{1.26}$$

where $A = \text{const}$.

EXAMPLE 1.6. Consider a mesh located as in Fig. 1.2. In this case, it follows from conditions (1.25) for the operator (1.26) that

$$\underset{+h}{D_2}(Ax^1) \cos\alpha - \underset{+h}{D_1}(Ax^1) \sin\alpha + \underset{+h}{D_2}(x^2) \sin\alpha + \underset{+h}{D_1}(x^2) \cos\alpha = 0,$$

which implies that

$$(A - 1) \sin\alpha \cos\alpha = 0.$$

In particular, the latter condition means that the orthogonal mesh preserves its orthogonality under the dilation transformations (1.26) with any A if it is parallel to the coordinate axes. Under the condition $\alpha \neq 0$, we obtain $A = 1$; i.e., the dilation group does not change the orthogonality of an "inclined" orthogonal mesh (Fig. 1.2) only if the dilations are *uniform*.

1.3.3. Invariant nonuniform meshes and invariance criterion

Now we assume that a nonuniform mesh $\underset{\hbar}{\omega}$ is given in $\underset{h}{\widetilde{Z}}$. At each point $z^i \in \underset{h}{\widetilde{Z}}$, there is a pair of given quantities, the right spacing h_+ and the left spacing h_-. (We first consider the case of a single independent variable x.) The operators $\underset{+h}{D}$ and $\underset{-h}{D}$ are no longer commutative and become "local," i.e., depend on the points x.

Let the right spacing h_+ be given as a sufficiently smooth function of x, $h_+ = \varphi(x)$. The left spacing is the right spacing at the point shifted by h_- to the left, $h_- = \varphi(x - h_-)$. Therefore, we can consider only h_+. Conversely, if the function $\varphi(x)$ and the point x_0 at which the discretization of the x-axis begins are given, then the points of the mesh $\underset{\hbar}{\omega}$ can be reconstructed uniquely.

Under the action of the group G_1, the quantity x and hence the variables h_+ and h_- vary. After the transformations of G_1, the new spacing h_+^* is, in general, some other function of x^*, $h_+^* = \widetilde{\varphi}(x^*)$.

DEFINITION. We say that a given nonuniform mesh $\underset{\hbar}{\omega}$ is invariant under transformations of G_1 in the space $\underset{h}{\widetilde{Z}}$ if the manifold

$$h_+ = \varphi(x) \tag{1.27}$$

is invariant, i.e., if the relation $h_+^* = \varphi(x^*)$ remains valid in the new variables.

The invariance criterion for the manifold (1.27) leads to the following assertion.

PROPOSITION 1.7. *For the difference mesh $\underset{\hbar}{\omega}$ given by Eq. (1.27) to be invariant under the transformations of G_1 with the operator (1.17), it is necessary and sufficient that the following condition be satisfied*:

$$\xi(\underset{+h}{S}(z)) - \xi(z)\left(1 + \frac{\partial\varphi}{\partial x}\right)\bigg|_{(1.27)} = 0. \tag{1.28}$$

Proof. Indeed, let us act by the operator (1.17) on the relation $h_+ = \varphi(x)$:

$$X(h_+ - \varphi(x)) = \xi(\underset{+h}{S}(z)) - \xi(z) - \xi(z)\frac{\partial\varphi}{\partial x} = 0.$$

Obtaining the manifold (1.27), we complete the proof. □

Note that condition (1.28) can be written in the equivalent form

$$\frac{\underset{+h}{D}(\xi)}{\xi} = \frac{\varphi'_x}{\varphi}.$$

EXAMPLE. Consider the transformations determined by the operator

$$X = x^2\frac{\partial}{\partial x}.$$

The criterion for preserving the uniformness is not satisfied, $\underset{+h-h}{D\,D}(x^2) = 2$, and therefore, we must consider the nonuniform mesh $h_+ = \varphi(x)$. Let us continue the operator to h_+ and h_-:

$$X = x^2\frac{\partial}{\partial x} + h_+(2x + h_+)\frac{\partial}{\partial h_+} + h_-(2x - h_-)\frac{\partial}{\partial h_-}. \tag{1.29}$$

An invariant mesh can readily be constructed starting from the invariants of the group G_1:

$$J_1 = x + \frac{x^2}{h_+}, \qquad J_2 = x - \frac{x^2}{h_-}.$$

For example, let us construct an invariant nonuniform mesh on an arbitrary interval $(0, L_0)$. By setting $J_1 = L_0$, we obtain the relation

$$h_+ = \frac{x^2}{L_0 - x}, \tag{1.30}$$

and the left spacing is determined by the equation $h_- = \varphi(x - h_-)$:

$$h_- = \frac{x^2}{L_0 + x}. \tag{1.31}$$

One can readily verify that relations (1.30) and (1.31) define an invariant manifold with respect to the operator (1.29).

We can see that the quantity

$$\frac{h_+}{h_-} = \frac{L_0 + x}{L_0 - x}$$

is also an invariant; i.e., *the transformations of G_1 preserving the mesh invariance preserve the difference stencil proportions.* This situation also takes place in the general case.

Indeed, let a nonuniform difference mesh be given,

$$h_+ = \varphi(x), \qquad h_- = \varphi(x - h_-). \tag{1.32}$$

If the mesh is invariant, then on (1.32) we have

$$(\xi^+ - \xi) = \xi\varphi_x, \qquad (\xi - \xi)^- = \xi^-\varphi_x^-. \tag{1.33}$$

The invariance of the manifold

$$\frac{h_+}{h_-} = \frac{\varphi(x)}{\varphi(x - h_-)}, \tag{1.34}$$

is determined by whether the following condition is satisfied on (1.34):

$$(\xi^+ - \xi)\varphi^- + \xi^-\varphi_x^- = (\xi - \xi^-)\varphi + h_-\xi\varphi_x,$$

which holds by virtue of (1.32) and (1.33).

Thus, we have the following assertion.

PROPOSITION 1.8. *Let a nonuniform mesh (1.27) be given in $\widetilde{Z}_h$. Then it follows from its invariance under the one-parameter group G_1 with the operator (1.17) that (1.34) is also invariant; i.e., the transformation group preserves the difference stencil proportions.*

EXAMPLE (exponential meshes). Consider the special case of a nonuniform one-dimensional mesh in which the spacings h^+ exponentially increase as $x \to \infty$,

$$qh^+ = h^-, \tag{1.35}$$

where $q = $ const, $0 < q < 1$. For example, relation (1.35) is satisfied for the mesh

$$\begin{cases} h^+ = xq^{-1} - x, \\ h^- = x - qx. \end{cases} \tag{1.36}$$

Acting by the operator (1.17), we readily obtain *necessary and sufficient conditions for the invariance of the exponential mesh*:

$$\left[q\xi^+ - (q+1)\xi + \xi^-\right]_{(1.35)} = 0,$$

where $\xi^+ = \underset{+h}{S}(\xi)$ and $\xi^- = \underset{-h}{S}(\xi)$. In the limit case $q \to 1$, the above conditions imply a criterion for the invariant uniformness of the mesh (1.18). One can readily verify that, in particular, this criterion is satisfied by the translation group ($\xi = $ const) and the dilation group ($\xi = A_0 x$).

In the special case (1.36), where the mesh depends only on the independent variable x, the action of the differentiation operators on $F(z) \in \underset{h}{A}$ can be written as

$$\underset{+h}{D}F(z)\big|_{(1.36)} = \frac{F(\underset{+h}{S}(z)) - F(z)}{x(q^{-1} - 1)} = D^-,$$

$$\underset{-h}{D}F(z)\big|_{(1.36)} = \frac{F(z) - F(\underset{-h}{S}(z))}{x(1 - q)} = D^+,$$

where the $\underset{\pm h}{S}$ are the shift operators along the mesh (1.36).

The discrete differentiation operators D^+ and D^- on the mesh (1.36) are used to write out the so-called "q-deformed" difference equations (e.g., see [55–57]).

1.3.4. Invariant meshes depending on the solution

Proposition 1.7 can be generalized to the case of a time-dependent mesh $\underset{\hbar}{\omega}$.

PROPOSITION 1.9. *Let a mesh be given by the equation* $h_+ = \varphi(z)$, *where* $\varphi(z) \in \underset{h}{A}$. *Then the invariance criterion for this mesh acquires the form*

$$\xi(\underset{+h}{S}(z)) - \xi(z) - X(\varphi(z))\big|_{h_+ = \varphi(z)} = 0, \qquad (1.37)$$

where X is an operator of the form (1.17).

In particular, if $h_+ = \varphi(x, u)$, *then the criterion* (1.37) *becomes*

$$\xi(\underset{+h}{S}(z)) - \xi(z)\left(1 + \frac{\partial\varphi}{\partial x}\right) - \eta(z)\frac{\partial\varphi}{\partial u}\bigg|_{h_+ = \varphi(x,u)} = 0.$$

This proposition can be proved by a straightforward application of the operator (1.17).

If the coordinate $\xi(x, u)$ of the operator of the group G_1 explicitly depends on the solution ($\xi_u \neq 0$),[1] then the invariance criterion for the mesh can be explicitly related to the solution of the corresponding invariant equation.

---

[1] The transformation groups with $\xi_u = 0$ were called the "x-autonomous" groups by Ovsyannikov [116].

EXAMPLE. Later on, we shall consider an example in which the invariance criteria for the difference mesh and the difference equation are related to each other, i.e., do not hold separately.

In particular, the second-order linear equation

$$\frac{d^2 u}{dx^2} = 0 \tag{1.38}$$

admits the operator

$$X = u\frac{\partial}{\partial x}. \tag{1.39}$$

In $\underset{h}{Z}$, consider the finite-difference equation

$$\underset{h}{v_{x\bar{x}}} = 0 \tag{1.40}$$

providing the second-order approximation to Eq. (1.38) on the uniform mesh. The operator

$$\underset{h}{X} = v\frac{\partial}{\partial x} - \underset{h}{v_x^2}\frac{\partial}{\partial v_x} + \underset{h}{v_{x\bar{x}}}(2h^- \underset{h}{v_{x\bar{x}}} - 3\underset{h}{v_x})\frac{\partial}{\partial \underset{h}{v_{x\bar{x}}}} \tag{1.41}$$

corresponding to (1.39) isomorphically represents the operator (1.39) in the mesh space. (The prolongation formulas are given below.) The operator (1.41) does not satisfy the criterion for preserving the mesh uniformness. But a uniform mesh can still be used.

Indeed, we write our manifold as the two relations

$$\underset{h}{v_{x\bar{x}}} = 0, \qquad h_+ = h_- \tag{1.42}$$

and prolong the operator (1.41) to h_+ and h_-,

$$\underset{h}{X} = v\frac{\partial}{\partial x} + \cdots + h_+ \underset{h}{v_x}\frac{\partial}{\partial h_+} + h_- \underset{h}{v_{\bar{x}}}\frac{\partial}{\partial h^-}, \tag{1.43}$$

where $\underset{h}{v_{\bar{x}}}$ is the first-order left difference derivative,

$$\underset{h}{v_{\bar{x}}} = \underset{h}{v_x} - h_- \underset{h}{v_{x\bar{x}}}.$$

We act by the operator (1.43) on the manifold (1.42):

$$(2h_- \underset{h}{v_{xx}} - 3\underset{h}{v_x})\underset{h}{v_{x\bar{x}}}\Big|_{(1.42)} = 0,$$

$$h_+ \underset{h}{v_x} - h_- \underset{h}{v_{\bar{x}}}\Big|_{(1.42)} = 0.$$

The invariance of the first equation is obvious; the invariance of the second equation follows from (1.42) and the relation

$$h_- \underset{h}{v_{x\bar{x}}} = \underset{h}{v_x} - \underset{h}{v_{\bar{x}}}.$$

Thus, the uniform mesh is not invariant in the entire space $\underset{h}{Z}$ but admits the operator (1.43) on the manifold (1.42).

In more detail, this situation can be considered together with the invariance criterion for difference equations (see Chapter 2).

1.3.5. Straightening of an invariant nonuniform difference mesh

The invariance property

$$h_+ = \varphi(x), \tag{1.44}$$

of an arbitrary mesh $\underset{h}{\omega}$ is independent of the choice of the coordinate system, because the invariance criterion for this mesh is a scalar expression. But an external point transformation can change the mesh *structure*.

It is well known (see [111]) that each *point* one-parameter group G_1 can be transformed by a change of variables into the translation group along the independent variable. In this case, the group operator in the new variables satisfies the invariant uniformness condition (1.18).

The following theorem solves the problem of the possibility of "straightening" an invariant nonuniform mesh (1.28), i.e., of making the mesh uniform.

PROPOSITION 1.10. *Let a smooth and locally invertible change of variables be given in* $\underset{h}{\tilde{Z}}$,

$$\bar{x} = f(x, u), \qquad \bar{u} = g(x, u). \tag{1.45}$$

Then the mesh $\underset{h}{\omega}$ *(1.44) invariant under the one-parameter group* G_1 *with the operator (1.17) becomes uniform after the change of variables (1.45) with the following condition on the function* f:

$$X\left\{f(x^+, u^+) - 2f(x, u) + f(x^-, u^-)\right\}\big|_{(1.44)} = 0, \tag{1.46}$$

where $z^+ = \underset{+h}{S}(z)$ *and* $z^- = \underset{-h}{S}(z)$.

Indeed, according to the transformation formulas for the coordinates of the operator of a point group (see [111]), the operator (1.17) in the new coordinate system acquires the form

$$\tilde{X} = X(f(x, u))\frac{\partial}{\partial \bar{x}} + X(g(x, u))\frac{\partial}{\partial \bar{u}} + \cdots$$
$$+ X(f(x^+, u^+) - f(x, u))\frac{\partial}{\partial \bar{h}_+} + X(f(x, u) - f(x^-, u^-))\frac{\partial}{\partial \bar{h}_-}, \tag{1.47}$$

where X is the operator (1.17) in the "old" variables (x, u) and $\bar{h}_+$ and $\bar{h}_-$ are the right and left spacings in the "new" variables. By applying the mesh uniformness condition to the operator (1.47), we obtain condition (1.46).

In particular, by solving the equations

$$X(f(x, u)) = 1, \qquad X(g(x, u)) = 0$$

for f and g, we obtain the change of variables (1.45), which transforms the group G_1 into the translation group $\widetilde{X} = \frac{\partial}{\partial \bar{x}}$.

A sufficient condition for "straightening" of the mesh (1.46) is simplified essentially if the change of variables concerns only the independent variable,

$$\bar{x} = f(x), \qquad \bar{u} = u.$$

In this case, the condition reads

$$\underset{+h}{S}(\xi f_x) - 2\xi f_x + \underset{-h}{S}(\xi f_x)\Big|_{\frac{\omega}{h}} = 0. \tag{1.48}$$

The new spacing has the form $\bar{h}_+ = \underset{+h}{S}(f(x)) - f(x) = f(x + h) - f(x)$.

Thus, *each invariant nonuniform mesh is uniform in some coordinate system.* Note that the difference equations written on this mesh are of course different.

EXAMPLE (of "straightening" of a mesh). We use the above condition for straightening of a mesh, which is invariant under the one-parameter projective group:

$$X = x^2 \frac{\partial}{\partial x} + h_+(2x + h_+)\frac{\partial}{\partial h_+} + h_-(2x - h_-)\frac{\partial}{\partial h_-}. \tag{1.49}$$

An invariant nonuniform mesh on the interval $(0, L_0)$ was already obtained:

$$h_+ = \frac{x^2}{L_0 - x}, \qquad h_- = \frac{x^2}{L_0 + x}.$$

One can readily verify that the change of variables $\bar{x} = -1/x$, $\bar{u} = u$ satisfies condition (1.48). The projective operator (1.49) in the new variables becomes the translation operator $\partial/\partial\bar{x}$, which does not change the mesh spacings under the transformations of the group $\bar{G}_1$. Let us verify that the mesh is uniform in the new variables:

$$\bar{h}_+ = -\frac{1}{x + \dfrac{x^2}{L_0 - x}} + \frac{1}{x} = \frac{1}{L_0},$$

$$\bar{h}_- = -\frac{1}{x} + \frac{1}{x - \dfrac{x^2}{L_0 + x}} = \frac{1}{L_0} = \bar{h}_+.$$

1.3.6. Invariant orthogonal nonuniform meshes on the plane

The above obtained invariance conditions for the geometric characteristics of difference meshes permit constructing their combinations. For example, consider the case of a two-dimensional orthogonal mesh that has an irregular structure along the x^1- and x^2-axes.

PROPOSITION 1.11. *Let there be given an orthogonal nonuniform rectangular mesh $\underset{h}{\omega}$,*

$$h^1_+ = \phi_1(x^1, x^2), \qquad h^2_+ = \phi_2(x^1, x^2). \tag{1.50}$$

For the mesh (1.50) to be invariant under the group G_1 with the operator (1.20), it is necessary and sufficient that the following conditions be satisfied:

$$\left. \underset{\pm h}{D}_1(\xi^2(z)) + \underset{\pm h}{D}_2(\xi^1(z)) \right|_{(1.50)} = 0,$$

$$\left. \xi^1(\underset{+h}{S}_1(z)) - \xi^1(z)\left(1 + \frac{\partial\phi_1}{\partial x^1}\right) - \xi^2(z)\frac{\partial\phi_1}{\partial x^2} \right|_{(1.50)} = 0,$$

$$\left. \xi^2(\underset{+h}{S}_2(z)) - \xi^2(z)\left(1 + \frac{\partial\phi_2}{\partial x^2}\right) - \xi^1(z)\frac{\partial\phi_2}{\partial x^1} \right|_{(1.50)} = 0.$$

Proof. The proof is by a straightforward application of the operator (1.20) with the above conditions for invariant orthogonality taken into account. □

1.3.7. Moving meshes preserving the flatness of time layers

Consider the situation in which the orthogonal mesh is not invariant for a given group G_1. In this case, it is sometimes important to preserve the plane structure of the mesh layers in any direction. This especially makes sense if evolution equations are considered. If we use a mesh whose time layers are straight lines parallel to the x-axis (see Fig. 1.3), then it is also important to preserve this property under the transformation group G_1. Otherwise, we meet the situation shown in Fig. 1.3, where, after the transformations of G_1, part of the space is in the "future," and another part is in the "past."

Let a mesh in the plane (t, x) be given by the relations

$$\tau_+ = \varphi(t), \qquad h_+ = \psi(t, x). \tag{1.51}$$

Then the mesh has flat time layers (the spacing τ^+ is independent of x).

We assume that the group G_1 is defined by its operator

$$X = \xi^t(z)\frac{\partial}{\partial t} + \xi^x(z)\frac{\partial}{\partial x} + \eta(z)\frac{\partial}{\partial u} + \cdots + [\xi^t(\underset{+\tau}{S}(z)) - \xi^t(z)]\frac{\partial}{\partial \tau_+}, \tag{1.52}$$

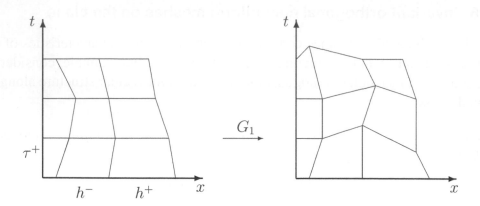

Figure 1.3

where

$$\underset{+\tau}{S} = \sum_{s\geq0}\frac{\tau_+^s}{s!}D_t^s$$

is the operator of shift to the subsequent time layer.

The new spacing τ_+^* is the same at each mesh node at a time layer if it is independent of the spatial variable x,

$$\underset{+h}{D}\tau_+^* = 0.$$

Accordingly, the infinitesimal characteristic of this is

$$\boxed{\underset{+h+\tau}{D\,D}(\xi^t(z)) = 0,}\qquad(1.53)$$

where

$$\underset{+\tau}{D} = \frac{\underset{+\tau}{S}-1}{\tau_+}.$$

PROPOSITION 1.12. *For a mesh with plane time layers to preserve this property under the transformations of the group G_1 with the operator (1.52), it is necessary and sufficient that condition (1.53) be satisfied at each mesh point.*

Note that Proposition 1.12 solves the problem of invariance of the time layer flatness (or the layer flatness with respect to any other coordinate) but does not solve the problem of invariance of the mesh (1.51). The application of the operator (1.52) to Eqs. (1.51) supplements conditions (1.53) with the following ones:

$$\xi^t(\underset{\tau}{S}(z)) - \xi^t(z)(1+\varphi_t)\big|_{(1.51)} = 0,$$

$$\xi^x(\underset{+h}{S}(z)) - \xi^x(z)(1+\psi_x) - \xi^t(z)\psi_t\big|_{(1.51)} = 0.$$

Note that the meshes thus constructed are moving in the originally chosen co-ordinate system. Several examples of meshes preserving the flatness of the time layers under the transformation group are considered in the next chapter.

Of course, we cannot consider all types of difference meshes and obtain conditions for their invariance. The above examples only show how closely the mesh geometric structure is related to transformation groups. In what follows, we propose a general method for constructing invariant meshes, which is based on the complete set of difference invariants.

1.4. Transformations Preserving the Geometric Meaning of Finite-Difference Derivatives; Prolongation Formulas

1. Consider a formal one-parameter transformation group G_1 in $\widetilde{Z}_h$:

$$
\begin{aligned}
x^* &= f(z, a), & u_1^* &= \varphi_1(z, a) \\
u^* &= g(z, a), & u_2^* &= \varphi_2(z, a) \\
u_1^* &= g_1(z, a), & &\dots\dots\dots\dots , \\
u_2^* &= g_2(z, a), & &\dots\dots\dots\dots , \\
&\dots\dots\dots\dots , & &\dots\dots\dots\dots ,
\end{aligned}
\tag{1.54}
$$

where f, g, and g_i are formal power series of the form (1.3).

The discrete shift operator $\underset{\pm h}{S}$ acts on all coordinates of the vector $(x, u, u_1, \dots)$ $\in \widetilde{Z}$, and therefore, it is natural to define the transformation of the spacing h_+ as the difference of two formal series,

$$
h_+{}^* = f(\underset{+h}{S}(z), a) - f(z, a),
$$

where

$$
f(\underset{+h}{S}(z), a) = \sum_{s \geq 0} \frac{a^s}{s!} A_s(\underset{+h}{S}(z))
$$

is a formal power series with "shifted" coefficients.

Likewise, $h_-{}^* = f(z, a) - f(\underset{-h}{S}(z), a)$.

The finite-difference derivatives were introduced in $\widetilde{Z}$ as formal power series of special form

$$
\underset{h}{u_1} = \underset{+h}{D}(u) = \sum_{s \geq 1} \frac{h_+^{s-1}}{s!} u_s, \qquad \underset{h}{u_2} = \underset{-h}{D}(\underset{h}{u_1}), \dots .
\tag{1.55}
$$

Transformations of an arbitrary formal group (1.54) may have the effect that the series $u_1^*, u_2^*, \ldots, h_+^*, h_-^*$ cannot be represented in the form (1.55) in the new variables.

We demand that the formal series (1.54) preserve the definitions of difference derivatives, i.e., that the representation (1.55) be invariant under (1.54). If the series (1.55) converge, then we can say that the geometric meaning of the difference derivatives is also preserved.

Thus, we define the transformations of $\underset{h}{u_k}$ as follows:

$$
\underset{h}{u_1^*} = \sum_{s \geq 1} \frac{(h_+^*)^{s-1}}{s!} g_s(z, a) = u_1^* + \frac{h_+^*}{2!} u_2^* + \cdots,
$$

$$
\underset{h}{u_2^*} = \sum_{l \geq 1} \sum_{s \geq 1} \frac{(-h_-^*)^{l-1}}{l!} \frac{(h_+^*)^{s-1}}{s!} g_{s+l}(z, a) = u_2^* + \frac{h_+^* - h_-^*}{2} u_3^* + \cdots, \tag{1.56}
$$

$$\cdots\cdots\cdots\cdots\cdots\cdots\cdots\cdots\cdots\cdots$$

From the definition of $\underset{\pm h}{S}$ and $\underset{h}{u_k}$, we have obtained the table of actions of $\underset{\pm h}{S}$ on the difference derivatives; in particular,

$$
\underset{+h}{S}(u) - u = \underset{h}{u_1} h_+. \tag{1.57}
$$

If the representation of $\underset{h}{u_k}$ in the form (1.56) is preserved, then so are naturally the relations in Table 1.1. In particular, relation (1.57) is preserved:

$$
g(\underset{+h}{S}(z), a) - g(z, a) = \underset{h}{u_1^*} h_+^*.
$$

By analogy with the tangent transformation groups [73, 107], we refer to relation (1.57) as the *discrete first-order tangency condition*.

If the transformations of G_1 preserve (1.57), then we say that they *preserve the meaning of the first difference derivative*, i.e., preserve the definition and the geometric meaning of the difference derivative.

Along with the preservation of the meaning of the first difference derivative, we can also say that the meaning (i.e., the definitions and the geometric meaning) of the continuous derivatives $u_1, u_2, \ldots$ in $\widetilde{Z}$ is preserved, i.e., that the following relations are invariant:

$$
du = u_1 \, dx, \quad du_1 = u_2 \, dx, \quad \ldots, \quad du_s = u_{s+1} \, dx, \quad \ldots. \tag{1.58}
$$

(For simplicity, we consider only the case of a single independent variable x.)

It is well known [73, 107] that system (1.58) is invariant under local Lie point and contact transformation groups, and for higher symmetries as well; i.e., system (1.58) preserves all continuous symmetries. The invariance of (1.58) implies the following chain of relations:

$$
\zeta_s = D^s(\eta - \xi u_1) + \xi u_{s+1}, \qquad s = 1, 2, \ldots, \tag{1.59}
$$

for the coordinates of the operator (1.54):

$$X = \xi \frac{\partial}{\partial x} + \eta \frac{\partial}{\partial u} + \sum_{s \geq 1} \zeta_s \frac{\partial}{\partial u_s} \tag{1.60}$$

of the group G_1, where $\zeta_s = \partial u_s^* / \partial a |_{a=0}$.

There is a natural question as to whether the preservation of the meaning of the difference derivatives can be combined with the preservation of the meaning of the "usual" derivatives, i.e., with the invariance of system (1.58).

The following statement answers this question.

THEOREM 1.13. *Let G_1 be a formal one-parameter group with the operator (1.60). Suppose that Eq. (1.57) is an invariant manifold of G_1 at each point of $\widetilde{Z}$. Then the coordinates of the operator (1.60) satisfy the chain of relations (1.59); i.e., system (1.58) is invariant. Conversely, it follows from the invariance of system (1.58) that the discrete first-order tangency is preserved, i.e., that relation (1.57) is invariant.*

Proof. In $\widetilde{Z}$, we calculate the coordinates of the operator (1.60) in the difference differential contained in (1.57):

$$\frac{\partial}{\partial a} \left[g(\underset{+h}{S}(z), a) - g(z, a) \right] \Big|_{a=0} = \sum_{s \geq 0} \frac{h_+^s}{s!} D^s(\eta) - \eta$$

$$= \sum_{s \geq 1} \frac{h_+^s}{s!} D^s(\eta) = h_+ \underset{+h}{D}(\eta). \tag{1.61}$$

Now from formulas (1.56) we obtain the coordinate of the operator (1.60), which determines the transformation $\underset{h}{u_1}$:

$$\underset{h}{\zeta_1} = \frac{\partial \underset{h}{u_1^*}}{\partial a} \Big|_{a=0} = \sum_{s \geq 1} \frac{h_+^{s-1}}{s!} \zeta_s + \sum_{s \geq 2} \frac{(s-1)h_+^{s-2}}{s!} u_s \sum_{l \geq 1} \frac{h_+^l}{l!} D^l(\xi), \tag{1.62}$$

where the functions ζ_s are not defined, because we do not assume the invariance of system (1.58). The invariance criterion (1.57) with formulas (1.61) and (1.62) taken into account has the form

$$\sum_{s \geq 1} \frac{h_+^s}{s!} D^s(\eta) - \sum_{s \geq 1} \frac{h_+^s}{s!} D^s(\xi) \sum_{l \geq 1} \frac{h_+^{l-1}}{l!} u_l$$

$$- h_+ \sum_{s \geq 1} \frac{h_+^{s-1}}{s!} \zeta_s - \sum_{s \geq 2} \frac{(s-1)h_+^{s-1}}{s!} u_s \sum_{l \geq 1} \frac{h_+^l}{l!} D^l(\xi) = 0. \tag{1.63}$$

Equation (1.63) means that the formal power series in h^+ is equal to zero. By equating the coefficients of powers of h^+ with zero, we obtain the chains of formulas (1.59).

Thus, *the preservation of the meaning (definition) of the first difference derivative under the transformations of G_1 automatically implies that the definition of all continuous derivatives $u_1, u_2, \ldots$ is preserved*, i.e., that the infinite system (1.58) is invariant. Note that when writing out the invariance criterion (1.57) for the first-order tangency we have not considered any specific difference mesh; i.e., the mesh invariance has not been assumed.

Conversely, assume that the chain of formulas (1.59) is given; then, by substituting them into criterion (1.63), we see that it is satisfied identically, which confirms the invariance of (1.57). The proof of the theorem is complete.   □

The invariance of (1.57) ensures that the meaning of the first difference derivative is preserved under the transformations of G_1 (1.54) and not only formulas (1.58) obtained in Theorem 1.13 but also the meaning of all "usual" derivatives $u_1, u_2, \ldots$ are preserved. Does such a group preserve the meaning of the second, third, and higher difference derivatives? The answer is given by the following assertion.

THEOREM 1.14. *Let a formal one-parameter group G_1 (1.54) with the operator (1.60) be given, and let relation (1.57) be invariant at each point of $\widetilde{Z}_h$ (at each node of the mesh ω_h). Then G_1 preserves the discrete tangency of any finite order.*

Proof. The discrete second-order tangency is defined to be the second relation in Table 1.1:

$$\underset{h}{u_1} - \underset{-h}{S}(\underset{h}{u_1}) = h\underset{h}{u_2}, \tag{1.64}$$

and the invariance of this relation is called the preservation of the meaning of the second difference derivative. We show that the invariance of (1.64) is a consequence of the invariance of discrete first-order tangency. Indeed, along with the preservation of condition (1.57),

$$\underset{+h}{S}(u) - u = h\underset{h}{u_1},$$

the invariance of the tangency at the neighboring point is also satisfied (by the assumtions of the theorem):

$$u - \underset{-h}{S}(u) = h\underset{-h}{S}(\underset{h}{u_1}). \tag{1.65}$$

By subtracting (1.65) from the preceding relation, we obtain

$$\underset{+h}{S}(u) - 2u + \underset{-h}{S}(u) = h(\underset{h}{u_1} - \underset{-h}{S}(\underset{h}{u_1})) = h^2\underset{h}{u_2}; \tag{1.66}$$

i.e., we obtain relation (1.61). Thus, the invariance of (1.61) is a consequence of the invariance of the terms in (1.57) and (1.65). Quite obviously, the proof of the theorem (for tangency of any order) can be completed by induction.   □

Thus, *the formal group G_1 (1.54) preserving the meaning of the first difference derivative is a group in $\widetilde{Z}$ preserving the meaning of all continuous derivatives and can be prolonged to $\underset{h}{\widetilde{Z}}$ with preserving the meaning of all difference derivatives of any finite order.*

Note the *nonlocality* of this interpretation of symmetry groups: two points on a smooth curve lying at a small but finite distance from each other are taken to two points on the image of this curve. (In the multidimensional case, the transformation take a neighborhood of a point z of a locally analytic manifold Φ to a neighborhood of a point z^* of the manifold Φ^*.)

We present the prolongation formulas for the finite-difference derivatives obtained by successive actions of the operator X (1.57) on the rows in Table 1.2:

$$
\begin{aligned}
\underset{h}{\zeta_1} &= \underset{+h}{D}(\eta) - u_1 \underset{h}{\underset{+h}{D}}(\xi), \\
\underset{h}{\zeta_2} &= \underset{-h}{D}(\underset{h}{\zeta_1}) - u_2 \underset{h}{\underset{-h}{D}}(\xi), \\
&\quad\quad\cdots\cdots\cdots\cdots\cdots, \\
\underset{h}{\zeta_{2k}} &= \underset{-h}{D}(\underset{h}{\zeta_{2k-1}}) - u_{2k}\underset{h}{\underset{-h}{D}}(\xi), \\
\underset{h}{\zeta_{2k+1}} &= \underset{+h}{D}(\underset{h}{\zeta_{2k}}) - u_{2k+1}\underset{h}{\underset{+h}{D}}(\xi), \qquad k = 1, 2, \dots.
\end{aligned}
\tag{1.67}
$$

Note that the recursive chain of formulas (1.67) as $h \to 0$ is formally taken to the formulas of Lie transformation group in the continuous case.

2. Consider how the operator of the group G_1 is extended in the two-dimensional case. We recall the notation of spaces in this case:

$$
\begin{aligned}
\underset{h}{Z} &= (x^1, x^2, u, \underset{h}{u_1}, \underset{h}{u_2}, \underset{h}{u_{12}}, \dots, h^1{}_+, h^2{}_-), \\
\widetilde{Z} &= (x^1, x^2, u, u_1, u_2, u_{12}, \dots), \\
\underset{h}{\widetilde{Z}} &= (x^1, x^2, u, u_1, u_2, \dots, \underset{h}{u_1}, \underset{h}{u_2}, \underset{h}{u_{12}}, \dots, h^1{}_+, h^2{}_-),
\end{aligned}
$$

where

$$
u_{ij} = \frac{\partial^2 u}{\partial x^i \partial x^j}, \qquad \underset{h}{u_{ij}} = \underset{+h}{D_j}\underset{+h}{D_i}(u), \qquad \dots, \qquad \underset{h}{\omega} = \underset{h}{\omega_1} \times \underset{h}{\omega_2},
$$

and $\underset{h}{\omega_i}$ is the difference mesh in the ith direction.

By $\underset{h}{\widetilde{Z}}$ we denote the space of sequences of formal power series with analytic coefficients,

$$
z^{j*} = \sum_{s \geq 0} A_s^j(z)a^s, \qquad A_0^j = z^j,
\tag{1.68}
$$

where z^j is a coordinate of the vector $(x, u, \underset{h}{u_1}, \underset{h}{u_2}, \ldots, \underset{h}{u_1}, \underset{h}{u_2}, \underset{h}{u_{12}}, \ldots)$.

We treat the sequence of series (1.68) as transformations in $\underset{h}{\widetilde{Z}}$. Among the series of the form (1.68), we are interested only in the series that produce formal one-parameter groups and are described by infinitesimal operators

$$X = \xi^1 \frac{\partial}{\partial x^1} + \xi^2 \frac{\partial}{\partial x^2} + \eta^k \frac{\partial}{\partial u^k} + \sum_{s \geq 1} \zeta_{i_1 \ldots i_s} \frac{\partial}{\partial u_{i_1 \ldots i_s}} + \sum_{l \geq 1} \zeta_{i_1 \ldots i_l} \frac{\partial}{\partial u_{i_1 \ldots i_l}}. \quad (1.69)$$

Supplementing $\underset{h}{\widetilde{Z}}$ with the variables h_+^1 and h_-^2, we prolong the operator (1.69):

$$X = \ldots + h_+^1 \underset{+h_1}{D}(\xi^1) \frac{\partial}{\partial h_+^1} + h_-^2 \underset{+h_2}{D}(\xi^2) \frac{\partial}{\partial h_-^2}. \quad (1.70)$$

We need to calculate the coordinates of the operator (1.69) for the difference derivatives.

To this end, in $\underset{h}{\widetilde{Z}}$ we consider the two-dimensional surface (the index k in u^k is still omitted)

$$u = \Psi(x^1, x^2). \quad (1.71)$$

We assume that in $\underset{h}{\widetilde{Z}}$ there acts a formal transformation group G_1,

$$x^1 = f^1(z, a), \qquad x^2 = f^2(z, a), \qquad u = g(z, a), \ldots,$$

whose tangent field is determined by the operator (1.69). Under the action of the group G_1, the manifold (1.71) is taken to $u^* = \Psi^*(x^{1*}, x^{2*})$, or

$$g(z, a) = \Psi^*(f^1(z, a), f^2(z, a)). \quad (1.72)$$

Let us apply the operator $\underset{+h_1}{S} - 1$ to relation (1.72) (see Fig. 1.4):

$$(\underset{+h_1}{S} - 1)g(z, a) = \Psi^*(\underset{+h_1}{S}(f^1), \underset{+h_1}{S}(f^2)) - \Psi^*(f^1, f^2)$$

$$= \frac{\Psi^*(\underset{+h_1}{S}(f^1), \underset{+h_1}{S}(f^2)) - \Psi^*(\underset{+h_1}{S^1}(f^1), f^2)}{(\underset{+h_!}{S} - 1)f^2}(\underset{+h_1}{S} - 1)f^2$$

$$+ \Psi^*(\underset{+h_1}{S}(f^1), f^2) - \Psi^*(f^1, f^2).$$

By applying the operation $\frac{\partial}{\partial a}\big|_{a=0}$ to the above relation, we obtain

$$(\underset{+h_1}{S} - 1)(\eta) = \underset{+h_1}{S}(u_2)(\underset{+h_1}{S} - 1)(\xi^2) + \underset{h}{\zeta_1}h_1 + \underset{h}{u_1}(\underset{+h_1}{S} - 1)(\xi^1),$$

which implies the expression for the desired coordinate:

$$\zeta_1 = \underset{h}{D}(\eta) - u_1 \underset{+h_1}{D}(\xi^1) - \underset{+h_1}{S}(u_2) \underset{+h_1}{D}(\xi^2), \quad (1.73)$$

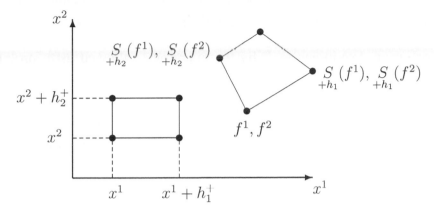

Figure 1.4

where $\underset{+h_1}{S}(u_2)$ is the "continuous" derivative $u_2 = \frac{\partial u}{\partial x^2}$ at the point shifted to the right by the spacing h^1_+ along the axis x^1. (About the discrete representation of "continuous" derivatives in $\underset{h}{Z}$, see below.)

In a similar way, we obtain the following prolongation formulas:

$$\underset{h}{\zeta_2} = \underset{+h_2}{D}(\eta) - \underset{+h_2}{S}(u_1)\underset{+h_2}{D}(\xi^1) - \underset{h}{u_2}\underset{+h_2}{D}(\xi^2),$$

$$\underset{h}{\zeta_{1\bar{1}}} = \underset{-h_1+h_1}{D\,D}(\eta) - 2\underset{h}{u_{1\bar{1}}}\underset{+h_1}{D}(\xi^1) - \frac{1}{h^1}\underset{+h_1}{S}(u_2)\underset{+h_1}{D}(\xi^2) + \frac{1}{h^1}\underset{-h_1}{S}(u_2)\underset{-h_1}{D}(\xi^2),$$

$$\underset{h}{\zeta_{2\bar{2}}} = \underset{-h_2+h_2}{D\,D}(\eta) - 2\underset{h}{u_{2\bar{2}}}\underset{+h_2}{D}(\xi^2) - \frac{1}{h^2}\underset{+h_2}{S}(u_1)\underset{+h_2}{D}(\xi^1) + \frac{1}{h^2}\underset{-h_2}{S}(u_1)\underset{-h_2}{D}(\xi^1), \qquad (1.74)$$

$$\underset{h}{\zeta_{12}} = \underset{+h_2+h_1}{D\,D}(\eta) - \underset{h}{u_{12}}\Big(\underset{+h_1}{D}(\xi^1) + \underset{+h_1}{D}(\xi^1)\Big) + \frac{1}{h^1}\underset{+h_2}{S}(u_1)\underset{+h_2}{D}(\xi^1)$$

$$+ \frac{1}{h^2}\underset{+h_1}{S}(u_2)\underset{+h_1}{D}(\xi^2),$$

$$\dotfill .$$

In formulas (1.73)–(1.74), it is not assumed that the corresponding mesh is invariantly uniform or invariantly orthogonal. If precisely such meshes are considered, then the prolongation formulas (1.73)–(1.74) must be supplemented with the corresponding formulas for invariant meshes.

Note that the prolongation formulas (1.73)–(1.74) can be obtained using the formulas of Lie transformation groups in the continuous space, just as in the one-dimensional case.

1.5. Newton's Group and Lagrange's Formula

The Taylor group determined in $\underset{h}{\widetilde{Z}}$ by the operator D allowed us to prolong the action of a formal group to the mesh variables $(\underset{h}{u_1}, \underset{h}{u_2}, \dots)$. In the theory of higher symmetry groups [73, 107], the Taylor group also plays a significant role. Its generalization, i.e., the group determined in $\widetilde{Z}$ by the operators $\xi^i D_i$ with arbitrary functions $\xi^i(z) \in A$, which is admitted by any differential equations, is used to pass to the quotient operators algebra. The representatives of this quotient algebra have independent variables as invariants, and the prolongation formulas for them have a simple convenient form.

In this and subsequent sections, in the simplest case of a single independent variable x and a uniform mesh $\underset{h}{\omega}$, we consider the difference analog of this construction. We construct the transformation group in the mesh space, i.e., the Newton group, which is isomorphic to the Taylor group. In the next section, we use the Newton group to construct an ideal of the Lie algebra of the set of all operators in the mesh space. The ideal thus constructed is used to factorize the set of operators of a formal group.

A Taylor group orbit, i.e., a one-parameter curve in $\widetilde{Z}$, obtained as the trajectory of an arbitrary point $(x, u, u_1, u_2, \dots)$ under the action of the operator $T_a = e^{aD}$, coincides at the points $a = \pm nh$, $n = 0, 1, 2, \dots$, with the points obtained by the action of the discrete shift operator $\underset{\pm h}{S^n}$. In other words, the Taylor group orbit is the "continuous shift" performed via the "discrete shift." This leads to the question as to whether this procedure is invertible, i.e., whether a continuous shift can be obtained via a discrete shift, or, in other words, to the *problem of the Taylor group representation in the mesh space* $\underset{h}{\widetilde{Z}}$.

The following heuristic considerations permits finding which power series must be used to obtain such a representation.

If the shift of the Taylor group orbit for $a = h$ gives the discrete shift $\underset{+h}{S}$, then, to obtain the shift by a certain quantity $a \neq nh$, we act by the operator $\underset{+h}{S}$ a "noninteger number of times" at a point in $\underset{h}{\widetilde{Z}}$; i.e., *we introduce the fractional power of the operator* $\underset{+h}{S}$. For the definition of the fractional power $\underset{+h}{S^{a/h}}$ of the shift operator we take the following operator series:

$$
\begin{aligned}
\underset{+h}{S^{a/h}} &\equiv (1 + h\underset{+h}{D})^{a/h} \equiv 1 + \frac{a}{h}h\underset{+h}{D} + \frac{a}{h}\left(\frac{a}{h} - 1\right)\frac{h^2}{2!}\underset{+h}{D^2} + \cdots \\
&= 1 + a\underset{+h}{D} + \frac{a(a - h)}{2!}\underset{+h}{D^2} + \frac{a(a - h)(a - 2h)}{3!}\underset{+h}{D^3} + \cdots \\
&= 1 + \sum_{s=1}^{\infty}\left(\prod_{k=0}^{s-1}(a - kh)\right)\frac{1}{s!}\underset{+h}{D^s}.
\end{aligned}
\tag{1.75}
$$

The quantity $a^{[s]} = \prod_{k=0}^{s-1}(a-kh)$ contained in (1.75) is called a *generalized power of the quantity a* [61]. Under the action of the operator series (1.75), the coordinate x becomes $x^* = x + a$, and the coordinate u becomes the series

$$u^* = u + a\underset{h}{u_1} + \frac{a(a-h)}{2!}(\underset{h}{u_2} + h\underset{h}{u_3}) + \cdots,$$

i.e., is the *expansion into the Newton series of the function $u = u(x)$ at the point $(x + a)$ on the uniform mesh* $x, (x+h), (x+2h), \ldots$.

In a similar way, we obtain the expansion into fractional power series of the left discrete shift operator $(a > 0)$,

$$\underset{-h}{S}^{a/h} \equiv (1 - h\underset{-h}{D})^{a/h} = 1 + \sum_{s=1}^{\infty}\left(\prod_{k=0}^{s-1}(kh - a)\right)\frac{1}{s!}\underset{-h}{D}^s. \tag{1.76}$$

The action of the series (1.76) on the coordinate u gives the expansion in the Newton series of the function $u = u(x)$ at the point $(x - a)$ on the uniform mesh $(x, x - h, x - 2h, \ldots)$.

The action of the operator series (1.75)–(1.76) on the point $(x, u, \underset{h}{u_1}, \underset{h}{u_2}, \ldots)$ coincides with the action of the Taylor group at the points $a = \pm nh$. In addition, note that the series (1.75)–(1.76) terminate at these points; i.e., they have finitely many terms $(s = 1, 2, \ldots, n)$:

$$\left[1 + \sum_{s=1}^{\infty} \frac{\prod_{k=0}^{s-1}(a-kh)}{s!}\underset{+h}{D}^s\right]_{a=nh} = \underset{+h}{S}^n, \qquad \left[1 + \sum_{s=1}^{\infty} \frac{\prod_{k=0}^{s-1}(kh-a)}{s!}\underset{-h}{D}^s\right]_{a=nh} = \underset{-h}{S}^n.$$

We regroup the formal operator power series in the parameter a,

$$N_a^+ = \sum_{s=0}^{\infty} \frac{a^s}{s!}\left(\sum_{n=1}^{\infty} \frac{(-h)^{n-1}}{n}\underset{+h}{D}^n\right)^s, \quad N_a^- = \sum_{s=0}^{\infty} \frac{(-a)^s}{s!}\left(\sum_{n=1}^{\infty} \frac{h^{n-1}}{n}\underset{-h}{D}^n\right)^s. \tag{1.77}$$

Thus, the operators (1.77) are defined in $\underset{h}{Z} = (x, u, \underset{h}{u_1}, \underset{h}{u_2}, \ldots)$ and can be represented in exponential form,

$$N_a = e^{a\underset{+h}{\widetilde{D}}}, \qquad N_a = e^{-a\underset{-h}{\widetilde{D}}}, \tag{1.78}$$

where

$$\underset{+h}{\widetilde{D}} = \sum_{n=1}^{\infty} \frac{(-h)^{n-1}}{n}\underset{+h}{D}^n, \qquad \underset{-h}{\widetilde{D}} = \sum_{n=1}^{\infty} \frac{h^{n-1}}{n}\underset{-h}{D}^n. \tag{1.79}$$

The exponential representation (1.78)–(1.79) means that the action of the operators N_a^+ and N_a^- at the point $(x, u, \underset{h}{u_1}, \underset{h}{u_2}, \ldots)$ forms the following pair of formal transformation groups in $\underset{h}{Z}$:

$$x^* = x \pm a,$$

$$u^* = u \pm a \underset{\pm h}{\widetilde{D}}(u) + \frac{a^2}{2!} \underset{\pm h}{\widetilde{D}}^2(u) \pm \cdots,$$

$$\underset{h}{u_1^*} = \underset{h}{u_1} \pm a \underset{\pm h}{\widetilde{D}}(\underset{h}{u_1}) + \frac{a^2}{2!} \underset{\pm h}{\widetilde{D}}^2(\underset{h}{u_1}) \pm \cdots, \qquad (1.80)$$

$$\cdots \cdots \cdots \cdots \cdots \cdots$$

$$\underset{h}{u_s^*} = \underset{h}{u_s} \pm a \underset{\pm h}{\widetilde{D}}(\underset{h}{u_s}) + \frac{a^2}{2!} \underset{\pm h}{\widetilde{D}}^2(\underset{h}{u_s}) \pm \cdots.$$

The second row of these transformations is a *formal (right and left) Newton series expansion of the function $u = u(x)$ at the point $(x \pm a)$* (e.g., see [61]). The other rows can be obtained by the termwise discrete differentiation, because the operators $\underset{+h}{D}, \underset{-h}{D}$ and $\underset{+h}{\widetilde{D}}, \underset{-h}{\widetilde{D}}$ commute. The group (1.80) is called the *Newton group* [29].

The action of N_a^+ and N_a^- for $a > 0$ can be treated as a formal (resp., right and left) interpolation in the sense of Newton on infinitely many equidistance nodes; for $a < 0$, N_a^+ and N_a^- provide the respective left and right extrapolations.

We calculate the tangent field of a pair of formal groups (1.80), i.e., the Newton groups:

$$\xi^{\pm} = \left. \frac{\partial x^*}{\partial a} \right|_{a=0} = \pm 1,$$

$$\eta^{\pm} = \left. \frac{\partial u^*}{\partial a} \right|_{a=0} = \pm \underset{\pm h}{\widetilde{D}}(u),$$

$$\underset{h1}{\zeta^{\pm}} = \left. \frac{\partial \underset{h}{u_1^*}}{\partial a} \right|_{a=0} = \pm \underset{\pm h}{\widetilde{D}}(\underset{h}{u_1}), \qquad (1.81)$$

$$\cdots \cdots \cdots \cdots \cdots \cdots .$$

Instead of the pair of tangent fields (1.81), consider the infinitesimal operators of the Newton group:

$$\underset{h}{D^+} = \frac{\partial}{\partial x} + \underset{+h}{\widetilde{D}}(u) \frac{\partial}{\partial u} + \underset{+h}{\widetilde{D}}(\underset{h}{u_1}) \frac{\partial}{\partial \underset{h}{u_1}} + \cdots,$$

$$\underset{h}{D^-} = -\frac{\partial}{\partial x} - \underset{-h}{\widetilde{D}}(u) \frac{\partial}{\partial u} - \underset{-h}{\widetilde{D}}(\underset{h}{u_1}) \frac{\partial}{\partial \underset{h}{u_1}} - \cdots.$$

In the operator $\underset{h}{D^-}$, we preserve the sign "$-$", because $\underset{h}{D^-}$ determines the left shift for a positive value of the parameter a.

Thus, using heuristic considerations, we have constructed a formal group in $\underset{h}{Z}$, i.e., the Newton group. Its orbit coincides with the Taylor group orbit at the points $a = nh$.

Now let us show that the Newton group (1.80) with the tangent field (1.81) is indeed a "discrete" representation of the Taylor group in $\underset{h}{Z}$.

It is well known that finite transformations of a continuous group are bijectively related to infinitesimal transformations. In the case of point groups, this relation is expressed by a finite system of Lie equations. In the case of a higher symmetry group, the corresponding relation is expressed by an infinite chain of Lie equations, whose solution is given by a unique recursive sequence of coefficients of formal series [73]. In both cases, the solution of this system can be represented as the exponential mapping. In the case of $\underset{h}{\widetilde{Z}}$, the finite transformation of any coordinate z^i is given by the formula

$$z^{i*} = \underset{a}{S}(z^i) \equiv e^{aX}(z^i) \equiv \sum_{s \geq 0} \frac{a^s}{s!} X^s(z^i). \tag{1.82}$$

The series (1.82) can be inverted; i.e., the infinitesimal transformation $aX(z^i)$ can be reconstructed from the finite transformation $\underset{a}{S}(z^i)$ as the logarithmic series (e.g., see [111])

$$aX(z^i) = \ln[1 + (\underset{a}{S} - 1)](z^i)$$

$$\equiv \left((\underset{a}{S} - 1) - \frac{1}{2}(\underset{a}{S} - 1)^2 + \cdots + \frac{(-1)^{n-1}}{n}(\underset{a}{S} - 1)^n + \cdots \right)(z^i)$$

$$= \sum_{s=1}^{\infty} \frac{(-1)^{s-1}}{s}(\underset{a}{S} - 1)^s(z^i), \qquad i = 1, 2, \ldots .$$

We apply the process of reconstruction of the tangent field X from the finite transformations to the Taylor group, taking the parameter value $a = h$:

$$e^{aD}\Big|_{a=h} = \underset{+h}{S} = 1 + h\underset{+h}{D},$$

$$hD = h\underset{+h}{D} - \frac{1}{2}(h\underset{+h}{D})^2 + \cdots + \frac{(-1)^{n-1}}{n}(h\underset{+h}{D})^n + \cdots ,$$

which implies that

$$D = \sum_{n=1}^{\infty} \frac{(-h)^{n-1}}{n} \underset{+h}{D^n}; \tag{1.83}$$

i.e., we obtain an expression coinciding with the operator $\underset{+h}{\widetilde{D}}$. In formulas (1.83), we omit the argument z^i under the action of the corresponding operator. If $z^i \in \underset{h}{Z}$, then we assume that the difference derivatives on the left-hand side of (1.83) are given by series; if $z^i \in \widetilde{Z}$, then the operator $\underset{+h}{D}$ must be expressed in terms of e^{hD}. Formula (1.83) gives the action of the tangent field of the Taylor group on the coordinate z^i. The infinitesimal operator of the Taylor group in $\underset{h}{Z}$ can be written as (note that $\underset{+h}{\widetilde{D}}(x) = 1$)

$$\underset{h}{D^+} = \frac{\partial}{\partial x} + \underset{+h}{\widetilde{D}}(u)\frac{\partial}{\partial u} + \underset{+h}{\widetilde{D}}(\underset{h}{u_1})\frac{\partial}{\partial \underset{h}{u_1}} + \cdots + \underset{+h}{\widetilde{D}}(\underset{h}{u_s})\frac{\partial}{\partial \underset{h}{u_s}} + \dots . \tag{1.84}$$

In a similar way, for $a = -h$ we obtain

$$\underset{h}{D^-} = \frac{\partial}{\partial x} + \underset{-h}{\widetilde{D}}(u)\frac{\partial}{\partial u} + \underset{-h}{\widetilde{D}}(\underset{h}{u_1})\frac{\partial}{\partial \underset{h}{u_1}} + \cdots + \underset{-h}{\widetilde{D}}(\underset{h}{u_s})\frac{\partial}{\partial \underset{h}{u_s}} + \cdots . \tag{1.85}$$

Thus, the Taylor group with tangent field

$$D = \frac{\partial}{\partial x} + u_1\frac{\partial}{\partial u} + u_2\frac{\partial}{\partial u_1} + \cdots + u_{s+1}\frac{\partial}{\partial u_s} + \cdots$$

in $\widetilde{Z}$ can be represented in $\underset{h}{Z}$ by the Newton group with pair of tangent fields (1.84)–(1.85); i.e., *the Taylor and Newton groups are isomorphic.* If a transformation of $\widetilde{Z}$ into $\underset{h}{Z}$ is given, then the coordinates of the infinitesimal operator of the Taylor group are changed by using the operator series $\underset{\pm h}{\widetilde{D}}$, which can be written as

$$D \Longleftrightarrow \begin{cases} \displaystyle\sum_{n=1}^{\infty} \frac{(-h)^{n-1}}{n} \underset{+h}{D^n}, \\ \displaystyle\sum_{n=1}^{\infty} \frac{(+h)^{n-1}}{n} \underset{-h}{D^n}. \end{cases} \tag{1.86}$$

The upper part of the formula uses the right Newton series; and the lower part, the left Newton series. Note that these representations are taken to each other by a discrete reflection group, which obviously admits a uniform mesh.

Formula (1.86) has been known for a long time (e.g., see [61]). Apparently, it was first obtained by Lagrange [81, 82]. Of course, the fact that (1.86) is a relation between the coordinates of infinitesimal operators of the corresponding groups was not known, because, at these times, the notion of the group had not yet been formulated. This is just the novelty of formula (1.86).

Remark. **1.** The discrete representation of the Taylor group was constructed by using the formal Newton series. This representation is also possible for a nonuniform mesh, but the tangent field of the Newton group for an arbitrary nonuniform mesh is very cumbersome. Considering only a specific case of such a representation (on an *invariant* nonuniform mesh), we can reduce the problem to the preceding one by using the theorem about the "straightening" of an invariant nonuniform mesh.

2. Here we do not consider the problems of convergence of the Newton series, because we are interested only in the algebraic aspects of the above constructions. Nevertheless, it is of interest to note that the Newton group, although it is an isomorphic representation of the Taylor group, has analytic properties that are significantly different from those of the Taylor group. For example, the domain of convergence of the Newton series is a half-plane in a complex domain to the right of the vertical line passing through the real number λ_0, which is called the *convergence asymptote*. Apparently, the first estimates of the asymptote λ_0 were obtained by Abel. The further history of such estimates and the study of problems of the Newton series convergence can be found in [61, 71, 105].

In the next section, we consider some structure properties of the Lie algebra of operators of a formal group in the simplest one-dimensional case and on uniform difference meshes.

1.6. Commutation Properties and Factorization of Group Operators on Uniform Difference Meshes

As was already noted, in the theory of group properties of differential equations there are two equivalent approaches to describing point symmetries, the classical approach based on operators of the form

$$X = \xi(x, u)\frac{\partial}{\partial x} + \eta(x, u)\frac{\partial}{\partial u} + \cdots$$

and the approach based on the use of the factorized form of operators or the evolution vector fields. Under the second approach, the independent variables are invariants, which, at first sight, is very attractive from the standpoint of difference models, because the problem of invariance of meshes has already been solved. But, as follows from the detailed considerations of this approach to difference models, the evolution vectors fields thus obtained have an extremely complicated form and use infinitely many mesh nodes rather then only the difference stencil nodes.

In this section, we consider the structure of the set of operators of a formal group in the simplest case of a single independent variable and a uniform mesh (see [29]). The Newton group is used to construct an ideal of the Lie algebra of the set of all operators in the mesh space. The ideal thus constructed is used to factorize the set of operators of the formal group. A difference mesh whose description is obtained only by using the independent variable is invariant under such an ideal.

We assume that, on the same uniform mesh $\underset{h}{\omega}$, a set of operators of a formal group G_1 preserving the discrete first-order tangency is given:

$$X_i = \xi^i \frac{\partial}{\partial x} + \eta^i \frac{\partial}{\partial u} + [\underset{+h}{D}(\eta^i) - u_1 \underset{h}{\underset{+h}{D}}(\xi^i)]\frac{\partial}{\partial \underset{h}{u_1}} + \cdots + h\underset{+h}{D}(\xi^i)\frac{\partial}{\partial \eta}, \qquad i = 1, 2, \ldots.$$

For any two operators X_1 and X_2, we introduce the multiplication (commutation) operation by the usual formula

$$[X_1, X_2] = X_1 X_2 - X_2 X_1.$$

The commutator $[X_1, X_2]$ contains differentiation of at most first order and hence is an operator of the formal group:

$$
\begin{aligned}
[X_1, X_2] &= (X_1(\xi^2) - X_2(\xi^1))\frac{\partial}{\partial x} + (X_1(\eta^2) - X_2(\eta^1))\frac{\partial}{\partial u} \\
&\quad + [X_1(\underset{+h}{D}(\eta^2) - u_1 \underset{+h}{D}(\xi^2)) - X_2(\underset{+h}{D}(\eta^1) - u_1 \underset{+h}{D}(\xi^1))]\frac{\partial}{\partial \underset{h}{u_1}} \\
&\quad + \cdots + [X_1(h\underset{+h}{D}(\xi^2)) - X_2(h\underset{+h}{D}(\xi^1))]\frac{\partial}{\partial h}. \quad (1.87)
\end{aligned}
$$

Is the commutator $[X_1, X_2]$ an operator preserving the meaning of the difference derivatives? To answer this question, it suffices to verify whether it preserves the "discrete tangency" of first order (i.e., whether it preserves the meaning of the first difference derivative at each point of $\underset{h}{\omega}$):

$$\underset{h}{du} = \underset{h}{u_1} h, \qquad \text{where} \quad \underset{h}{d} = \underset{+h}{S} - 1. \qquad (1.88)$$

We extend the operator (1.87) to the variable $\underset{h}{du}$ by the formula (see [21])

$$\left[\frac{\partial}{\partial a}(\underset{h}{du^*})\right]_{a=0} = h\underset{+h}{D}(X_1(\eta^2) - X_2(\eta^1));$$

acting by this operator on (1.88), we hence obtain the condition

$$\underset{+h}{D}X_1(\eta^2) - \underset{+h}{D}X_2(\eta^1) - X_1\underset{+h}{D}(\eta^2) + X_2\underset{+h}{D}(\eta^1) - \underset{+h}{D}(\xi^1)\underset{+h}{D}(\eta^2) + \underset{+h}{D}(\xi^2)\underset{+h}{D}(\eta^1) = 0. \qquad (1.89)$$

To prove relation (1.89), we need to compute the commutator $[X, \underset{+h}{D}] \equiv X\underset{+h}{D} - \underset{+h}{D}X$. The expression $X(\eta)$ is a function in $\underset{h}{A}$, i.e., an analytic function of finitely many variables in $\underset{h}{Z}$ if $\eta \in \underset{h}{A}$. By the definition of discrete differentiation of a function in $\underset{h}{A}$, we have

$$\underset{+h}{D}X(\eta) = \frac{1}{h}(X(\underset{+h}{S}(\eta)) - X(\eta))$$

and hence

$$[X, \underset{+h}{D}] = -\underset{+h}{D}(\xi)\underset{+h}{D}. \qquad (1.90)$$

The substitution of formula (1.90) into (1.89) takes the latter into an identity. Thus, the commutator (1.87) preserves discrete first-order tangency. Since any operator of a formal group preserving discrete first-order tangency at each point of $\underset{h}{\omega}$ also preserves tangency of any finite order, it follows that the commutator $[X_1, X_2]$ preserves any finite discrete tangency.

Thus, we have the following assertion.

THEOREM 1.15. *The set of all operators of a formal group given on the same uniform mesh $\underset{h}{\omega}$ forms a Lie algebra with multiplication*

$$[X_1, X_2] = X_1 X_2 - X_2 X_1.$$

Now consider the tangent field of the Newton group, i.e., the pair of operators

$$
\begin{aligned}
\underset{h}{D^+} &= \frac{\partial}{\partial x} + \underset{+h}{\widetilde{D}}(u)\frac{\partial}{\partial u} + \underset{+h}{\widetilde{D}}(\underset{h}{u_1})\frac{\partial}{\partial \underset{h}{u_1}} + \cdots, \\
\underset{h}{D^-} &= \frac{\partial}{\partial x} + \underset{-h}{\widetilde{D}}(u)\frac{\partial}{\partial u} + \underset{-h}{\widetilde{D}}(\underset{h}{u_1})\frac{\partial}{\partial \underset{h}{u_1}} \cdots,
\end{aligned}
\tag{1.91}
$$

where

$$\underset{\pm h}{\widetilde{D}} = \sum_{n=1}^{\infty} \frac{(\mp h)^{n-1}}{n}\underset{\pm h}{D^n}.$$

Note that one tangent field of the Taylor group in $\widetilde{Z}$,

$$D = \frac{\partial}{\partial x} + u_1\frac{\partial}{\partial u} + \cdots + u_{s+1}\frac{\partial}{\partial u_s} + \cdots,$$

is associated with the pair of fields (1.91) in the mesh space $\underset{h}{Z}$. This doubling of objects, i.e., the appearance of the "right" and "left" objects, is a typical feature of mesh spaces and concerns not only the operators of the Newton group (1.91) but also the discrete shift $\underset{\pm h}{S} = e^{\pm hD}$, the discrete differentiation $\underset{\pm h}{D}$, etc. In the case of a *uniform* difference mesh, this doubling is related to the existence of a specific discrete group, namely, the reflection group: $x \to -x$, which changes the sign of the mesh spacing h, $h \to -h$. Therefore, instead of the pair of Newton groups with operators (1.91), we can consider one group, which means factorization with respect to the reflection roup. Thus, *in the one-dimensional case with a uniform mesh $\underset{h}{\omega}$, the transition from the Taylor group in $\widetilde{Z}$ to the Newton group in $\underset{h}{\bar{Z}}$ is the transition to an isomorphic continuous group with addition of a discrete reflection group.*

Consider the following formal group operator:

$$X = \xi\frac{\partial}{\partial x} + \eta\frac{\partial}{\partial u} + \underset{h}{\zeta_1}\frac{\partial}{\partial \underset{+h}{u_1}} + \underset{h}{\zeta_2}\frac{\partial}{\partial \underset{+h}{u_2}} + \cdots + h\underset{+h}{D}(\xi)\frac{\partial}{\partial h},$$

where

$$\zeta_1 = D_{\substack{+h}}(\eta) - u_1 D_{\substack{+h}}(\xi), \qquad \zeta_2 = D(\zeta_1) - u_2 D(\xi), \ \dots . \tag{1.92}$$

Formulas (1.92) for $\zeta_1, \zeta_2, \dots$ ensure the preservation of the meaning of the finite-difference derivatives under the formal group transformations.

One can readily see that the operators (1.91) satisfy relations (1.92); i.e., the *Newton group preserves the meaning of finite-difference derivatives of any finite order.*

Multiplying the operator (1.92) by a certain function $\widetilde{\xi}(z) \in A_h$ on the left, we generally take it out of the set of operators preserving discrete tangency. Let us introduce a special operation of left multiplication of an operator of a formal group by an arbitrary analytic function $\widetilde{\xi}(z) \in A_h$: $\widetilde{\xi} * X$. In the operator $\widetilde{\xi} * X$, the first coordinates are multiplied by $\widetilde{\xi}$,

$$\widetilde{\xi} * X = \widetilde{\xi}\xi \frac{\partial}{\partial x} + \widetilde{\xi}\eta \frac{\partial}{\partial u} + \cdots,$$

and the other coordinates are constructed so that the operator determines a group preserving the finite-difference derivative of first order (and hence any difference derivative of finite order). Thus, $\widetilde{\xi} * X$ must satisfy formulas (1.92),

$$\widetilde{\xi} * X = \widetilde{\xi}\xi \frac{\partial}{\partial x} + \widetilde{\xi}\eta \frac{\partial}{\partial u} + [D_{\substack{+h}}(\widetilde{\xi}\eta) - u_1 D_{\substack{+h}}(\widetilde{\xi}\xi)]\frac{\partial}{\partial u_1} + \cdots + h D_{\substack{+h}}(\widetilde{\xi}\xi)\frac{\partial}{\partial h}, \tag{1.93}$$

and does not coincide with the operator $\widetilde{\xi}X$.

The same operation of left multiplication by $\widetilde{\xi}(z)$ can be introduced in the "continuous" space $\widetilde{Z}$ with the requirement to *preserve the infinite-order tangency.*

Suppose that the following formal group operator is given in $\widetilde{Z} = (x, u, u_1, \dots)$:

$$X = \xi \frac{\partial}{\partial x} + \eta \frac{\partial}{\partial u} + \sum_{s \geq 0} \zeta_s \frac{\partial}{\partial u_s},$$

where $\zeta_s = D^s(\eta - \xi u_1) + \xi u_{s+1} = D(\zeta_{s-1}) - u_s D(\xi)$.

Then the multiplication operation $(*)$ implies

$$\widetilde{\xi} * X = \widetilde{\xi}\xi \frac{\partial}{\partial x} + \widetilde{\xi}\eta \frac{\partial}{\partial u} + [D(\widetilde{\xi}\eta) - u_1 D(\widetilde{\xi}\xi)]\frac{\partial}{\partial u_1} + \cdots .$$

One can readily see that

$$\widetilde{\xi} * X = \widetilde{\xi}X + \sum_{s \geq 0}\sum_{n=1}^{s} C_s^n D^n(\widetilde{\xi}) D^{(s-n)}(\eta - \xi u_1)\frac{\partial}{\partial u_s};$$

i.e., for a formal group operator to be multiplied on the left by an arbitrary function $\widetilde{\xi} \in A$ so that it remains an operator preserving the infinite-order tangency condition, it is necessary and sufficient that $\eta = \xi u_1$. This condition is satisfied by the coordinates of the operator ξD of the Taylor group. Thus, the operator D is the only operator that can be multiplied on the left by $\widetilde{\xi}(z) \in A$ "without penalty."

This situation does not hold in the mesh space $\underset{h}{Z}$; namely, the tangent field of the Newton group $\underset{h}{D^{\pm}}$ cannot be multiplied on the left by $\widetilde{\xi}(z) \in \underset{h}{A}$, which is related to the specific features of the difference Leibniz rule. Therefore, it is necessary to use formula (1.93), i.e., construct an operator of the form

$$\widetilde{\xi} * \underset{h}{D^{\pm}} = \pm \widetilde{\xi}\frac{\partial}{\partial x} \pm \widetilde{\xi}\underset{\pm h}{\widetilde{D}}(u)\frac{\partial}{\partial u} \pm [\underset{+h}{D}(\widetilde{\xi}\underset{\pm h}{D}(u)) - u_1\underset{h}{D}(\widetilde{\xi})]\frac{\partial}{\partial \underset{h}{u_1}} + \cdots \pm h\underset{+h}{D}(\widetilde{\xi})\frac{\partial}{\partial h}.$$
$$(1.94)$$

Note that the coordinate $h\underset{+h}{D}(\widetilde{\xi})$ appears in the operator (1.94); this coordinate determines the deformation of the spacing of the mesh $\underset{h}{\omega}$ and is zero for the operator $\underset{h}{D^{\pm}}$ of the Newton group.

Consider the commutation properties of the Lie–Bäcklund operators X, $\underset{h}{D^{\pm}}$, and $\xi * \underset{h}{D^{\pm}}$ in the mesh space $\underset{h}{Z}$.

LEMMA 1.16. *For the formal group operators X and $\underset{h}{D^{\pm}}$ defined on the same uniform mesh $\underset{h}{\omega}$, the following relation holds*:

$$[X, \underset{h}{D^{\pm}}] = -(\underset{h}{D^{\pm}}(\xi)) * \underset{h}{D^{\pm}}. \qquad (1.95)$$

Proof. To prove relation (1.95), consider only the "right" operator $\underset{h}{D^+}$ of the Newton group. Let us write out the left- and right-hand sides of (1.95):

$$-\underset{h}{D^+}(\xi)\frac{\partial}{\partial x} - \underset{h}{D^+}(\xi)\underset{+h}{\widetilde{D}}(u)\frac{\partial}{\partial u} + \cdots + h\underset{+h}{D}(\underset{h}{D^+}(\xi))\frac{\partial}{\partial h}$$
$$= -\underset{h}{D^+}(\xi)\frac{\partial}{\partial x} + [X(\underset{+h}{\widetilde{D}}(u)) - \underset{h}{D^+}(\eta)]\frac{\partial}{\partial u} + \cdots + h\underset{+h}{D^+}(\underset{h}{D^+}(\xi))\frac{\partial}{\partial h}.$$

The coordinates of $\partial/\partial x$ and $\partial/\partial h$ coincide, because $\underset{h}{D^+}$ and $\underset{h}{\widetilde{D}}$ commute. The simplest way to prove that the coordinates of $\partial/\partial u$ coincide in relation (1.95) is to use the continuous representation of the coefficients, i.e., reflect them from $\underset{h}{Z}$ into $\widetilde{Z}$, and then vice versa:

$$X(\underset{+h}{\widetilde{D}}(u)) - \underset{h}{D^+}(\eta) \longleftrightarrow X(u_1) - D(\eta) = D(\eta) - u_1 D(\xi) - D(\eta)$$

$$= -D(\xi)u_1 \longleftrightarrow -\underset{h}{D^+}(\xi)\underset{+h}{\widetilde{D}}(u).$$

The coincidence of the other coordinates, i.e., of $\partial/\partial u_s$, $s = 1, 2, \ldots$, is ensured by Theorem 1.15 and the multiplication $(*)$ introduced above, because the left- and right-hand sides in relation (1.95) contain formal group operators whose coordinates of $\partial/\partial x$ and $\partial/\partial u$ coincide, while the other coordinates are obtained by the same prolongation formulas. The proof of the lemma is complete. $\square$

LEMMA 1.17. *The following commutation relation holds for the formal group operators* X, $\widetilde{\xi} * D_h^\pm$, *and* $\widetilde{\xi}(z) \in A_h$ *defined on the same uniform mesh* ω_h:

$$[\widetilde{\xi} * D_h^\pm, X] = \left(\widetilde{\xi} * D_h^\pm(\xi) - X(\widetilde{\xi})\right) * D_h^\pm.$$

Proof. The proof of Lemma 1.17 is quite similar to that of the preceding lemma; namely, one should establish the coincidences of the coordinates of the operators for $\partial/\partial x$ and $\partial/\partial u$ (see [29]). $\square$

The multiplication $(*)$ introduced above and Lemma 1.17 imply the following assertion.

THEOREM 1.18. *The set of operators of the form*

$$\widetilde{\xi} * D_h^\pm = \widetilde{\xi}\frac{\partial}{\partial x} + \widetilde{\xi} * \underset{\pm h}{\widetilde{D}}(u)\frac{\partial}{\partial u} + \cdots \qquad (1.96)$$

with arbitrary analytic functions $\widetilde{\xi}(z) \in A_h$ *is an ideal in the Lie algebra of all formal group operators on the same uniform mesh* ω_h.

Therefore, instead of the Lie algebra of the operators

$$X = \xi\frac{\partial}{\partial x} + \eta\frac{\partial}{\partial u} + [\underset{+h}{D}(\eta) - u_1\underset{h}{}\underset{+h}{D}(\xi)]\frac{\partial}{\partial u_1} + \cdots + h\underset{+h}{D}(\xi)\frac{\partial}{\partial h}$$

we can consider the quotient algebra by the ideal (1.96).

For representatives of the above quotient algebra we take the operators with coordinate $\xi \equiv 0$,

$$\bar{X} = \bar{\eta}\frac{\partial}{\partial u} + \underset{h}{\zeta_1}\frac{\partial}{\partial u_1} + \underset{h}{\zeta_2}\frac{\partial}{\partial u_2} + \cdots, \qquad (1.97)$$

where $\bar{\eta} = \eta - \xi\underset{\pm h}{\widetilde{D}}(u)$.

The operators (1.97) are called *canonical operators*, just as in the continuous case (see [73]). For these operators, the prolongation formulas have the simple form

$$\underset{h}{\zeta_1} = \underset{+h}{D}(\bar{\eta}), \qquad \underset{h}{\zeta_2} = \underset{h}{D}\underset{-h+h}{D}(\bar{\eta}), \ldots.$$

Note that the independent variable for the canonical operator $\bar{X}$ is an invariant, and hence so is the mesh spacing. (The coordinate of $\partial/\partial h$ in the operator $\bar{X}$ is zero.) In Lemma 1.16, the canonical operators $\bar{X}$ commute with the operators $\underset{h}{D^+}$ and $\underset{h}{D^-}$.

Another approach to the factorization of formal group operators is as follows. Consider the space $(\dots, x^-, u^-, x, u, x^+, u^+, \dots)$ of discrete variables without introducing the finite-difference derivatives, i.e., the space of sequences obtained by successive actions of the shift operators on the coordinates (x, u).

Consider the tangent field of the Newton group in such a space:

$$
\begin{aligned}
\underset{h}{D^+} &= \frac{\partial}{\partial x} + \frac{\partial}{\partial x^+} + \underset{+h}{\tilde{D}}(u)\frac{\partial}{\partial u} + \underset{+h}{\tilde{D}}(u^+)\frac{\partial}{\partial u^+} + \cdots, \\
\underset{h}{D^-} &= \frac{\partial}{\partial x} + \frac{\partial}{\partial x^-} + \underset{-h}{\tilde{D}}(u)\frac{\partial}{\partial u} + \underset{-h}{\tilde{D}}(u^-)\frac{\partial}{\partial u^-} \cdots, \\
\underset{\pm h}{\tilde{D}} &= \sum_{n=1}^{\infty} \frac{(\mp h)^{n-1}}{n} \underset{\pm h}{D^n}.
\end{aligned}
\tag{1.98}
$$

We introduce the *usual* left multiplication of an operator of the Newton group by a function $\tilde{\xi}$,

$$
\begin{aligned}
\tilde{\xi}\underset{h}{D^+} &= \tilde{\xi}\frac{\partial}{\partial x} + \tilde{\xi}\frac{\partial}{\partial x^+} + \tilde{\xi}\underset{+h}{\tilde{D}}(u)\frac{\partial}{\partial u} + \tilde{\xi}\underset{+h}{\tilde{D}}(u^+)\frac{\partial}{\partial u^+} + \cdots, \\
\tilde{\xi}\underset{h}{D} &= \tilde{\xi}\frac{\partial}{\partial x} + \tilde{\xi}\frac{\partial}{\partial x^-} + \tilde{\xi}\underset{h}{\tilde{D}}(u)\frac{\partial}{\partial u} + \tilde{\xi}\underset{-h}{\tilde{D}}(u^-)\frac{\partial}{\partial u^-} \cdots.
\end{aligned}
$$

Note that the action of the operator $\tilde{\xi}\underset{h}{D^\pm}$ on the mesh spacing $h^+ = x^+ - x$ is zero.

Lemma 1.19. *Any formal group operators X, $\tilde{\xi}\underset{h}{D^\pm}$, and $\underset{h}{\tilde{\xi}}(z) \in A$ given on the same uniform mesh $\underset{h}{\omega}$ satisfy the commutation relation*

$$
[\tilde{\xi}\underset{h}{D^\pm}, X] = \left(\tilde{\xi}\underset{h}{D^\pm}(\xi) - X(\tilde{\xi})\right)\underset{h}{D^\pm}.
\tag{1.99}
$$

Lemma 1.19 can be proved by straightforward computations.   □

In fact, relation (1.99) is completely similar to the corresponding relation for the Lie–Bäcklund groups [73], which is extended to the "shifted" points x^-, u^-, x^+, $u^+, \dots$. Note that the difference derivatives and the mesh spacings are not contained in this relation.

In the space $(\dots, x^-, u^-, x, u, x^+, u^+, \dots)$, consider the point group operator

$$
X = \xi\frac{\partial}{\partial x} + \xi^+\frac{\partial}{\partial x^+} + \xi^-\frac{\partial}{\partial x^-} + \cdots + \eta\frac{\partial}{\partial u} + \eta^+\frac{\partial}{\partial u^+} + \eta^-\frac{\partial}{\partial u^-} + \cdots.
\tag{1.100}
$$

Suppose that the operators of the form (1.100) on a uniform mesh form a Lie algebra. Then Lemma 1.19 implies the following assertion.

THEOREM 1.20. *The set of operators of the form*

$$\widetilde{\xi} \underset{h}{D^{\pm}} = \widetilde{\xi} \frac{\partial}{\partial x} + \widetilde{\xi} \frac{\partial}{\partial x^{+}} + \widetilde{\xi} \underset{\pm h}{\widetilde{D}}(u) \frac{\partial}{\partial u} + \widetilde{\xi} \underset{\pm h}{\widetilde{D}}(u^{+}) \frac{\partial}{\partial u^{+}} + \cdots \qquad (1.101)$$

with arbitrary analytic functions $\widetilde{\xi}(z) \in \underset{h}{A}$ *is an ideal in the Lie algebra of all formal group operators on the same uniform mesh* $\underset{h}{\omega}$.

Therefore, instead of the Lie algebra of operators (1.100), we can consider the quotient algebra by the ideal (1.101). For the representatives of this quotient algebra we can take the operators

$$\bar{X} = (\xi^{+} - \xi) \frac{\partial}{\partial x^{+}} + (\xi^{-} - \xi) \frac{\partial}{\partial x^{-}} + \cdots + (\eta - \xi \underset{\pm h}{\widetilde{D}}(u)) \frac{\partial}{\partial u}$$

$$+ (\eta^{+} - \xi \underset{\pm h}{\widetilde{D}}(u^{+})) \frac{\partial}{\partial u^{+}} + \cdots + (\xi^{+} - \xi) \frac{\partial}{\partial h_{+}} + (\xi - \xi^{-}) \frac{\partial}{\partial h_{-}} + \cdots .$$

Note that, in contrast to the preceding version of factorization, the operator $\bar{X}$ transforms the mesh spacings.

Here we have considered only the simplest case of a Lie algebra of operators on a uniform mesh. The transition to a nonuniform mesh changes the tangent vector field of the group but does not vary the structure properties of the set of operators, which are scalar quantities and are independent of coordinate system. In particular, if the nonuniform mesh is *invariant*, then the theorem about the "straightening" of such a mesh reduces the problem to the already studied problem.

Thus, both versions of factorization in the difference case have the consequence that the coefficients of the operators become formal series defined on infinitely many points of the difference mesh. This fact significantly complicates their practical use. An example of application of factorized symmetry operators will be considered in a subsequent chapter.

1.7. Finite-Difference Integration and Prolongation of the Mesh Space to Nonlocal Variables

The above-introduced discrete (finite-difference) variables were obtained by using a pair of discrete shift operators $\underset{\pm h}{S}$ and the differentiation operator $\underset{\pm h}{D}$. Since the operator $\underset{\pm h}{S}$ is an operator of *finite transformations of the group* for a fixed value of the parameter $a = \pm h$, it is invertible, just as any action of the Taylor group. The inverse of the shift operator is the operator of finite transformations of the Taylor group with opposite sign of the group parameter; i.e.,

$$\left(\underset{+h}{S}\right)^{-1} = \underset{-h}{S}, \qquad \left(\underset{-h}{S}\right)^{-1} = \underset{+h}{S},$$

or, otherwise, $\underset{+h-h}{S\,S} = \underset{-h+h}{S\,S} = 1$.

It is of interest to find the operator of *discrete integration* as the inverse of the pair $\underset{+h}{D}, \underset{-h}{D}$.

We consider only the simplest case of a single independent variable and a uniform mesh with spacing h.

The *operator of discrete integration on a uniform mesh* $\underset{h}{\omega}$ *is defined to be a pair of linear operators* $\underset{+h}{D^{-1}}, \underset{-h}{D^{-1}}$ *commuting with the pair* $\underset{+h}{D}, \underset{-h}{D}$ *and satisfying*

$$\underset{+h}{D^{-1}}\underset{+h}{D} = \underset{+h+h}{D\,D^{-1}} = 1, \qquad \underset{-h}{D^{-1}}\underset{-h}{D} = \underset{-h-h}{D\,D^{-1}} = 1, \qquad (1.102)$$

Starting from this definition, one can readily obtain the "mixed" actions of the operators $\underset{\pm h}{D}$ and $\underset{\pm h}{D^{-1}}$:

$$\underset{-h+h}{D\,D}{}^{1} = \underset{+h}{D}{}^{1}\underset{-h}{D} = \underset{-h}{S}, \qquad \underset{+h-h}{D\,D^{-1}} = \underset{-h}{D^{-1}}\underset{+h}{D} = \underset{+h}{S}.$$

Definition (1.102) readily permits obtaining the table of action of the operator $\underset{\pm h}{D^{-1}}$ on the finite-difference derivatives $\underset{h}{u_s}$ (see Table 1.3). Here we use the following notation for difference derivatives: $\underset{h}{u_1} = \underset{h}{u_x}$, $\underset{h}{u_2} = \underset{h}{u_{x\bar{x}}}$, etc.

Starting from the definition of the operators $\underset{\pm h}{D}{}^{-1}$, one can readily obtain the discrete rule of integration by parts on a uniform mesh $\underset{h}{\omega}$:

$$\boxed{\begin{array}{l} \underset{+h}{D^{-1}}(\underset{h}{u_1}v) = uv - \underset{+h}{D^{-1}}(\underset{h}{uv_1}) - h\underset{+h}{D^{-1}}(\underset{h}{u_1}\underset{h}{v_1}) = uv - \underset{+h}{D^{-1}}(u^{+}\underset{h}{v_1}), \\ \underset{-h}{D^{-1}}(\underset{h}{u_{\bar{1}}}v) = uv - \underset{-h}{D^{-1}}(\underset{h}{uv_{\bar{1}}}) - h\underset{-h}{D^{-1}}(\underset{h}{u_{\bar{1}}}\underset{h}{v_{\bar{1}}}) = uv - \underset{-h}{D^{-1}}(u^{-}\underset{h}{v_{\bar{1}}}), \end{array}} \qquad (1.103)$$

where $f^{+} = \underset{+h}{S}(f)$, $f^{-} = \underset{-h}{S}(f)$, and $\underset{h}{u_{\bar{1}}}$ and $\underset{h}{v_{\bar{1}}}$ are the left difference derivatives of first order.

Formulas (1.103), which are called the "Abel transformations" in the old literature (e.g., see [61]), can readily be verified by the action of the operators $\underset{\pm h}{D}$ with the use of the difference Leibniz rule.

Note that the above-introduced discrete integration does not permit closing its action in $\underset{h}{Z} = (x, u, \underset{h}{u_1}, \underset{h}{u_2}, \underset{h}{u_3}, \ldots)$. It is not difficult to extend the action of $\underset{\pm h}{D^{-1}}$ to x:

$$\underset{+h}{D^{-1}}(x) = \frac{x\check{x}}{2}, \qquad \underset{-h}{D^{-1}}(x) = \frac{x\hat{x}}{2}.$$

But the action of $\underset{\pm h}{D^{-1}}$ on u takes it outside $\underset{h}{Z}$. To make the action of $\underset{\pm h}{D^{-1}}$ in $\underset{h}{Z}$ well defined, it is necessary to extend it to the *nonlocal* variables, which can be

Table 1.3: Action of the discrete integration operators on the finite-difference derivatives $\left(u_1, u_2, u_3, \ldots\right)$.
 $\underset{h}{}\underset{h}{}\underset{h}{}$

"left" $\underset{-h}{D^{-1}}$-integration	"right" $\underset{+h}{D^{-1}}$-integration
$\underset{-h}{D^{-1}}(\underset{h}{u_1}) = u + h\underset{h}{u_1}$	$\underset{+h}{D^{-1}}(\underset{h}{u_1}) = u$
$\underset{-h}{D^{-1}}(\underset{h}{u_2}) = \underset{h}{u_1}$	$\underset{+h}{D^{-1}}(\underset{h}{u_2}) = \underset{h}{u_1} - h\underset{h}{u_2}$
$\underset{-h}{D^{-1}}(\underset{h}{u_3}) = \underset{h}{u_2} + h\underset{h}{u_3}$	$\underset{+h}{D^{-1}}(\underset{h}{u_3}) = \underset{h}{u_2}$
$\cdots\cdots\cdots\cdots\cdots$	$\cdots\cdots\cdots\cdots\cdots$
$\underset{-h}{D^{-1}}(\underset{h}{u_{2k+1}}) = \underset{h}{u_{2k}} + h\underset{h}{u_{2k+1}}$	$\underset{+h}{D^{-1}}\underset{h}{u_{2k+1}} = \underset{h}{u_{2k}}$
$\underset{-h}{D^{-1}}(\underset{h}{u_{2k+2}}) = \underset{h}{u_{2k+1}}$	$\underset{+h}{D^{-1}}(\underset{h}{u_{2k+2}}) = \underset{h}{u_{2k+1}} - h\underset{h}{u_{2k+2}}$
$\cdots\cdots\cdots\cdots\cdots$	$\cdots\cdots\cdots\cdots\cdots$

introduced as follows:

$$\underset{h}{u_{-1}} = \underset{-h}{D^{-1}}(u), \qquad\qquad \underset{h}{u_{-2}} = \underset{+h}{D^{-1}}(\underset{h}{u_{-1}}),$$

$$\cdots\cdots\cdots\cdots \qquad\qquad \cdots\cdots\cdots\cdots$$

$$\underset{h}{u_{-2k}} = \underset{+h}{D^{-1}}(\underset{h}{u_{-2k+1}}), \qquad \underset{h}{u_{-2k-1}} = \underset{-h}{D^{-1}}(\underset{h}{u_{-2k}}),$$

$$\cdots\cdots\cdots\cdots \qquad\qquad \cdots\cdots\cdots\cdots$$

In the space $\underset{h}{Z}$ supplemented with the nonlocal variables $\underset{h}{u_{-s}}$, the action of $\underset{\pm h}{D^{-1}}$ is closed (see Table 1.4, which includes Table 1.3).

It is of interest to learn to express the operators $\underset{\pm h}{D^{-1}}$ in terms of the discrete shift operators $\underset{\pm h}{S}$, i.e., to relate them to the Taylor group.

LEMMA 1.21. *Equation* $\underset{\pm h}{D^{-1}}\underset{\pm h}{D} = 1$ *is solvable in the class of formal operator series of the form*

$$\underset{+h}{D^{-1}} = -h\sum_{\alpha=0}^{\infty}\underset{+h}{S}^{\alpha} = -h(1 + \underset{+h}{S} + \underset{+h}{S}^2 + \cdots) \equiv -h\sum_{\alpha=0}^{\infty}e^{+\alpha hD},$$

$$\underset{-h}{D^{-1}} = +h\sum_{\alpha=0}^{\infty}\underset{-h}{S}^{\alpha} = h(1 + \underset{-h}{S} + \underset{-h}{S}^2 + \cdots) \equiv h\sum_{\alpha=0}^{\infty}e^{-\alpha hD}. \tag{1.104}$$

Indeed,

$$\underset{+h}{D^{-1}}\underset{+h}{D} = \underset{+h+h}{D\,D^{-1}} = \frac{1}{h}(\underset{+h}{S} - 1)[-h(1 + \underset{+h}{S} + \underset{+h}{S}^2 + \cdots)]$$

$$= 1 + \underset{+h}{S} + \underset{+h}{S}^2 + \cdots - \underset{+h}{S} - \underset{+h}{S}^2 - \cdots = 1.$$

Table 1.4: Discrete integration in the extended space
$\widetilde{Z}_{h} = (\ldots, u_{-2}, u_{-1}, x, u, u_1, u_2, \ldots)$.

"left" $\underset{-h}{D^{-1}}$-integration	"right" $\underset{+h}{D^{-1}}$-integration
$\underset{-h}{D^{-1}}(\underset{h}{u_{-2k}}) = \underset{h}{u_{-2k-1}}$	$\underset{+h}{D^{-1}}(\underset{h}{u_{-2k}}) = \underset{h}{u_{-2k-1}} - \underset{h}{hu_{-2k}}$
$\underset{-h}{D^{-1}}(\underset{h}{u_{-2k+1}}) = \underset{h}{u_{-2k}} + \underset{h}{hu_{-2k+1}}$	$\underset{+h}{D^{-1}}(\underset{h}{u_{-2k+1}}) = \underset{h}{u_{-2k}}$
$\cdots\cdots\cdots\cdots\cdots\cdots\cdots$	$\cdots\cdots\cdots\cdots\cdots\cdots\cdots$
$\underset{-h}{D^{-1}}(\underset{h}{u_2}) = \underset{h}{u_3}$	$\underset{+h}{D^{-1}})\underset{h}{u_2} = \underset{h}{u_3} - \underset{h}{hu_2}$
$\underset{-h}{D^{-1}}(\underset{h}{u_1}) = \underset{h}{u_2} + \underset{h}{hu_1}$	$\underset{+h}{D^{-1}}(\underset{h}{u_1}) = \underset{h}{u_2}$
$\underset{-h}{D^{-1}}(u) = \underset{h}{u_1}$	$\underset{+h}{D^{-1}}(u) = \underset{h}{u_1} - hu$
$\underset{-h}{D^{-1}}(\underset{h}{u_1}) = u + \underset{h}{hu_1}$	$\underset{+h}{D^{-1}}(\underset{h}{u_1}) = u$
$\underset{-h}{D^{-1}}(\underset{h}{u_2}) = \underset{h}{u_1}$	$\underset{+h}{D^{-1}}(\underset{h}{u_2}) = \underset{h}{u_1} - \underset{h}{hu_2}$
$\underset{-h}{D^{-1}}(\underset{h}{u_3}) = \underset{h}{u_2} + \underset{h}{hu_3}$	$\underset{+h}{D^{-1}}(\underset{h}{u_3}) = \underset{h}{u_2}$
$\cdots\cdots\cdots\cdots\cdots\cdots\cdots$	$\cdots\cdots\cdots\cdots\cdots\cdots\cdots$
$\underset{-h}{D^{-1}}(\underset{h}{u_{2k+1}}) = \underset{h}{u_{2k}} + \underset{h}{hu_{2k+1}}$	$\underset{+h}{D^{-1}}(\underset{h}{u_{2k+1}}) = \underset{h}{u_{2k}}$
$\underset{-h}{D^{-1}}(\underset{h}{u_{2k+2}}) = \underset{h}{u_{2k+1}}$	$\underset{+h}{D^{-1}}(\underset{h}{u_{2k+2}}) = \underset{h}{u_{2k+1}} - \underset{h}{hu_{2k+2}}$
$\cdots\cdots\cdots\cdots\cdots\cdots\cdots$	$\cdots\cdots\cdots\cdots\cdots\cdots\cdots$

The representation (1.104) permits extending the action of $\underset{\pm h}{D^{-1}}$ to analytic functions $F(z) \in \underset{h}{A}$ of finitely many variables:

$$\underset{\pm h}{D^{-1}}F(z) = \mp h[F(z) + F(\underset{\pm h}{S}(z)) + F(\underset{\pm h}{S}^2(z)) + \cdots].$$

The series on the right-hand side converges if $F(z)$ decreases sufficiently rapidly at infinity. (It follows from the analysis that this series converges, for example, simultaneously together with the corresponding improper integral.)

By using the operators $\underset{\pm h}{D^{-1}}$, one can introduce the *inner product*. For example, the analog of the inner product in L_2 acquires the form

$$(u, v) = (\underset{-h}{D^{-1}} - \underset{-h}{D^{-1}} - h)(uv). \qquad (1.105)$$

One can show that the *discrete shift operator* $\underset{\pm h}{S}$ is *unitary* with respect to the inner product (1.105):

$$\underset{\pm h}{S}^* = (\underset{\pm h}{S})^{-1} = \underset{\mp h}{S}.$$

Indeed, it follows from (1.104) and (1.105) that

$$(\underset{+h}{S}(u), v) = h(\dots + \underset{+h}{S}(u)v + u\underset{-h}{S}(v) + \underset{-h}{S}(u)\underset{-h}{S}^2(v) + \cdots),$$

$$(u, (\underset{+h}{S})^{-1}(v)) = h(\dots + \underset{+h}{S}(u)v + u\underset{-h}{S}(v) + \underset{-h}{S}(u)\underset{-h}{S}^2(v) + \cdots);$$

i.e.,

$$(\underset{+h}{S}(u), v) = (u, \underset{-h}{S}(v)).$$

In conclusion, note that the problems of convergence of the above-considered series in the shift operator are not the object of our attention and do not affect the algebraic aspects of studying the difference forms and equations.

1.8. Change of Variables in the Mesh Space

In this section, we present some formulas of the change of variables in $\underset{h}{Z}$, which will be used in what follows to study group properties of difference equations. We consider only the case of a single independent variable.

1. Suppose that a formal one-parameter group G_1 acts in $\widetilde{Z}$ and $\underset{h}{Z}$:

$$
\begin{aligned}
x* &= f(z, a), & h^{+*} &= \underset{+h}{S}(f(z, a)) - f(z, a), \\
u^* &= g(z, a), & \underset{h}{u_1^*} &= \varphi_1(z, a), \\
\underset{h}{u_1^*} &= g_1(z, a), & \underset{h}{u_2^*} &= \varphi_2(z, a), \\
& \cdots\cdots\cdots & & \cdots\cdots\cdots
\end{aligned}
\tag{1.106}
$$

with the operator

$$X = \xi\frac{\partial}{\partial x} + \eta\frac{\partial}{\partial u} + \sum_{i\geq 1}\zeta_i\frac{\partial}{\partial u_i} + \sum_{s\geq 1}\underset{h}{\zeta_s}\frac{\partial}{\partial \underset{h}{u_s}} + (\xi^+ - \xi)\frac{\partial}{\partial h^+} + (\xi - \xi^-)\frac{\partial}{\partial h^-}. \tag{1.107}$$

The scalar function $\mathcal{F}(z) \in \underset{h}{A}$ at the transformed point $z^* \in \underset{h}{Z}$ is the formal power series

$$\mathcal{F}(z^*) = \mathcal{F}(z) + a\left(\xi\frac{\partial\mathcal{F}}{\partial x} + \eta\frac{\partial\mathcal{F}}{\partial u} + \cdots\right) + \cdots = \sum_{s=0}^{\infty}\frac{a^s}{s!}X^{(s)}\mathcal{F}(z). \tag{1.108}$$

In particular, if G_1 (1.106) is the Taylor group with operator D, then for $a = \pm h$ formula (1.108) gives

$$\underset{\pm h}{S}(\mathcal{F}(z)) = \mathcal{F}(\underset{\pm h}{S}(z));$$

i.e., the shift operators $\underset{\pm h}{S}$ commute with any function in $\underset{h}{A}$.

Now consider an "external" point transformation (not necessarily comprising a group!):

$$\bar{x} = F(x, u), \qquad \bar{u} = G(x, u). \tag{1.109}$$

In the change of variables (1.109), the coordinates are changed in $\widetilde{Z}$ and $\underset{h}{Z}$. The changes of differential variables in $\widetilde{Z}$ are well known (see [107, 111, 113]). Therefore, we consider the change of the difference variables. The points (x^+, u^+) and (x^-, u^-) pass respectively into

$$\bar{x}^+ = F(x^+, u^+), \quad \bar{u}^+ = G(x^+, u^+), \qquad \bar{x}^- = F(x^-, u^-), \quad \bar{u}^- = G(x^-, u^-);$$

i.e.,

$$\bar{h}^+ = F(x^+, u^+) - F(x, u) = h^+ \underset{+h}{D}(F), \quad \bar{h}^- = F(x, u) - F(x^-, u^-) = h^- \underset{-h}{D}(F).$$

We introduce the shift operators $\underset{\pm h}{\bar{S}}$ in the new coordinate system:

$$\underset{+h}{\bar{S}} = \sum_{s=0}^{\infty} \frac{(\bar{h}^+)^s}{s!} \bar{D}^s, \qquad \underset{-h}{\bar{S}} = \sum_{s=0}^{\infty} \frac{(-\bar{h}^-)^s}{s!} \bar{D}^s,$$

where $\bar{D}$ is the differentiation operator in the new coordinate system.

The discrete differentiation operator in the new coordinate system becomes

$$\underset{\pm h}{\bar{D}} = \frac{1}{\underset{\pm h}{D}(F(x, u))} \underset{\pm h}{D}. \tag{1.110}$$

In particular, it follows from (1.110) that

$$\underset{h}{\bar{u}_x} = \frac{\underset{+h}{D}(G)}{\underset{+h}{D}(F)}, \qquad \underset{h}{\bar{u}_{x\bar{x}}} = \frac{\underset{+h}{D}(G)\underset{-h}{D}(F) - \underset{+h}{D}(F)\underset{-h}{D}(G)}{h^- \underset{+h}{D}(F)[\underset{-h}{D}(F)]^2}, \qquad \dots . \tag{1.111}$$

Let us find how the coefficients of the operator (1.107) vary in the change (1.109). The well-known formulas of group analysis (see [107, 111, 113]) give the coefficients of the operator (1.107) in $\widetilde{Z}$:

$$\bar{X} = X(F)\frac{\partial}{\partial \bar{x}} + X(G)\frac{\partial}{\partial \bar{u}} + X\left(\frac{D(G)}{D(F)}\right)\frac{\partial}{\partial \bar{u}_x} + \cdots . \tag{1.112}$$

Thus, the operator (1.107) is an operator of a formal group preserving "continuous" tangency of any order and "discrete" tangency of any finite order. This property is independent of the coordinate system (the operator (1.107) is a scalar expression), and hence the other coordinates in (1.112) can be obtained by the pro-longation formulas:

$$\bar{X} = \cdots + [\underset{+h}{\bar{D}}(X(G)) - \underset{h}{\bar{u}_x}\underset{+h}{\bar{D}}(X(F))]\frac{\partial}{\partial \underset{h}{\bar{u}_x}} + \cdots$$

$$+ X(F(x^+, u^+) - F(x, u))\frac{\partial}{\partial \bar{h}^+} + X(F(x, u) - F(x^-, u^-))\frac{\partial}{\partial \bar{h}^-}. \quad (1.113)$$

Indeed, for example, $\bar{h}^+ = F(x^+, u^+) - F(x, u)$ under the action of $\bar{G}_1$ becomes the new spacing:

$$(\bar{h}^+)^* = F(x^{+*}, u^{+*}) - F(x^*, u^*).$$

By differentiating the last relation with respect to a and by equating a with zero, we obtain the desired coefficient of $\frac{\partial}{\partial \bar{h}^+}$ in (1.113).

In a similar way, we obtain the following expression from (1.111):

$$\underset{h}{\bar{u}_x^*} = \frac{G(x^{+*}, u^{+*}) - G(x^*, u^*)}{F(x^{+*}, u^{+*}) - F(x^*, u^*)};$$

by applying it to some $\partial/\partial a|_{a=0}$, we obtain the coefficient of $\partial/\partial \underset{h}{\bar{u}_x}$ in (1.113). Thus, the group $\bar{G}_1$, similar to the group G_1, is determined in $\underset{h}{Z}$ by an operator that can be written in the unique form

$$\bar{X} = X(\bar{x})\frac{\partial}{\partial \bar{x}} + X(\bar{u})\frac{\partial}{\partial \bar{u}} + \sum_{s=1}^{\infty} X(\underset{h}{\bar{u}_s})\frac{\partial}{\partial \underset{h}{\bar{u}_s}} + X(\bar{h}^+)\frac{\partial}{\partial \bar{h}^+} + X(\bar{h}^-)\frac{\partial}{\partial \bar{h}^-}. \quad (1.114)$$

2. In what follows, we need to apply the operator (1.107) to the functions $F(z) \in \underset{h}{A}$, which contain the neighboring points of a given point or some of their coordinates. To obtain the desired extension of the coordinates in the operator (1.107), note that the operator (1.107) is the inner product (in $\underset{h}{Z}$) of the vectors

$$(\xi, \eta, \underset{h}{\zeta_1}, \underset{h}{\zeta_2}, \ldots, h^+), \qquad \left(\frac{\partial}{\partial x}; \frac{\partial}{\partial u}; \frac{\partial}{\partial \underset{h}{u_x}}, \frac{\partial}{\partial \underset{h}{u_{x\bar{x}}}}, \ldots, \frac{\partial}{\partial h^+}\right). \quad (1.115)$$

To obtain the product of vectors at the points $z^+ = \underset{+h}{S}(z)$ and $z^- = \underset{-h}{S}(z)$, we write both vectors (1.115) at the points z^+ and z^- and take their inner product:

$$X^+ = \xi^+\frac{\partial}{\partial x^+} + \eta^+\frac{\partial}{\partial u^+} + \underset{h}{\zeta_1^+}\frac{\partial}{\partial \underset{h}{u^+_x}} + \cdots + h^{++}\frac{\partial}{\partial h^{++}},$$

$$X^- = \xi^-\frac{\partial}{\partial x^-} + \eta^-\frac{\partial}{\partial u^-} + \underset{h}{\zeta_1^-}\frac{\partial}{\partial \underset{h}{u^-_x}} + \cdots + h^{--}\frac{\partial}{\partial h^{--}}.$$

$$(1.116)$$

The desired coefficients in (1.116) can be obtained in a different way. Formula (1.114) permits extending the action of G_1 to any new variables. For example, by introducing the new variables

$$\bar{x} = x^+ = x + h^+,$$
$$\bar{u} = u^+ = u + \underset{h}{h^+ u_x},$$

$$\cdots \cdots \cdots \cdots$$

according to (1.114), we obtain

$$\bar{X} = X(x + h^+)\frac{\partial}{\partial x^+} + X(u + \underset{h}{h^+ u_x})\frac{\partial}{\partial u^+} + \cdots = \xi^+\frac{\partial}{\partial x^+} + \eta^+\frac{\partial}{\partial u^+} + \cdots,$$

i.e., formulas (1.116).

We use the formulas obtained above to prolong the group G_1 to the neighboring points of a given point, i.e., to the set of points of the difference mesh that comprise the *difference stencil*. Their generalization to the multidimensional case is quite obvious.

Chapter 2

Invariance of Finite-Difference Models

In this chapter, we use the mathematical technique developed in the preceding chapter to study the symmetry properties of finite-difference models, i.e., difference equations considered together with difference meshes. The main theorem proved in this chapter deals with necessary and sufficient conditions for the invariance of difference equations and meshes. We develop a simple algorithm, called *the method of finite-difference invariants*, for constructing invariant difference model from a given differential equation and admitted transformation group. We also consider several examples of constructing finite-difference models, where we completely preserve the symmetry of the original differential equations. We show that symmetries of difference models permits symmetry reduction by means of subgroups (just as in the case of differential equations). In addition, we present an example where the group admitted by a difference model is calculated.

We should mention the pioneering papers on the invariance of difference equations [95–97].

2.1. An Invariance Criterion for Finite-Difference Equations on the Difference Mesh

1. For simplicity, we consider the case of a single independent variable. Let $\underset{h}{Z}$ be the space of sequences of mesh variables $(x, u, \underset{h}{u_1}, \ldots, h^+)$, and let $\underset{h}{A}$ be the space of analytic functions of finitely many coordinates z^i of a vector $z \in \underset{h}{Z}$. Then each finite-difference equation on the mesh $\underset{h}{\omega}$ can be written as

$$F(z) = 0, \tag{2.1}$$

where $F \in \underset{h}{A}$. This equation is written on finitely many points of the difference mesh $\underset{h}{\omega}$, which may be uniform or nonuniform. We assume that the mesh is determined by the equation

$$\Omega(z) = 0, \tag{2.2}$$

where $\Omega \in \underset{h}{A}$. The function Ω is uniquely determined by the "discretization" of the space of independent variables, which is obtained by the action of the operator

$S^\alpha_{\pm h}$ on the "starting point" point x_0. Thus, we initially assume that Eq. (2.2) is invariant under the discrete shift operator $S_{\pm h}$ in Z_h. A difference equation (2.1) that admits $S_{\pm h}$ is said to be *homogeneous* ([122]). The function $\Omega(z)$ depends on finitely many variables in Z_h and explicitly depends on the mesh spacing h^+. In the continuum limit, the mesh equation $\Omega(z) = 0$ degenerates into an identity (for example, $0 = 0$).

If a transformation group acts on Z_h, then, in contrast to the continuous case, one should also include Eq. (2.2) in the invariance condition, because an arbitrary mesh ω_h does not necessarily admit this group.

PROPOSITION 2.1. *Let G_1 be a one-parameter group in Z_h with operator*

$$X = \xi\frac{\partial}{\partial x} + \eta\frac{\partial}{\partial u} + \cdots + h^+ \underset{+h}{D}(\xi)\frac{\partial}{\partial h^+}. \tag{2.3}$$

For the difference equation (2.1) to admit the group G_1 with operator (2.3) on the mesh (2.2), it is necessary and sufficient that the following condition be satisfied:

$$XF(z)\big|_{(2.1),(2.2)} = 0, \qquad X\Omega(z)\big|_{(2.1),(2.2)} = 0. \tag{2.4}$$

Proof. In the proof, we restrict ourselves to the one-dimensional case,

$$\widetilde{Z}_h = (x, u, u_1, u_2, \ldots, \underset{h}{u_1}, \underset{h}{u_2}, \ldots, h^+).$$

We assume that the finite-difference equation (2.1) and Eq. (2.2) are invariant in Z_h; i.e.,

$$F(z^*) = 0, \qquad \Omega(z^*) = 0$$

for each point z of the manifold (2.1), (2.2). By expanding the functions $F, \Omega \in A_h$ in power series in a, we obtain

$$F(z^*) = F(z) + a\left[\xi(z)\frac{\partial F(z)}{\partial x} + \eta(z)\frac{\partial F(z)}{\partial u} + \cdots\right] + a^2 N(z, a),$$
$$\Omega(z^*) = \Omega(z) + a\left[\xi(z)\frac{\partial\Omega(z)}{\partial x} + \eta(z)\frac{\partial\Omega(z)}{\partial u} + \cdots\right] + a^2 M(z, a), \tag{2.5}$$

where $N(z, a)$ and $M(z, a)$ are formal power series in the parameter a as well. In particular, the fact that the formal series (2.5) are zero *for the points of the manifold* (2.1), (2.2) implies that the first coefficients are zero; i.e., (2.4) is satisfied.

Now let us assume that the manifold (2.1), (2.2) satisfies conditions (2.4). The group G_1 with operator (2.3) takes each point $z = (x, u, u_1, u_2, \ldots, \underset{h}{u_1}, \underset{h}{u_2}, \ldots, h^+)$ to the point z^* with coordinates given by the formal power series

$$z^{i*} = f^i(z, a) = \sum_{k=0}^{\infty} A^i_k(z, a)a^k. \tag{2.6}$$

Since the series (2.6) form a group, it follows that they can be represented in exponential form (see Chapter I),

$$z^{i*} = e^{aX}(z^i) \equiv \sum_{s=0}^{\infty} \frac{a^s}{s!} X^{(s)}(z^i). \tag{2.7}$$

Consider $F(z^*)$ and $\Omega(z^*)$, i.e., a superposition of series of the form (2.6) and (2.7). Let us find the derivative of $F(z^*)$ and $\Omega(z^*)$ with respect to the parameter a:

$$\frac{dF(z^*)}{da} = \xi(z^*)\frac{\partial F(z^*)}{\partial x^*} + \eta(z^*)\frac{\partial F(z^*)}{\partial u^*} + \cdots = \sum_{s=1}^{\infty} \frac{a^{s-1}}{(s-1)!} X^{(s)} F(z),$$

$$\frac{d\Omega(z^*)}{da} = \xi(z^*)\frac{\partial \Omega(z^*)}{\partial x^*} + \eta(z^*)\frac{\partial \Omega(z^*)}{\partial u^*} + \cdots = \sum_{s=1}^{\infty} \frac{a^{s-1}}{(s-1)!} X^{(s)} \Omega(z). \tag{2.8}$$

Equations (2.8) are the Lie equations for the formal series $F(z^*)$ and $\Omega(z^*)$ with the initial data $F(z) = 0$ and $\Omega(z) = 0$. By assumption, the manifold $F(z^*) = 0$, $\Omega(z^*) = 0$ satisfies Lie equations (2.8) and the initial conditions. Since Eqs. (2.8) determine the solutions as unique recursion relations for the coefficients of the formal series $F(z^*)$ and $\Omega(z^*)$, it follows that the solution of the system $F(z^*) = 0$, $\Omega(z^*) = 0$ is unique. Hence the manifold (2.1), (2.2) is invariant under the formal group G_1 with operator (2.3). □

Note that the only essentially new fact in considering the invariance of finite-difference equations for the formal one-parameter group G_1 is the appearance of Eq. (2.2) for the difference mesh, which is not contained in the differential setting.

Remark. If the equation $\Omega(x, h^+) = 0$ is independent of the variables $u, \underset{h}{u_1}, \underset{h}{u_2}, \ldots$ (i.e., of the solution), then the mesh invariance condition

$$\left[\xi(x, u)\frac{\partial \Omega}{\partial x} + \eta(x, u)\frac{\partial \Omega}{\partial u} + \cdots + (\underset{+h}{S}(\xi) - \xi)\frac{\partial \Omega}{\partial h^+}\right]\Bigg|_{(2.1),(2.2)} = 0$$

is not related to the invariance criterion for the difference equation $F(z) = 0$ for groups with $\xi_u = 0$ (so-called x-autonomous groups):

$$\left[\xi(x)\frac{\partial \Omega}{\partial x} + (\xi(x + h^+) - \xi(x))\frac{\partial \Omega}{\partial h^+}\right]\Bigg|_{(2.2)} = 0.$$

These groups include shifts and dilations of independent variables, rotations, the Lorentz group, and some other widely used symmetries of mathematical models in physics. This fact significantly simplifies the construction of invariant difference equations for specific mathematical models of physics. Indeed, in this case the mesh invariance criterion can be separated from the invariance criterion for the difference equations, and the difference mesh can be constructed independently.

2. We see that criterion (2.4) implies necessary and sufficient conditions for the invariance of finite-difference equations. Conditions (2.4) are *linear partial differ- ence equations for the coefficients* $\xi^i(x, u)$ and $\eta^k(x, u)$ of the infinitesimal opera- tor (2.3). If the problem is to find all operators of the form (2.3) admitted by a given finite-difference equation (system), then it always reduces to the solution of the *lin- ear* system (2.4) regardless of whether the original system (2.1), (2.2) is linear. Thus, the problem of finding the group admitted by given finite-difference equa- tions is always linear. But even such a linear problem is intractable, because the problem of exact integration of partial finite-difference equations is in fact open. It is significantly simpler to solve the inverse problem, i.e., construct difference equations and meshes admitting a given transformation group. For example, if one needs to *preserve the symmetry group of the original model in finite-difference modeling*, then one can use the fact that conditions (2.4) are *sufficient* for difference equations and meshes to be invariant.

Consider several examples of difference equations and meshes that preserve the symmetry group of the original model.

EXAMPLE 2.2. The ordinary differential equation

$$\frac{d^2 u}{dx^2} = e^u \tag{2.9}$$

admits the two-parameter point transformation group generated by the operators

$$X_1 = \frac{\partial}{\partial x}, \qquad X_2 = x\frac{\partial}{\partial x} - 2\frac{\partial}{\partial u}. \tag{2.10}$$

Consider the finite-difference equation

$$u_{x\bar{x}} = e^u \tag{2.11}$$

in $\underset{h}{Z}$. Let us represent the operators (2.10) in $\underset{h}{Z}$ by extending X_2 to h^+ and h^- as follows (the corresponding coordinates in X_1 are zero):

$$X_1 = \frac{\partial}{\partial x}, \quad X_2 = x\frac{\partial}{\partial x} - 2\frac{\partial}{\partial u} - u_x\frac{\partial}{\partial u_x} - 2u_{x\bar{x}}\frac{\partial}{\partial u_{x\bar{x}}} + h^+\frac{\partial}{\partial h^+} + h^-\frac{\partial}{\partial h^-}. \tag{2.12}$$

Both operators satisfy the criterion for the preservation of the mesh uniformity (see Chapter 1); therefore, the uniform mesh $(h^+ = h^-)$ is invariant.

We check the invariance criterion (2.4) for Eq. (2.11) and the operators (2.12):

$$X_1(u_{x\bar{x}} - e^u)\big|_{(2.11)} = 0, \qquad X_2(u_{x\bar{x}} - e^u)\big|_{(2.11)} = -2(u_{x\bar{x}} - e^u)\big|_{(2.11)} = 0.$$

Note that we have first constructed a difference equation and then verified whether the invariance criterion (2.4) is satisfied.

Is Eq. (2.11) unique in the sense that it is the only equation that admits the group (2.12) and approximates (2.9) modulo $O(h^2)$? The following example gives another difference equation with the same properties:

$$u_{x\bar{x}} = e^u + h^2 e^{2u}.$$

By way of example, consider the following equation, which approximates (2.9) to the second order but does not admit the same group as the original equation:

$$u_{x\bar{x}} = e^u + h^2 e^u. \tag{2.13}$$

It is obvious that Eq. (2.13) admits X_1 but does not admit X_2,

$$X_2(u_{x\bar{x}} - e^u(1 + h^2))\big|_{(2.13)} \neq 0.$$

EXAMPLE 2.3. The nonlinear heat equation

$$u_t = (u^\sigma u_x)_x, \qquad \sigma > 0 \tag{2.14}$$

admits (see [112]) the four-parameter group with the operators

$$X_1 = \frac{\partial}{\partial t}, \qquad X_2 = \frac{\partial}{\partial x}, \qquad X_3 = 2t\frac{\partial}{\partial t} + x\frac{\partial}{\partial x}, \qquad X_4 = \sigma t\frac{\partial}{\partial t} - u\frac{\partial}{\partial u}.$$

The operators in $\underset{h}{\tilde{Z}}$ in extended form become

$$X_1 = \frac{\partial}{\partial t}, \qquad X_2 = \frac{\partial}{\partial x},$$

$$X_3 = 2t\frac{\partial}{\partial t} + x\frac{\partial}{\partial x} - u_x\frac{\partial}{\partial u_x} - 2u_t\frac{\partial}{\partial u_t} - 2u_{x\bar{x}}\frac{\partial}{\partial u_{x\bar{x}}} + 2\tau\frac{\partial}{\partial \tau} + h\frac{\partial}{\partial h}, \tag{2.15}$$

$$X_4 = \sigma t\frac{\partial}{\partial t} - u\frac{\partial}{\partial u} - u_x\frac{\partial}{\partial u_x} - (\sigma + 1)u_t\frac{\partial}{\partial u_t} - u_{x\bar{x}}\frac{\partial}{\partial u_{x\bar{x}}} + \sigma\tau\frac{\partial}{\partial \tau}.$$

Note that all four operators generate x-autonomous subgroups and preserve the orthogonality and uniformity of the mesh $\underset{h}{\omega} \times \underset{\tau}{\omega}$. (The operators X_1 and X_2 do not change the mesh spacings h and τ, and the operators X_3 and X_4 dilate them uniformly.) Therefore, we choose just such an invariant mesh.

Consider a difference mesh approximating Eq. (2.14) in $\underset{h}{Z}$ to the order of $\tau + h^2$:

$$u_t = \underset{+h}{D}(k(u)\hat{u}_{\bar{x}}) = k(u)\hat{u}_{x\bar{x}} + \underset{+h}{D}(k(u))\hat{u}_x, \tag{2.16}$$

where $\hat{z} = \underset{+\tau}{S}(z)$ is the value of z at the "upper layer" in t and $k(u)$ is the difference approximation to the thermal conductivity coefficient u^σ. The implicit scheme (2.16) is based on a six-point stencil and is divergence free; i.e., it can be written in the form of a difference conservation law. We choose the following approximation to the coefficient $k = u^\sigma$:

$$k(u) = \frac{1}{2}(u^\sigma + \underset{-h}{S}(u^\sigma));$$

then $\underset{+h}{D}(k(u)) = \frac{1}{2h}(\underset{+h}{S}(u^\sigma) - \underset{-h}{S}(u^\sigma))$, and the scheme (2.16) becomes

$$u_t = \frac{1}{2}(u^\sigma + \underset{-h}{S}(u^\sigma))\hat{u}_{x\bar{x}} + \frac{1}{2h}(\underset{+h}{S}(u^\sigma) - \underset{-h}{S}(u^\sigma))\hat{u}_x. \qquad (2.17)$$

We extend the operators (2.15) to the variables $\underset{h}{\hat{u}_{\bar{x}}}, \underset{h}{\hat{u}_{x\bar{x}}}, \underset{+h}{S}(u^\sigma), \underset{-h}{S}(u^\sigma)$ as follows:

$$X_3 = \cdots - \underset{h}{\hat{u}_x}\frac{\partial}{\partial\underset{h}{\hat{u}_x}} - 2\underset{h}{\hat{u}_{x\bar{x}}}\frac{\partial}{\partial\underset{h}{\hat{u}_{x\bar{x}}}};$$

$$X_4 = \cdots - \underset{+h}{S}(u)\frac{\partial}{\partial(\underset{+h}{S}(u))} - \underset{-h}{S}(u)\frac{\partial}{\partial(\underset{-h}{S}(u))} - \underset{h}{\hat{u}_x}\frac{\partial}{\partial\underset{h}{\hat{u}_x}} - \underset{h}{\hat{u}_{x\bar{x}}}\frac{\partial}{\partial\underset{h}{\hat{u}_{x\bar{x}}}}.$$

(The operators X_1 and X_2 are "nonextensible"; i.e., the coordinates of any additional variables are zero.)

Obviously, the difference equation (2.17) admits X_1 and X_2. Let us verify whether it is invariant under X_3 and X_4:

$$X_3\left[u_t - \frac{1}{2}[u^\sigma + \underset{-h}{S}(u^\sigma)]\underset{h}{\hat{u}_{x\bar{x}}} - \frac{1}{2h}[\underset{+h}{S}(u^\sigma) - \underset{-h}{S}(u^\sigma)]\underset{h}{\hat{u}_x}\right]\Bigg|_{(2.17)}$$

$$= -2\left[u_t - \frac{1}{2}[u^\sigma + \underset{-h}{S}(u^\sigma)]\underset{h}{\hat{u}_{x\bar{x}}} - \frac{1}{2h}[\underset{+h}{S}(u^\sigma) - \underset{-h}{S}(u^\sigma)]\underset{h}{\hat{u}_x}\right]\Bigg|_{(2.17)} = 0,$$

$$X_4\left[u_t - \frac{1}{2}[u^\sigma + \underset{-h}{S}(u^\sigma)]\underset{h}{\hat{u}_{x\bar{x}}} - \frac{1}{2h}[\underset{+h}{S}(u^\sigma) - \underset{-h}{S}(u^\sigma)]\underset{h}{\hat{u}_x}\right]\Bigg|_{(2.17)}$$

$$= -(\sigma + 1)\left[\frac{1}{2}[u^\sigma + \underset{-h}{S}(u^\sigma)]\underset{h}{\hat{u}_{x\bar{x}}} - \frac{1}{2h}[\underset{+h}{S}(u^\sigma) - \underset{-h}{S}(u^\sigma)]\underset{h}{\hat{u}_x}\right]\Bigg|_{(2.17)} = 0.$$

Here we have used the fact that the operators $\underset{\pm h}{S}$ commute with any function in $\underset{h}{A}$.

Thus, the scheme (2.17) is invariant on a uniform orthogonal mesh.

3. The most famous applications of the theory of group properties of differential equations are methods for obtaining group-invariant solutions [111], of which self-similar solutions are used most widely. The method for constructing such solutions is based on the so-called "π-theorem."

If an equation (a system) admits an r-parameter transformation group G_r, then the solutions invariant under a subgroup of G_r can be obtained by integrating an equation (a system) of smaller dimension. If the necessary conditions [111] are satisfied, then invariant solutions can be constructed on subgroups of various dimensions; they are classified according to their rank, invariance defects, and other properties.

Just as in the differential case, the invariance of a finite-difference equation (system) together with the difference mesh permits constructing invariant solutions. But discrete equations are nonlocal, and so this procedure has several specific features. A difference equation is defined on a difference stencil, which is a finite set of difference mesh points, and hence the mapping into the space of the group invariants should match the difference mesh and the stencil in the original space with those in the space of invariants. In other words, the projection onto the space of invariants should take all points of the original difference stencil to the points of the stencil of the reduced model. Let us illustrate this by examples.

Symmetry reduction by means of subgroups of the admissible group

Consider several typical invariant solutions of the difference nonlinear heat equation (2.17),

$$u_t = \frac{1}{2}(u^\sigma + \underset{-h}{S}(u^\sigma))\hat{u}_{x\bar{x}} + \frac{1}{2h}(\underset{+h}{S}(u^\sigma) - \underset{-h}{S}(u^\sigma))\hat{u}_x, \qquad \sigma > 0, \qquad (2.18)$$

which admits the group G_4 with operators (2.15) on a uniform orthogonal mesh.

Translation groups

Consider the stationary solution of Eq. (2.18). In the space (t, x, u), the operator X_1 has two invariants, x and u. The invariant solution $u(x,t) = \tilde{u}(x)$ is the stationary solution determined by the ordinary difference equation obtained by substituting $u(x,t) = \tilde{u}(x)$ into the difference equation:

$$k(\tilde{u})\tilde{u}_{x\bar{x}} + \underset{+h}{D}(k(\tilde{u}))\tilde{u}_x = 0, \qquad k(\tilde{u}) = \frac{1}{2}(\tilde{u}^\sigma + \underset{-h}{S}(\tilde{u}^\sigma)),$$

and the mesh spacing h must remain the same as in the original space.

A homogeneous solution on the operator X_2 can be obtained in a similar way.

A solution invariant under the operator

$$X_1 + \alpha X_2 = \frac{\partial}{\partial t} + \alpha\frac{\partial}{\partial x}, \qquad \alpha = \text{const},$$

with invariants

$$J_1 = u, \qquad J_2 = x - \alpha t,$$

is a difference traveling wave,

$$u(x,t) = \tilde{u}(\lambda), \qquad \lambda = x - \alpha t, \qquad (2.19)$$

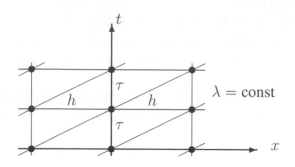

Figure 2.1

propagating for $\alpha > 0$ to the right with respect to x at velocity α. The conditions necessary for the existence of a traveling wave type solution are satisfied,

$$R\left(\left\|\frac{\partial\lambda}{\partial u},\frac{\partial\tilde{u}}{\partial u}\right\|\right) = 1, \qquad R\left(\|1,\alpha,0\|\right) = 1 < N = m + n = 3.$$

But this is insufficient for the desired solution of Eq. (2.18) on the difference stencil in the subspace (x, t) to be taken to the difference stencil of the mesh in the invariant subspace (λ).

Additional specific difference conditions for the existence of an invariant solution relate the original stencil to the stencil in the space of invariants (the "multipoint" mapping conditions).

The difference mesh spacing $\Delta\lambda$ along the λ-axis should be matched with the original mesh spacings h and τ and the wave velocity α,

$$\alpha = h/\tau, \qquad \Delta\lambda = h. \tag{2.20}$$

These relations mean that the lines $\lambda = \text{const}$ pass through the nodes of the original mesh in the plane (x, t) (see Fig 2.1).

The difference stencils can be matched if $\alpha = kh/\tau$, where k is any rational number, i.e., if the traveling wave velocity α is a multiple of the "difference velocity" h/τ. Difference traveling waves of the heat equation were originally obtained in [126].

Under conditions (2.20), we can substitute the invariant representation (2.19) of a solution into the invariant difference equation (2.18), thus obtaining the following ordinary difference equation for the function $\tilde{u}(\lambda)$:

$$\alpha\tilde{u}_\lambda + \frac{1}{2}[\tilde{u}^\sigma + \check{\tilde{u}}^\sigma\tilde{u}_{\bar{\lambda}}]_\lambda = 0, \tag{2.21}$$

where $\check{\tilde{u}} = \tilde{u}(\lambda - \Delta\lambda)$.

Just as in the differential case, the difference equation (2.21) has a first integral. By applying the right integration operator to (2.21), we obtain

$$\alpha\tilde{u} + \frac{1}{2}(\tilde{u}^\sigma + \check{\tilde{u}}^\sigma)\tilde{u}_{\bar{\lambda}} = \text{const.}$$

The dilation group

Consider a self-similar solution of Eq. (2.18) invariant under the one-parameter dilation group and corresponding to X_4. The invariants are x and $ut^{1/\sigma}$.

We seek the invariant solution in the form

$$u(x,t) = \tilde{u}(x)t^{-\frac{1}{\sigma}}.$$

By substituting the invariant representation of $u(x,t)$ into Eq. (2.18), we obtain the following equation for $\tilde{u}(x)$ at the $(n+1)$st t-layer:

$$(\underset{+h}{S}(\tilde{u}^\sigma) + \tilde{u}^\sigma)\underset{h}{\tilde{u}_x} - (\tilde{u}^\sigma + \underset{-h}{S}(\tilde{u}^\sigma))\underset{h}{\tilde{u}_x} + 2nh(\sqrt{n^2 + n} - 1)\tilde{u} = 0. \qquad (2.22)$$

In this equation, the mesh spacing h coincides with the original one. Having solved Eq. (2.22), we can find the solution of the original equation by the formula

$$u(x,t) = \frac{\tilde{u}(x)}{[\tau(n+1)]^{1/\sigma}}, \qquad n = 0, 1, 2, \ldots .$$

Thus, finding an invariant solution of the scheme (2.18) is reduced to finding a solution of an ordinary difference equation, i.e., solving the problem at each time layer reduced to a one-time process.

4. Consider an example where the invariance criterion is applied to a difference equation for factorized (canonical) symmetry operators.

Consider the ordinary differential equation

$$u_{xx} = u^2. \qquad (2.23)$$

Equation (2.23) admits the two-parameter point transformation group generated by the operators

$$X_1 = \frac{\partial}{\partial x}, \qquad X_2 = x\frac{\partial}{\partial x} - 2u\frac{\partial}{\partial u}. \qquad (2.24)$$

Consider the finite-difference equation

$$\frac{u^+ - 2u + u^-}{h^2} = u^2, \qquad (2.25)$$

where $u^+ = \underset{+h}{S}(u)$ and $u^- = \underset{-h}{S}(u)$, on the uniform difference mesh

$$h^+ = h^-. \qquad (2.26)$$

Equations (2.25)–(2.26) use the three-point stencil $(x, x^+, x^-, u, u^+, u^-)$. We extend the operators (2.24) to the points of this stencil by setting

$$X_1 = \frac{\partial}{\partial x} + \frac{\partial}{\partial x^+} + \frac{\partial}{\partial x^-},$$

$$X_2 = x\frac{\partial}{\partial x} + x^+\frac{\partial}{\partial x^+} + x^-\frac{\partial}{\partial x^-} - 2u\frac{\partial}{\partial u} - 2u^+\frac{\partial}{\partial u^+} - 2u^-\frac{\partial}{\partial u^-}$$
$$+ h^+\frac{\partial}{\partial h^+} + h^-\frac{\partial}{\partial h^-}.$$

One can readily verify the invariance of Eqs. (2.25)–(2.26):

$$X_2\left[\frac{u^+ - 2u + u^-}{h^2} - u^2\right]\Bigg|_{(2.25),(2.26)} = 0, \quad X_2\left(h^+ - h^-\right)\Big|_{(2.25),(2.26)} = 0. \quad (2.27)$$

(The operator X_1 does not change Eqs. (2.25)–(2.26).)

It follows that the difference model (2.25)–(2.26) admits the same transformation group as the original differential equation (2.23). Note that the invariance conditions (2.27) split into two independent equations.

Now consider the invariance criterion for the difference model (2.25)–(2.26) in the case of canonical operators. Without loss of generality, consider the representation of the operator D^+ for the right half-line,

$$D^+ = \frac{\partial}{\partial x} + \frac{\partial}{\partial x^+} + \frac{\partial}{\partial x^-} + u_x\frac{\partial}{\partial u} + u_x^+\frac{\partial}{\partial u^+} + u_x^-\frac{\partial}{\partial u^-},$$

where

$$u_x \equiv \sum_{n=1}^{\infty}\frac{(-h)^{n-1}}{n}\mathop{D^n}_{+h}(u), \qquad u_x^+ \equiv \sum_{n=1}^{\infty}\frac{(-h)^{n-1}}{n}\mathop{D^n}_{+h}(u^+),$$

$$u_x^- \equiv \sum_{n=1}^{\infty}\frac{(-h)^{n-1}}{n}\mathop{D^n}_{+h}(u^-)$$

are the difference representations of continuous derivatives at different points of the stencil.

The canonical operators acquire the form

$$\mathop{\bar{X}_1}_{+h} = -X_1 + D^+ = u_x\frac{\partial}{\partial u} + u_x^+\frac{\partial}{\partial u^+} + u_x^-\frac{\partial}{\partial u^-},$$

$$\mathop{\bar{X}_2}_{+h} = -X_2 + xD^+ = -h^+\frac{\partial}{\partial h^+} - h^-\frac{\partial}{\partial h^-} + (2u + xu_x)\frac{\partial}{\partial u} \qquad (2.28)$$

$$+ (2u^+ + xu_x^+)\frac{\partial}{\partial u^+} + (2u^- + xu_x^-)\frac{\partial}{\partial u^-}.$$

It is obvious that the mesh equation (2.26) admits the operators (2.28).

The action of the operators (2.28) on Eq. (2.25), considered on the equation itself, leads to

$$u_x + 2u_x^+ + u_x^- = 2uu_x h^2.$$

The last equation is a differential corollary of Eq. (2.25) based on the use of all points of the right half-line.

One can see that the use of factorized (canonical) operators in the difference case leads to significant technical difficulties even on uniform meshes. If it is required to *construct* a difference model, then it is impossible to use canonical operators, because no methods for constructing invariants have been developed for such operators.

2.2. Symmetry Preservation in Difference Modeling: Method of Finite-Difference Invariants

In the preceding section, the invariance criterion (2.4) was used to *verify the invariance* of finite-difference equations. In this section, we consider the problem of *how to construct* difference models for a given differential equation if the symmetry group of this equation is known.

One can readily construct an *arbitrary* invariant difference equation for a given group G_r. For simplicity, consider the case of a single independent variable and first-order difference equations. We assume that there is given a point group G_r that acts in the space of a single independent and m dependent variables. We extend it to the first difference derivatives, i.e., to the space with $m + 1$ variables, which is supplemented with the mesh spacing and m first difference derivatives; i.e., the total number of variables is $2m + 2$. Just as in the differential case, to find the invariants, one has to solve the linear partial differential system

$$X_\alpha(I^k) = 0, \qquad \alpha = 1, 2, \ldots, r.$$

If there exists a nontrivial solution of this system, then, according to the classical Lie theory, for a group with a set of operators of rank R we have $\lambda = 2m + 2 - R$ independent invariants. The invariants that are not invariants in the original $(m+1)$-dimensional space are naturally called *first-order finite-difference invariants*. By repeating this process, we can construct finite-difference invariants of any order, and the number of them increases together with the number of repetitions, just as in the continuous case of extension to higher-order derivatives (cf. [111]).

Now it is clear how to construct an arbitrary invariant difference equation for a given group G_r. *Any* equations relating (in a sufficiently smooth way) these λ independent invariants,

$$\Phi_k(I^1(z), I^2(z), \ldots, I^\lambda(z)) = 0, \qquad k = 1, 2, \ldots, s, \qquad (2.29)$$

are first-order finite-difference equations admitting the given group. The number s of equations must ensure the solvability of system (2.29) for the dependent variables; moreover, one of the equations in the system should determine the difference mesh. As the group is extended to the second and higher difference derivatives, the number of difference invariants increases, and the invariant difference equations of the second-, third-, and higher-orders can be written in a completely similar way.

Note, however, that these equations have nothing in common with any differential model. Consider the following problem: of equations of the form (2.29), choose equations that *approximate* a given equation (system) to a given order $O(h^k)$. Let us illustrate this by an example.

EXAMPLE. We return to the example of the equation

$$u'' = e^u.$$

For this second-order equation, we should have at least a three-point difference stencil, on which we should construct (to be definite) a second-order approximation to the original equation. The group G_2 with operators

$$X_1 = \frac{\partial}{\partial x}, \qquad X_2 = x\frac{\partial}{\partial x} - 2\frac{\partial}{\partial u} - \underset{h}{u_x}\frac{\partial}{\partial \underset{h}{u_x}} - 2\underset{h}{u_{x\bar{x}}}\frac{\partial}{\partial \underset{h}{u_{x\bar{x}}}} + h^+\frac{\partial}{\partial h^+} + h^-\frac{\partial}{\partial h^-}$$

extended to the space $(x, u, \underset{h}{u_x}, \underset{h}{u_{x\bar{x}}}, h^+, h^-)$ has $6 - 2 = 4$ independent invariants, which can be obtained by solving the corresponding linear system (which we omit here). The solution of this system has a functional ambiguity in the choice of the solution. (Any function of invariants is again an invariant.) For example, we can choose the following difference invariants:

$$J_1 = (h^+)^2 e^u, \qquad J_2 = \underset{h}{u_x}e^{-u/2}, \qquad J_3 = \underset{h}{u_{x\bar{x}}}e^{-u}, \qquad J_4 = \frac{h^+}{h^-}.$$

We should write out two invariant difference equations of the form

$$\Phi_1(I^1(z), I^2(z), \ldots, I^4(z)) = 0, \qquad \Phi_2(I^1(z), I^2(z), \ldots, I^4(z)) = 0,$$

one of which should determine the difference mesh. A differential equation contains no "traces" of a difference mesh, and hence it is only the developer of the difference scheme who can choose it. However, if the problem is to preserve the symmetry of the original differential equation in the difference model, then the choice is significantly narrower, because the equation generating the mesh should be written in terms of difference invariants. Note that it is possible to choose a mesh independent of the solution in our example:

$$J_4 = \frac{h^+}{h^-} = 1.$$

This choice of the simplest invariant mesh seems to be quite logical if there is no special modeling problem. On such a uniform mesh, as is easily seen, the third difference invariant gives a second-order approximation to the differential invariant $u''e^{-u}$ modulo $O(h^2)$. Therefore, for the invariant difference equation we take

$$J_3 = u_{x\bar{x}} e^{-u} = 1.$$

Thus, we have constructed a difference model completely preserving the two-parameter symmetry group of the original differential equation:

$$u_{x\bar{x}} = e^u, \qquad h^+ = h^-.$$

Is the constructed invariant model unique? Of course, not. The equation for invariant mesh generation provides infinitely many possibilities. In addition, even on any chosen invariant mesh, an infinite choice is still possible. For example, on the above-chosen invariant uniform mesh, another invariant approximation can be taken. The equation

$$u_{x\bar{x}} = e^u + \left(h^+\right)^2 e^{2u},$$

can also be represented in the invariant form

$$J_3 = 1 + J_1, \tag{2.30}$$

and it provides a second-order approximation to the differential equation. But in this case the equation

$$u_{x\bar{x}} = e^u + \left(h^+\right)^2 e^u$$

has the second order of approximation as well but cannot be written in invariant form.

Now consider the general case. Let there be given a system of m differential equations

$$\mathcal{F}_\alpha(x, u, u_1, u_2) = 0, \qquad \alpha = 1, \ldots, m, \tag{2.31}$$

that admits a known transformation group G_r. Without loss of generality, we assume that (2.31) is a second-order system in the highest-order derivatives.

We should construct a system of difference equations

$$F_\alpha(x, h^+, h^-, u, u^+, u^-) = 0, \qquad \alpha = 1, \ldots, m,$$

where u^+ and u^- are the values of the desired function at the neighbors of a given point x in all n directions, and a system of equations for constructing the difference mesh,

$$\Omega_\beta(x, h^+, h^-, u, u^+, u^-) = 0, \qquad \beta = 1, \ldots, n, \tag{2.32}$$

which admit the same group G_r in Z and provide a second-order approximation to the original system (2.31).

Note that the order of Eq. (2.32) may be less than 2.

Since system (2.31) admits the group G_r (which is assumed to have invariants), it follows that, after the complete set of τ functionally independent invariants

$$(J^1(z), J^2(z), \ldots, J^\tau(z)), \qquad J^\alpha \in A,$$

of order k is constructed, system (2.31) can be rewritten in the invariant representation

$$\Phi_\alpha(J^1(z), J^2(z), \ldots, J^\tau(z)) = 0, \qquad \alpha = 1, \ldots, m.$$

(It is assumed that system (2.31) is nondegenerate.)

The next step is to construct finite-difference invariants of the group G_r. The set of difference invariants can be found by solving the following standard linear problem of group analysis:

$$X_\gamma(I) = 0, \qquad \gamma = 1, 2, \ldots, r.$$

The group G_r represented in $\underset{h}{Z}$ has $\tau + nm$ functionally independent difference invariants

$$I^1(z), I^2(z), \ldots, I^{\tau+nm}(z), \qquad I^\alpha(z) \in \underset{h}{A},$$

because in the difference space we have two sets (right and left) of first difference derivatives. (This situation is preserved for any difference derivatives of odd order.)

Now one has to choose an invariant mesh from the general equation for invariant mesh generation:

$$\omega_\beta(I^1(z), I^2(z), \ldots, I^{\tau+nm}(z)) = 0,$$

where the ω_β are arbitrary smooth functions. This choice can be made from different standpoints. In our examples, we usually take an invariant mesh that is, in a sense, simplest. At this step of choice, a preliminary analysis of possible meshes, which was developed in the preceding chapter, may be of great help.

The next step is to form a set of τ invariants with the desired approximation property; i.e., for each invariant I^α representable in $\tilde{Z}$ by the Taylor group, we have the relation

$$I^\alpha(z) = J^\alpha(z) + O(h^2), \qquad \alpha = 1, \ldots, \tau, \quad i = 1, \ldots, n, \qquad (2.33)$$

which holds on the mesh chosen above. In practice, as a rule, it suffices to have less than τ of such invariants.

Finally, we write out the difference equations

$$\Phi_\alpha(I^1(z), I^2(z), \ldots, I^\tau(z)) = 0, \quad \omega_\beta(I^1(z), I^2(z), \ldots, I^{\tau+nm}(z)) = 0, \quad (2.34)$$

where $\omega_\beta = 0$ is the above-chosen invariant representation of the finite-difference mesh.

Thus, roughly speaking, the algorithm is that we substitute the *difference* invariants for the *differential* ones into the invariant representation of the original system.

Since the functions $\Phi_\alpha(I^1, I^2, \ldots, I^\tau) = 0$ are assumed to be locally analytic in their arguments, it follows from (2.33) that the difference equations $\Phi_\alpha = 0$ model the corresponding differential equations with second-order approximation.

Obviously, the system of difference equations (2.34) thus constructed admits the complete group G_r.

The above method for constructing invariant difference equations was called the *method of finite-difference invariants* in [29].

Of course, this method is not unique. More possibilities can be obtained by increasing the number of points in the difference stencil, i.e., by increasing the number of variables in the mesh space, by increasing the number of difference invariants, and accordingly, by increasing the possibilities for obtaining relations (2.33). Moreover, the finite-difference invariants can be used to approximate the original differential equation itself rather than its invariant representation. But in all cases the main tool is the set of finite-difference invariants. As we shall see later, when constructing an invariant model of the linear heat equation, it may happen that the differential equation is degenerate in the sense that there are no differential invariants, while simultaneously there exist difference invariants that can be used to approximate the heat equation by a nondegenerate difference model. (Thus, the degeneration occurs in the continuum limit.)

In the next section, we consider examples of applications of the above algorithm.

2.3. Examples of Construction of Difference Models Preserving the Symmetry of the Original Continuous Models

2.3.1. Invariant difference model for the equation $u_{xx} = u^{-3}$

Consider a more complicated situation in which the invariance conditions for the mesh and the difference equations are related to each other and do not hold separately.

The ordinary differential equation

$$u_{xx} = u^{-3} \tag{2.35}$$

admits the three-parameter point group associated with the Lie algebra spanned by the operators

$$X_1 = \frac{\partial}{\partial x}, \qquad X_2 = 2x\frac{\partial}{\partial x} + u\frac{\partial}{\partial u}, \qquad X_3 = x^2\frac{\partial}{\partial x} + xu\frac{\partial}{\partial u}. \tag{2.36}$$

Equation (2.35) is a special case of the Ermakov–Pinney equation (see also [26, 27]). It can be treated as the equation of one-dimensional motion of a particle in a field with potential $U = u^{-2}$ (where u is the coordinate and x is time). Equation (2.35) (and its three-dimensional analog) are in a sense unique: it is only for the quadratic potential that there exists an additional projection symmetry and a variational symmetry corresponding to *all* symmetries of (2.36). This results in additional conservation laws for the particle motion in such a field. In the case of a quadratic potential, the Boltzmann equation and the hydrodynamic equations of a polytropic gas for an appropriate value of the adiabatic constant have additional symmetry and the corresponding conservation laws as well. (See the discussion of this topic in [16].) Thus, the ordinary differential equation (2.35) is the simplest model with additional symmetry in this hierarchy.

To approximate Eq. (2.35), we need a difference stencil with at least three points, which is associated with the subspace $(x, h^+, h^-, u, u^+, u^-)$.

The operator X_3 violates the uniformity invariance condition (see Chapter 1), and hence the uniform mesh is not invariant. Although the group G_3 corresponding to the algebra (2.36) is autonomous in x, it has no invariants in the subspace (x, h^+, h^-). This means that it is impossible to construct an invariant mesh independent of the solution u. On the above-chosen stencil of the nonuniform mesh, we have three difference invariants, for which we can take the following ones:

$$J_1 = \frac{h^+}{uu^+}, \qquad J_2 = \frac{h^-}{uu^-}, \qquad J_3 = u^2 u \frac{\underset{h}{u_x} - \underset{h}{u_{\bar{x}}}}{h^-},$$

where

$$\underset{h}{u_x} = \frac{u^+ - u}{h^+}, \qquad \underset{h}{u_{\bar{x}}} = \frac{u - u^-}{h^-}.$$

Note that there is only one invariant, $I_1 = u^3 u_{xx}$, in the original space of differential variables (x, u, u_x, u_{xx}).

The general equation determining the family of difference meshes invariant under (2.36) can be written as

$$F\left(\frac{h^+}{uu^+}, \frac{h^-}{uu^-}, u^2 u \frac{\underset{h}{u_x} - \underset{h}{u_{\bar{x}}}}{h^-} \right) = 0, \tag{2.37}$$

where F is an arbitrary function of its arguments. We choose the simplest version

$$\frac{h^+}{uu^+} = \frac{h^-}{uu^-} \tag{2.38}$$

of Eq. (2.37). This mesh has the obvious integral

$$\frac{h^+}{uu^+} = \epsilon, \qquad \frac{h^-}{uu^-} = \epsilon, \qquad \epsilon = \text{const}, \quad 0 < \epsilon \ll 1, \tag{2.39}$$

where the constant ϵ characterizes the mesh spacing smallness. Using Taylor series expansions, we can obtain the relation

$$h^+ = h^- + O(h^2) \tag{2.40}$$

for the spacings of the invariant mesh (2.38). On the mesh (2.38), the third difference invariant can be rewritten as

$$J_3 = u^2 u^- \frac{\frac{u_x - u_{\bar{x}}}{h}}{h^-} = 2u^2 u^+ u^- \frac{\frac{u_x - u_{\bar{x}}}{h}}{h^- u^+ + h^+ u^-}, \tag{2.41}$$

and the differential invariant can be estimated as

$$J_3 = u^3 u_{xx} + O(h^2).$$

Thus, for the difference equation approximating Eq. (2.35) we can take

$$J_3 = 1, \qquad \text{or} \qquad u_{x\bar{x}} \equiv \frac{\frac{u_x - u_{\bar{x}}}{h}}{h^-} = \frac{1}{u^2 u^-}. \tag{2.42}$$

Since our model (2.38)–(2.42) is constructed from difference invariants of the group (2.36), it is completely invariant.

The difference mesh (2.38) depends on the solution u, u^+, u^-, which must be found from Eq. (2.42) containing the mesh spacings as well. At first glance, the scheme (2.38)–(2.42) is implicit, and it is not clear in general how to perform the actual computations for a specific boundary-value problem on this mesh.

Let us make several transformations of (2.38)–(2.42). By substituting the mesh spacings (2.39) into Eq. (2.35), we obtain the one-dimensional mapping

$$u^+ u^- (2 - \epsilon^2) = u(u^+ + u^-), \tag{2.43}$$

which uniquely determines the unknown value u^+ from the two known values u and u^-. Thus, the scheme (2.38)–(2.42) is in fact an *explicit scheme* for solving the boundary value problem: first, we calculate the sequence of values u from the mapping (2.43), and then we use formulas (2.39) to calculate the mesh spacings h^+ and h^-, i.e., the values of x at the corresponding mesh points.

As will be shown in Chapter 7, no numerical computations at all are required for the scheme (2.38)–(2.42). It turns out that the scheme (2.38)–(2.42) is completely integrable, which is related to the fact that it has three first integrals. (A method for constructing these integrals is considered in detail in Chapter 7.)

2.3.2. Invariant difference model of the sine–Gordon equation.

It is well known (e.g., see [74]) that the equation

$$u_{xy} = \sin u \tag{2.44}$$

admits the three-parameter point transformation group generated by the operators

$$X_1 = \frac{\partial}{\partial x}, \qquad X_2 = \frac{\partial}{\partial y}, \qquad X_3 = x\frac{\partial}{\partial x} - y\frac{\partial}{\partial y}. \qquad (2.45)$$

Note that Eq. (2.44) also admits an infinite series of higher-order symmetries and nonlocal symmetries (see [74]), which we do not consider here. Now Eq. (2.44) can be rewritten in the different form

$$v_{tt} - v_{zz} = \sin v. \qquad (2.46)$$

The symmetry of Eq. (2.46) is described by the operators

$$X_1 = \frac{\partial}{\partial t}, \qquad X_2 = \frac{\partial}{\partial z}, \qquad X_3 = z\frac{\partial}{\partial t} + t\frac{\partial}{\partial x}. \qquad (2.47)$$

Equations (2.44) and (2.46) are related to each other by the point transformation

$$t = x + y, \qquad z = x - y, \qquad u(x,y) = v(t,z), \qquad (2.48)$$

which of course takes the symmetry (2.45) to (2.47).

Let us construct a finite-difference model of Eq. (2.44) preserving the symmetry of (2.45). The first problem is the problem of choosing the mesh on which an approximation to this equation is possible so that it is invariant under the operators (2.45). Note that in this case the transformation (2.45) does not affect the dependent variable (u is an invariant), and therefore, the invariance conditions for the mesh can be considered independently of the invariance conditions for the difference equation approximating (2.44).

In this case, it is easily seen that one can use the simplest orthogonal mesh uniform in both directions. Indeed, all three operators (2.45) satisfy the mesh invariant orthogonality conditions

$$\underset{\pm h}{D_i}(\xi^j) = -\underset{\pm h}{D_j}(\xi^i), \qquad i = 1,2, \quad i \neq j,$$

developed in Chapter 1 and the mesh invariant uniformness conditions

$$\underset{+h}{D_i}\underset{-h}{D_i}(\xi^i) = 0, \qquad i = 1,2.$$

These conditions are satisfied in the entire space $\underset{h}{Z}$, in particular, on solutions of the difference equation approximating Eq. (2.44). On the orthogonal difference mesh, we should approximate the derivative u_{xy}. We introduce the notation for the variable u at different points of the mesh according to Fig. 2.2.

All nine variables $u, u^+, u^-, \hat{u}, \hat{u}^+, \hat{u}^-, \check{u}, \check{u}^-, \check{u}^+$ are invariants of the operators (2.45); hence one can use any of these, say, $\sin u$, to approximate the right-hand

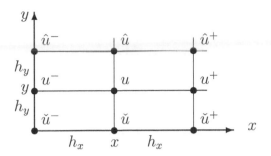

Figure 2.2

side of (2.44) at (x, y, u). The space $(x, y, h_x, h_y, u, u^+, u^-, \hat{u}, \hat{u}^+, \hat{u}^-, \breve{u}, \breve{u}^-, \breve{u}^+)$ contains one more invariant

$$J_{10} = h_x h_y.$$

Thus the invariant approximation problem is very simple in this case, and it only remains to construct second-order approximations to the mixed derivative u_{xy}.

For example, this can be realized as follows:

$$u_{xy} \approx \left(\frac{\hat{u}^+ - \hat{u}^-}{2h_x} - \frac{\breve{u}^+ - \breve{u}^-}{2h_x} \right) \frac{1}{2h_y} + O(h^2),$$

which finally implies the equation

$$\frac{\hat{u}^- - \hat{u}^- - \breve{u}^+ + \breve{u}^-}{4h_x h_y} = \sin u. \tag{2.49}$$

Equation (2.49) is based on the use of a five-point stencil. In particular, we can set $h_x = h_y$ in (2.49). In this case, the x- and y-directions become equivalent, just as in the original equation. The central point can be eliminated by using another approximation to the right-hand side,

$$\frac{\hat{u}^+ - \hat{u}^- - \breve{u}^+ + \breve{u}^-}{4h_x h_y} = \sin \frac{\hat{u}^+ + \hat{u}^- + \breve{u}^+ + \breve{u}^-}{4}.$$

By using Taylor expansions, one can readily estimate the order of approximation of the difference equation (2.49) on the uniform orthogonal mesh:

$$\frac{\hat{u}^+ - \hat{u}^- - \breve{u}^+ + \breve{u}^-}{4h_x h_y} - \sin u = u_{xy} - \sin u + O(h_x^2 + h_y^2).$$

Thus, the difference equation (2.49) on an orthogonal mesh provides a second-order approximation to the differential equation (2.44) and admits the entire symmetry (2.47) of the original equation.

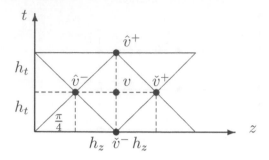

Figure 2.3

The simplest way to obtain an invariant scheme and a mesh for Eq. (2.46) is to use the transformation (2.48). This transformation takes Eq. (2.49) to the equation

$$\frac{\hat{v}^+ - \hat{v}^- - \check{v}^+ + \check{v}^-}{4h_z h_t} = \sin v, \qquad (2.50)$$

where h_z and h_t are the spacings of the mesh shown in Fig. 2.3.

Under the action of the transformations (2.48), the orthogonal mesh shown in Fig. 2.2 becomes the diagonal orthogonal mesh shown in Fig. 2.3. Of course, we could obtain this invariantly orthogonal mesh directly, by using the result obtained in Chapter 1: the Lorentz transformations (2.47) leave a mesh orthogonal only if the mesh makes an angle of $45°$ with the coordinate axes.

Equation (2.50) on the mesh in Fig. 2.3 admits the complete group (2.47), because the point transformation (2.48) preserves the symmetry group of the original equation by taking the operators (2.45) to the operators (2.47).

In the subsequent chapters, we present numerous examples of construction of other finite-difference models preserving the symmetry of the original differential equation.

2.3.3. An example of solving the determining equations for a difference model

In this section, we use the invariance criterion for difference models to *calculate the symmetry group*. For the difference model, we consider the following difference equation and mesh:

$$\frac{u^+ - 2u + u^-}{h^+ h^-} = e^u, \qquad h^+ = h^-. \qquad (2.51)$$

Here, as usual, $h^+ = x^+ - x$ and $h^- = x - x^-$. This model is known to give a second-order approximation to the ordinary differential equation

$$\frac{d^2 u}{dx^2} = e^u \qquad (2.52)$$

on the uniform mesh.

We seek the point symmetry admitted by system (2.51) in the operator form

$$X = \xi(x, u)\frac{\partial}{\partial x} + \eta(x, u)\frac{\partial}{\partial u}. \tag{2.53}$$

The equation and the mesh (2.51) are written on the three-point difference stencil $(x^-, x, x^+, u^-, u, u^+)$, to which we extend the operator (2.53) by setting

$$X = \xi(x, u)\frac{\partial}{\partial x} + \eta(x, u)\frac{\partial}{\partial u} + \xi(x^-, u^-)\frac{\partial}{\partial x^-} + \eta(x^-, u^-)\frac{\partial}{\partial u^-}$$
$$+ \xi(x^+, u^+)\frac{\partial}{\partial x^+} + \eta(x^+, u^+)\frac{\partial}{\partial u^+}. \tag{2.54}$$

Acting on the difference equation (2.51) by the operator (2.54), we obtain the linear difference equations

$$\eta^+ - 2\eta + \eta^- = e^u \eta h^- h^+ + e^u h^+(\xi - \xi^-) + e^u h^-(\xi^+ - \xi),$$
$$\xi^+ - 2\xi + \xi^- = 0 \tag{2.55}$$

for the desired functions ξ and η, where the notation $f^+ = f(x^+, u^+)$ and $f^- = f(x^-, u^-)$ is used. The equation for the difference mesh has the obvious integral

$$h^+ = h^- = \text{const} = h,$$

which we take into account in the subsequent calculations.

The determining equations (2.55) should be considered on the solutions of system (2.51). In the case of differential equations, the standard method is to eliminate the second derivative from the determining equation and use the splitting procedure for the first derivative. In the case of the difference system (2.55), we can, say, eliminate u^+ by using the first equation in (2.51):

$$u^+ = h^2 e^u + 2u - u^-.$$

Consider the equation for the mesh with the above elimination taken into account:

$$\xi^+(x^+, (h^2 e^u + 2u - u^-)) - 2\xi(x, u) + \xi^-(x^-, u^-) = 0. \tag{2.56}$$

Now this equation should be satisfied identically at the remaining *two* values u, u^-. We differentiate Eq. (2.56) with respect to the variable u^-,

$$\xi^+{}_{u^+} = \xi^-{}_{u^-},$$

which implies that

$$\xi^+{}_{u^+} = \xi^-{}_{u^-} = A = \text{const},$$

and the constant cannot depend on x. Thus, we have specified the form of the desired coordinate,

$$\xi(x, u) = Au + B(x), \quad A = \text{const.}$$

By substituting this expression into (2.56), we obtain

$$Ah^2 e^u + B(x^+) - 2B(x) + B(x^-) = 0.$$

By splitting the last equation with respect to u, we obtain

$$A = 0, \quad B(x^+) - 2B(x) + B(x^-) = 0.$$

The last linear equation has the obvious solution $B(x) = \alpha x + \beta$, where $\alpha, \beta = \text{const.}$ Thus, the desired coordinate has the form

$$\xi = \alpha x + \beta. \tag{2.57}$$

Taking into account this form of the desired function $\xi(x)$, let us now consider the second determining equation on the solution of the original equation,

$$\eta^+(x^+, (h^2 e^u + 2u - u^-) - 2\eta(x, u) + \eta^-(x^-, u^-) = h^2 e^u(\eta(x, u) + 2\alpha). \tag{2.58}$$

In a similar way, we differentiate Eq. (2.58) with respect to the variable u^-,

$$\eta^+{}_{u^+} = \eta^-{}_{u^-},$$

which implies that

$$\eta^+{}_{u^+} = \eta^-{}_{u^-} = \gamma = \text{const;}$$

namely,

$$\eta(x, u) = \gamma u + \delta(x).$$

We substitute this specified form of the desired function into (2.58):

$$\gamma h^2 e^u + \delta(x^+) - 2\delta(x) + \delta(x^-) = h^2 e^u(2\alpha + \gamma u + \delta(x)).$$

The splitting of this equation with respect to u and e^u yields

$$\gamma = 0, \quad \delta = -2\alpha, \quad \delta(x^+) - 2\delta(x) + \delta(x^-) = 0. \tag{2.59}$$

Since $\delta = -2\alpha = \text{const}$, it follows that the last equation in (2.59) is satisfied identically.

Thus, we have obtained the following solution of the determining system (2.55):

$$\xi = \alpha x + \beta, \quad \eta = -2\alpha. \tag{2.60}$$

By choosing a basis in the two-parameter family, we obtain

$$X_1 = \frac{\partial}{\partial x}, \qquad X_2 = x\frac{\partial}{\partial x} - 2\frac{\partial}{\partial u}; \tag{2.61}$$

i.e., the difference equation admits the same transformation group as the differential equation (2.52). In the three-point stencil extended to all variables, the operators (2.61) can be written as

$$X_1 = \frac{\partial}{\partial x} + \frac{\partial}{\partial x^+} + \frac{\partial}{\partial x^-}, \quad X_2 = x\frac{\partial}{\partial x} + x^+\frac{\partial}{\partial x^+} + x^-\frac{\partial}{\partial x^-} - 2\frac{\partial}{\partial u} - 2\frac{\partial}{\partial u^+} - 2\frac{\partial}{\partial u^-}.$$

Some other examples of solving the determining equations for difference models can be found in [85, 89].

An alternative approach to symmetry preservation in difference equations can be found in [108–110].

Chapter 3

Invariant Difference Models of Ordinary Differential Equations

3.1. First-Order Invariant Difference Equations and Lattices

It is well known that every ordinary differential equation (ODE) of first order admits an infinite point group (e.g., see [107, 111]), but it is precisely in the case of first-order equations that a search for this group is ineffective. If the admissible symmetry is known, then an integrating factor for the ODE can be found explicitly (see the Introduction). Since the knowledge of the symmetry of a first-order ODE permits finding its general solution, we see that the construction of an invariant scheme does not make any practical sense. But it is of interest to find out what the relation between the symmetry and integrability is in the case of ordinary difference equations (see [120]).

Consider a first-order ODE

$$y' = f(x, y). \tag{3.1}$$

We assume that its symmetry is known, which permits rewriting this equation as a total differential equation

$$A(x, y)dx + B(x, y)dy = 0, \tag{3.2}$$

where $A_y = B_x$, $A(x, y) = V_x(x, y)$, $B(x, y) = V_y(x, y)$, and $V(x, y)$ is a function implicitly determining the general solution of Eq. (3.1) by the formula

$$V(x, y) = c = \text{const.} \tag{3.3}$$

Equation (3.2) admits the one-parameter group generated by the operator

$$X = B(x, y)\frac{\partial}{\partial x} - A(x, y)\frac{\partial}{\partial y}. \tag{3.4}$$

In the (x, y)-space, this operator has the only invariant

$$J_1 = V(x, y).$$

The situation is different in the case of difference equations. To approximate the ODE (3.1), we need at least a two-point stencil on a difference mesh; i.e., we should consider the subspace (x, y, x_+, y_+) of difference variables, where the operator (3.4) already has three difference invariants. To find the invariants of the operator extended to the difference variables,

$$X = B(x,y)\frac{\partial}{\partial x} - A(x,y)\frac{\partial}{\partial y} + B(x_+, y_+)\frac{\partial}{\partial x_+} - A(x_+, y_+)\frac{\partial}{\partial y_+},$$

one has to solve the characteristic system

$$\frac{dx}{V_y(x,y)} = -\frac{dy}{V_x(x,y)} = \frac{dx_+}{V_{y_+}(x_+, y_+)} = -\frac{dy_+}{V_{x_+}(x_+, y_+)}.$$

This gives the following three invariants:

$$I_1 = V(x,y), \qquad I_2 = V(x_+, y_+),$$

$$I_3 = \int \frac{dx_+}{V_{y_+}(x_+, y_+(x_+, I_2))} - \int \frac{dx}{V_y(x, y(x, I_1))},$$

where the third invariant contains the expressions of y and y_+ via the first two invariants I_1 and I_2.

The two-point difference scheme can be written out using I_1 and I_2 as

$$V(x_+, y_+) - V(x,y) = 0, \tag{3.5}$$

and this scheme is *exact on any difference mesh*, for example, on the uniform mesh $h_+ = h_-$. But the uniform mesh is not invariant under an arbitrary group. The general equation for invariant meshes can be written out in terms of the difference invariants,

$$F(I_1, I_2, I_3) = 0. \tag{3.6}$$

The difference scheme (3.5), (3.6) is *invariant and exact*. Its general solution is implicitly given by formula (3.3), the same formula as for the original equation.

Consider an example of constructing an invariant difference model for a first-order ODE.

EXAMPLE. Consider the ODE

$$y' = -2\frac{y}{x}.$$

This equation can be rewritten as the total differential equation

$$2xy\,dx + x^2\,dy = 0, \tag{3.7}$$

Equation (3.7) admits the operator

$$X = x^2 \frac{\partial}{\partial x} - 2xy \frac{\partial}{\partial y} \qquad (3.8)$$

and has the first integral

$$V(x, y) = x^2 y = C_0 = \text{const},$$

which is an invariant of the operator (3.8). In the space of the difference variables (x, y, x_+, y_+), the extended operator (3.8) given by

$$X = x^2 \frac{\partial}{\partial x} - 2xy \frac{\partial}{\partial y} + x_+{}^2 \frac{\partial}{\partial x_+} - 2x_+ y_+ \frac{\partial}{\partial y_+}$$

has the following three difference invariants:

$$I_1 = x^2 y, \qquad I_2 = x_+{}^2 y_+, \qquad I_3 = \frac{1}{x} - \frac{1}{x_+}.$$

An exact difference equation can be constructed from the first two invariants as follows:

$$I_2 - I_1 = 0 \quad \Longrightarrow \quad x_+{}^2 y_+ - x^2 y = 0. \qquad (3.9)$$

An invariant mesh is determined by any equation of the form

$$F(I_1, I_2, I_3) = 0.$$

By way of example, consider the mesh

$$I_3 = \frac{1}{x} - \frac{1}{x_+} = \varepsilon \ll 1,$$

which we rewrite explicitly for $h_+ = x_+ - x$ as

$$h_+ = \frac{\varepsilon x^2}{1 - \varepsilon x}, \qquad 0 < x < \frac{1}{\varepsilon}. \qquad (3.10)$$

The invariant equation (3.9) can be rewritten as

$$\frac{y_+ - y}{h_+} = -\frac{(x + x_+)y_+}{x^2}. \qquad (3.11)$$

Note that the exact invariant scheme (3.10), (3.11) is implicit. A series of examples of invariant schemes for first-order ODE can be found in [120].

Thus, if the symmetry of the first-order ODE is known, then it is always possible to write out a family of exact difference schemes containing a family of invariant schemes. On the other hand, it is always possible (by using the difference invariants) to write out an invariant scheme that is not exact. This fact is of general character, because the order of approximation (the accuracy) and the invariance are distinct properties of difference equations.

3.2. Invariant Second-Order Difference Equations and Lattices

In this section, we consider the more general problem of how to list all difference models invariant under a given group, i.e., how to solve the group classification problem for difference schemes by analogy with the same problem for the differential equations. When solving this problem, it is of interest to compare the group classifications of difference schemes and of differential equations. *Is the list of invariant schemes wider or narrower than the corresponding list of differential equations?*

In this section, by way of example, we solve the group classification problem for second-order ordinary difference equations on the corresponding difference meshes [46].[1]

In the classical papers [90–93], S. Lie obtained the group classification of second-order ordinary differential equations. In particular, he showed that the dimension n of the point transformation group under which the solution set of a second-order ODE is invariant can be $0, 1, 2, 3$, or 8. Moreover, he showed that each equation that admits the transformation group of maximum dimension $n = 8$ can be transformed into the simplest equation $y'' = 0$. Lie's classification is based on the list of all finite-dimensional Lie algebras realizable by vector fields on the two-dimensional space [92], i.e., by vector fields of the form

$$X = \xi(x, y) \frac{\partial}{\partial x} + \eta(x, y) \frac{\partial}{\partial y}.$$

Lie used vector fields over the field of complex numbers, using the list of all finite-dimensional subalgebras of infinite-dimensional Lie algebras. This classification was obtained up to arbitrary locally invertible transformations of the complex plane $\mathbb{C}^2$.

Much later, a classification over the field of real numbers was obtained [63], and here we follow this classification.

In our classification of difference models, we restrict ourselves to the minimal three-point difference stencil needed to approximate second-order difference equations. But we do not specify and restrict the types of meshes in advance.

Let x be an independent variable, and let y be a dependent variable. To consider a second-order equation and a difference mesh in the x-direction, we need the three-point stencil corresponding to the subspace $(x, x_-, x_+, y, y_-, y_+)$ (see Fig. 3.1).

The difference model under study consists of two difference equations

$$F(x, x_-, x_+, y, y_-, y_+) = 0, \qquad \Omega(x, x_-, x_+, y, y_-, y_+) = 0. \tag{3.12}$$

Here the first equation is a second-order difference equation and should become a second-order ODE at the point (x, y, y', y'') in the continuum limit. The second

---

[1]This research was carried out jointly with P. Winternitz and R. Kozlov.

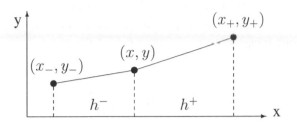

Figure 3.1

equation provides a difference mesh on which the first equation will be considered. The second equation need not be a second-order difference equation; for example, we can treat a uniform mesh as the equation $h_+ = h_-$, which "disappears" in the continuum limit.

In some cases, we use the following notation for the difference models:

$$y_+ = f(x, x_-, y, y_-), \qquad x_+ = g(x, x_-, y, y_-),$$

or

$$y_x = f(x, h_-, y, y_{\bar{x}}), \qquad h_+ = g(x, h_-, y, y_x).$$

We use the following notation for the mesh spacings and difference derivatives:

$$h_+ = x_+ - x, \quad h_- = x - x_-, \quad y_x = \frac{y_+ - y}{h_+}, \quad y_{\bar{x}} = \frac{y - y_-}{h_-}, \quad y_{x\bar{x}} = \frac{2(y_x - y_{\bar{x}})}{h_+ + h_-},$$

where y_x and $y_{\bar{x}}$ are the first right and left difference derivatives and $y_{x\bar{x}}$ is the second difference derivative. The continuous derivatives are denoted by y' and y''.

We must prolong the group operator of the continuous symmetry group to all points of the difference stencil. The corresponding coefficients of the operator are obtained by shifting the arguments to the difference stencil points adjacent to (x, y):

$$\tilde{X}_\alpha = X_\alpha + \xi_\alpha(x_-, y_-)\frac{\partial}{\partial x_-} + \xi_\alpha(x_+, y_+)\frac{\partial}{\partial x_+}$$

$$+ \eta_\alpha(x_-, y_-)\frac{\partial}{\partial y_-} + \eta_\alpha(x_+, y_+)\frac{\partial}{\partial y_+}. \quad (3.13)$$

The difference invariants are obtained as the solutions of the system of first-order partial differential equations:

$$\tilde{X}_\alpha \Phi = 0, \qquad \alpha = 1, \ldots, n,$$

where the function Φ depends on $(x, x_-, x_+, y, y_-, y_+)$. We see that the group G has $k = \dim M - \operatorname{rank} Z$ functionally independent invariants, where M is the manifold (3.12) and Z is the coefficient matrix of the operator (3.13).

Note that $\dim M = 4$ in the differential case and $\dim M = 6$ in the difference case. Hence we can expect that there are two more invariants in the difference case than in the continuous case.

In the continuous case, an invariant second-order ODE can be written as

$$E(I_1, \ldots, I_k) = 0, \qquad \frac{\partial E}{\partial y''} \neq 0,$$

while in the difference case we need two equations,

$$
\begin{aligned}
F(I_1, \ldots, I_k) = 0, \qquad & \frac{\partial F}{\partial y_-} \neq 0, \qquad \frac{\partial F}{\partial y} \neq 0, \qquad \frac{\partial F}{\partial y_+} \neq 0, \\
G(I_1, \ldots, I_k) = 0, \qquad & \frac{\partial G}{\partial x_-} \neq 0, \qquad \frac{\partial G}{\partial x} \neq 0, \qquad \frac{\partial G}{\partial x_+} \neq 0.
\end{aligned}
\tag{3.14}
$$

The conditions on F and G guarantee that Eqs. (3.14) determine an ordinary difference equation and a mesh.

3.2.1. Difference models invariant under one- and two-dimensional groups

We start from the simplest one-dimensional point transformation group, whose Lie algebra can always be reduced to the form

$$X_1 = \frac{\partial}{\partial y}$$

by an appropriate change of variables. The most general second-order ordinary differential equation invariant under the one-dimensional group has the form

$$y'' = F(x, y'),
\tag{3.15}$$

where F is an arbitrary function of its arguments.

To write out the invariant difference model, we should compute the basis of finite-difference invariants of the operator X_1 in the space $(x, x_-, x_+, y, y_-, y_+)$. Computations give the following five difference invariants:

$$\left\{ y_{x\bar{x}}, \quad \frac{y_x + y_{\bar{x}}}{2}, \quad x, \quad h_-, \quad h_+ \right\}.$$

Hence the general invariant difference model can be written as

$$y_{x\bar{x}} = f\left(x, \frac{y_x + y_{\bar{x}}}{2}, h_-\right), \qquad h_+ = h_- g\left(x, \frac{y_x + y_{\bar{x}}}{2}, h_-\right),
\tag{3.16}$$

where f and g are arbitrary functions of their arguments. Throughout the following, we assume that the functions occurring in the definition of difference models do not have any singularities as $h_+ \to 0$ and $h_- \to 0$. The form in which the equation for the mesh is written assumes that the equation "disappears" in the continuum limit.

In the family of difference models (3.16), the simplest difference scheme approximating the ODE (3.15) can be distinguished by taking a function f independent of h_- and the function $g \equiv 1$,

$$y_{x\bar{x}} = F\left(x, \frac{y_x + y_{\bar{x}}}{2}\right), \qquad h_- = h_+. \tag{3.17}$$

We point out that the difference scheme (3.17) is only a special case of the difference model (3.16) containing two arbitrary functions. In other words, the family of invariant difference models is much more ambiguous than the corresponding invariant differential equation.

Now consider two-dimensional algebras.

1. In the continuous case, the Abelian Lie algebra with two unconnected elements

$$X_1 = \frac{\partial}{\partial x}, \qquad X_2 = \frac{\partial}{\partial y}$$

leaves invariant the solution set of the family of second-order ODE

$$y'' = F(y'), \tag{3.18}$$

where F is an arbitrary function. In the difference case, the standard computational procedure gives the following set of functionally independent invariants:

$$\left\{ h_+, \quad h_-, \quad \frac{y_x + y_{\bar{x}}}{2}, \quad y_{x\bar{x}} \right\}.$$

Accordingly, the general invariant difference model can be represented as

$$y_{x\bar{x}} = f\left(\frac{y_x + y_{\bar{x}}}{2}, h_-\right), \qquad h_+ = h_- g\left(\frac{y_x + y_{\bar{x}}}{2}, h_-\right). \tag{3.19}$$

The simplest difference scheme approximating (3.18) can again be obtained by restricting f and by choosing $g = 1$:

$$y_{x\bar{x}} = F\left(\frac{y_x + y_{\bar{x}}}{2}\right), \qquad h_- = h_+. \tag{3.20}$$

2. The Abelian Lie algebra with two connected elements

$$X_1 = \frac{\partial}{\partial y}, \qquad X_2 = x\frac{\partial}{\partial y} \tag{3.21}$$

gives the invariant equation

$$y'' = F(x). \tag{3.22}$$

This equation can be transformed into the simplest linear equation $u'' = 0$ by the change of variables

$$u = y - W(x),$$

where $W(x)$ is an arbitrary solution of Eq. (3.22). The basis

$$\{y_{x\bar{x}}, \quad x, \quad h_+, \quad h_-\}$$

of finite-difference invariants of the algebra (3.21) permits writing out the following family of invariant models:

$$y_{x\bar{x}} = f(x, h_-), \qquad h_+ = h_- g(x, h_-). \tag{3.23}$$

By restricting f and by aking $g = 1$, we obtain the following simplest scheme on a uniform mesh, which approximates Eq. (3.22):

$$y_{x\bar{x}} = F(x), \qquad h_- = h_+.$$

Just as in the continuous case, the difference model (3.23) can be reduced to the form

$$u_{x\bar{x}} = 0, \qquad h_+ = h_- g(x, h_-) \tag{3.24}$$

by the change of variables

$$u = y - W(x, h_-, h_+), \qquad W_{x\bar{x}} = f(x, h_-),$$

where W is an arbitrary solution of the difference system (3.23).

3. The non-Abelian Lie algebra with two unconnected elements

$$X_1 = \frac{\partial}{\partial y}, \qquad X_2 = x\frac{\partial}{\partial x} + y\frac{\partial}{\partial y}$$

gives the invariant ODE

$$y'' = \frac{1}{x}F(y'). \tag{3.25}$$

For this algebra, computations give the following basis set of difference invariants:

$$\left\{ xy_{x\bar{x}}, \quad \frac{y_x + y_{\bar{x}}}{2}, \quad \frac{h_+}{h_-}, \quad \frac{h_-}{x} \right\}.$$

The invariant model has the general form

$$y_{x\bar{x}} = \frac{1}{x}f\left(\frac{y_x + y_{\bar{x}}}{2}, \frac{h_-}{x}\right), \qquad h_+ = h_- g\left(\frac{y_x + y_{\bar{x}}}{2}, \frac{h_-}{x}\right). \tag{3.26}$$

By restricting f and by taking $g = 1$, we obtain the invariant equation and mesh

$$y_{x\bar{x}} = \frac{1}{x}F\left(\frac{y_x + y_{\bar{x}}}{2}\right), \qquad h_- = h_+,$$

for which Eq. (3.25) is the continuum limit.

4. The non-Abelian Lie algebra with two linearly connected elements

$$X_1 = \frac{\partial}{\partial y}, \qquad X_2 = y \frac{\partial}{\partial y}$$

leads to the invariant ODE

$$y'' = F(x)y'. \qquad (3.27)$$

This equation can be reduced to the linear form $v'' = 0$ as well by the change of variables $t = g(x)$, $v(t) = y(x)$, where $g(x)$ is an arbitrary solution of Eq. (3.27). The complete set

$$\left\{ \frac{y_{x\bar{x}}}{y_x + y_{\bar{x}}}, \quad x, \quad h_-, \quad h_+ \right\}$$

of finite-difference invariants permits writing out the invariant difference model as

$$y_{x\bar{x}} = \frac{y_x + y_{\bar{x}}}{2} f(x, h_-), \qquad h_+ = h_- g(x, h_-). \qquad (3.28)$$

The special case of this family for $f(x, h_-) = F(x)$, $g(x, h_-) = 1$ is the scheme

$$y_{x\bar{x}} = \frac{y_x + y_{\bar{x}}}{2} F(x), \qquad h_- = h_+$$

approximating the corresponding ODE (3.27). The difference family (3.28) can be transformed into the scheme (3.24) by using any solution $\phi(x)$ of system (3.28) and the transformation

$$(x, y) \longmapsto (t = \phi(x), u(t) = y(x)) \qquad (3.29)$$

of the independent variable.

The results obtained for two-dimensional Lie algebras can be summarized as follows.

THEOREM 3.1. *Two subalgebras with linearly unconnected elements give families of invariant schemes* (3.19) *and* (3.26) *containing two arbitrary functions of two variables. Two subalgebras with linearly connected elements permit writing out the families of invariant models* (3.23) *and* (3.28), *which can be reduced to the form* (3.24) *if at least one solution of the original family is known.*

Remark. Theorem 3.1 is an analog of the well-known result obtained by S. Lie for second-order ODE in the cases where the equations with the same two-dimensional Lie symmetry algebras can be reduced to the linear equation $y'' = 0$ [90, 93].

3.2.2. Difference equations invariant under three-dimensional transformation groups

First, let us study three-dimensional Lie algebras containing two-dimensional subalgebras with linearly connected elements.

1. The three-dimensional algebra

$$X_1 = \frac{\partial}{\partial x}, \qquad X_2 = \frac{\partial}{\partial y}, \qquad X_3 = x\frac{\partial}{\partial y}$$

contains commuting linearly connected operators X_2 and X_3. The invariant differential equation $y'' = C$ is equivalent to $y'' = 0$. The set

$$\{y_{x\bar{x}}, \quad h_-, \quad h_+\}$$

of independent finite-difference invariants permits representing the family of invariant difference models as

$$y_{x\bar{x}} = f(h_-), \qquad h_+ = h_- g(h_-). \tag{3.30}$$

The equation $y'' = C$ is the continuum limit of (3.30) if $f = C$ and $g = 1$. System (3.30) is equivalent to (3.24) if the function g is independent of x.

2. The three-dimensional algebra

$$X_1 = \frac{\partial}{\partial y}, \qquad X_2 = x\frac{\partial}{\partial y}, \qquad X_3 = \frac{\partial}{\partial x} + y\frac{\partial}{\partial y}$$

contains two linearly connected operators X_1 and X_2. The invariant ODE has the form

$$y'' = C\exp(x). \tag{3.31}$$

The basis of difference invariants can be represented as

$$\{y_{x\bar{x}}\exp(-x), \quad h_-, \quad h_+\}.$$

The general form

$$y_{x\bar{x}} = f(h_-)\exp(x), \qquad h_+ = h_- g(h_-) \tag{3.32}$$

of the family of invariant models in particular contains the simplest scheme approximating Eq. (3.31) for $f = C$ and $g = 1$. The scheme (3.32) can be transformed into (3.24) for g independent of x.

3. The operator algebra

$$X_1 = \frac{\partial}{\partial x}, \qquad X_2 = \frac{\partial}{\partial y}, \qquad X_3 = y\frac{\partial}{\partial y}$$

leaves invariant the equation

$$y'' = Cy'. \tag{3.33}$$

The complete set of difference invariants can be chosen as

$$\left\{ \frac{y_{x\bar{x}}}{y_x}, \quad \frac{y_x + y_{\bar{x}}}{y_x}, \quad h_- \right\}.$$

The general invariant difference model can be written as

$$y_{x\bar{x}} = \frac{y_x + y_{\bar{x}}}{2} f(h_-), \qquad h_+ = h_- g(h_-). \tag{3.34}$$

By choosing $f = C$ and $g = 1$, we transform system (3.34) into a scheme approximating Eq. (3.33). The transformation (3.29) with g independent of x takes (3.34) to (3.24).

4. Consider the three-dimensional Lie algebra spanned by the pairwise linearly connected operators

$$X_1 = \frac{\partial}{\partial y}, \qquad X_2 = x\frac{\partial}{\partial y}, \qquad X_3 = y\frac{\partial}{\partial y}.$$

The only continuous invariant of this algebra is the function x, but $y'' = 0$ is an invariant manifold, and hence the equation $y'' = 0$ is an invariant ODE. The only difference invariants are $\{x_-, x, x_+\}$, but the difference equation $y_{x\bar{x}} = 0$ is an invariant manifold as well, and hence the difference scheme

$$y_{x\bar{x}} = 0, \qquad h_+ = h_- q(x, h_-)$$

is invariant. If we set $g = 1$ in this scheme, then we obtain the simplest approximation to the equation $y'' = 0$.

5. The operator algebra

$$X_1 = \frac{\partial}{\partial y}, \qquad X_2 = x\frac{\partial}{\partial y}, \qquad X_3 = (1-a)x\frac{\partial}{\partial x} + y\frac{\partial}{\partial y}, \qquad a \neq 1,$$

has the invariant equation

$$y'' = Cx^{\frac{2a-1}{1-a}}, \qquad a \neq 1. \tag{3.35}$$

Computations give the following set of difference invariants:

$$\left\{ y_{x\bar{x}} x^{\frac{2a-1}{a-1}}, \quad \frac{h_-}{x}, \quad \frac{h_+}{x} \right\},$$

according to which we obtain the invariant form

$$y_{x\bar{x}} = x^{\frac{2a-1}{1-a}} f\left(\frac{h_-}{x}\right), \qquad h_+ = h_- g\left(\frac{h_-}{x}\right) \tag{3.36}$$

of the difference model. An approximation to Eq. (3.35) is obtained for $f = C$ and $g = 1$. The difference scheme (3.36) can also be transformed into (3.24).

6. The operator algebra

$$X_1 = \frac{\partial}{\partial y}, \qquad X_2 = x\frac{\partial}{\partial y}, \qquad X_3 = (1+x^2)\frac{\partial}{\partial x} + (x+b)y\frac{\partial}{\partial y}$$

is associated with the invariant ODE

$$y'' = C(1+x^2)^{-3/2}\exp(b\arctan(x)), \tag{3.37}$$

which can be transformed to $y'' = 0$ by a change of variables. The complete set

$$\left\{ \frac{h_+}{1+xx_+}, \quad \frac{h_-}{1+xx_-}, \quad (y_x - y_{\bar{x}})\sqrt{1+x^2}\exp(-b\arctan(x)), \right\}$$

of difference invariants permits writing out the invariant difference model as

$$y_{x\bar{x}} = \frac{\exp(b\arctan(x))}{(h_- + h_+)\sqrt{1+x^2}}\left(\frac{h_+}{1+xx_+} + \frac{h_-}{1+xx_-} \right) f\left(\frac{h_-}{1+xx_-} \right),$$

$$h_+ = h_-\frac{1+xx_+}{1+xx_-}g\left(\frac{h_-}{1+xx_-} \right).$$

By setting $f = g = 1$, we obtain a difference approximation to the ODE (3.37).

7. The operator algebra

$$X_1 = \frac{\partial}{\partial x}, \qquad X_2 = \frac{\partial}{\partial y}, \qquad X_3 = x\frac{\partial}{\partial x} + y\frac{\partial}{\partial y}$$

has the only invariant y', but $y'' = 0$ is an invariant manifold. By using the basis set

$$\left\{ h_- y_{x\bar{x}}, \quad \frac{y_x + y_{\bar{x}}}{2}, \quad \frac{h_+}{h_-}, \right\}$$

of difference invariants, we can readily write out the invariant difference model

$$h_- y_{x\bar{x}} = f\left(\frac{y_x + y_{\bar{x}}}{2} \right), \qquad h_+ = h_- g\left(\frac{y_x + y_{\bar{x}}}{2} \right), \tag{3.38}$$

which does not have a continuum limit in general. This limit only exists for $f = 0$. The equation $y'' = 0$ is approximated by system (3.38) for $f = 0$ and $g = 1$.

Thus, the difference models invariant under three-dimensional algebras considered so far are equivalent to special cases of the system

$$y_{x\bar{x}} = 0, \qquad g(x, h_-, h_+) = 0. \tag{3.39}$$

The difference models considered below cannot be reduced to the form (3.39).

8. The operator algebra

$$X_1 = \frac{\partial}{\partial x} \qquad X_2 = \frac{\partial}{\partial y} \qquad X_3 = x\frac{\partial}{\partial x} + (x+y)\frac{\partial}{\partial y}$$

gives the invariant ODE

$$y'' = \exp(-y'). \tag{3.40}$$

Its general solution

$$y = -x + (x + B)\ln(x + B) + A$$

contains constants A and B of integration. The basis

$$\left\{ \frac{h_+}{h_-}, \quad h_+ \exp\left(-y_x\right), \quad h_- \exp\left(-y_{\bar{x}}\right) \right\}$$

of finite-difference invariants permits writing out the general invariant difference system

$$\frac{2\left(\exp\left(y_x\right) - \exp\left(y_{\bar{x}}\right)\right)}{h_- + h_+} = f\left(h_- \exp\left(-y_{\bar{x}}\right)\right), \qquad h_+ = h_- g\left(h_- \exp\left(-y_{\bar{x}}\right)\right).$$

An invariant approximation to the ODE (3.40) is obtained for $f = g = 1$.

9. The three-dimensional operator algebra

$$X_1 = \frac{\partial}{\partial x}, \quad X_2 = \frac{\partial}{\partial y}, \quad X_3 = x\frac{\partial}{\partial x} + ky\frac{\partial}{\partial y}, \qquad k \neq 0, 1,$$

gives the family of invariant ODE

$$y'' = y'^{\frac{k-2}{k-1}}, \tag{3.41}$$

where $k = \text{const}$. The general solution is

$$y = \left(\frac{1}{k-1}\right)^{k-1} \frac{1}{k}(x - x_0)^k + y_0$$

By using the basis set

$$\left\{ \frac{h_+}{h_-}, \quad y_x h_+^{(1-k)}, \quad y_{\bar{x}} h_-^{(1-k)} \right\},$$

of difference invariants, we can write out the invariant model as

$$\frac{2(k-1)}{h_- + h_+}\left((y_x)^{\frac{1}{k-1}} - (y_{\bar{x}})^{\frac{1}{k-1}}\right) = f\left(y_{\bar{x}} h_-^{(1-k)}\right), \qquad h_+ = h_- g\left(y_{\bar{x}} h_-^{(1-k)}\right).$$

The simplest approximation to the ODE (3.41) is obtained for $f = g = 1$.

10. The Lie algebra of linearly unconnected operators

$$X_1 = \frac{\partial}{\partial x}, \qquad X_2 = \frac{\partial}{\partial y}, \qquad X_3 = (kx + y)\frac{\partial}{\partial x} + (ky - x)\frac{\partial}{\partial y}$$

gives the invariant ODE

$$y'' = (1 + (y')^2)^{3/2} \exp(k \arctan(y')) \tag{3.42}$$

with the general solution

$$\exp\left(2k \arctan\left(\frac{-(x - x_0) + k(y - y_0)}{k(x - x_0) + (y - y_0)}\right)\right) ((x - x_0)^2 + (y - y_0)^2) = \frac{1}{1 + k^2}.$$

Computations give the basis

$$\left\{h_+\sqrt{1 + y_x^2}\exp\left(k \arctan y_x\right), \quad h_-\sqrt{1 + y_{\bar{x}}^2}\exp\left(k \arctan y_{\bar{x}}\right), \quad \frac{y_x - y_{\bar{x}}}{1 + y_x y_{\bar{x}}}\right\}$$

of finite-difference invariants. This implies the general form

$$y_{x\bar{x}} = \frac{1 + y_x y_{\bar{x}}}{h_- + h_+}\left(h_+\sqrt{1 + y_x^2}\exp(k \arctan y_x) + h_-\sqrt{1 + y_{\bar{x}}^2}\exp(k \arctan y_{\bar{x}})\right)$$

$$\times f\left(h_-\sqrt{1 + y_{\bar{x}}^2}\exp(k \arctan y_{\bar{x}})\right),$$

$$h_+\sqrt{1 + y_x^2}\exp(k \arctan y_x) = h_-\sqrt{1 + y_{\bar{x}}^2}\exp(k \arctan y_{\bar{x}})$$

$$+ g\left(h_-\sqrt{1 + y_{\bar{x}}^2}\exp(k \arctan y_{\bar{x}})\right)$$

of the invariant difference model, which has the ODE (3.42) as the continuum limit provided that $f = 1$ and $g = 0$.

11. The operator algebra

$$X_1 = \frac{\partial}{\partial x}, \qquad X_2 = 2x\frac{\partial}{\partial x} + y\frac{\partial}{\partial y}, \qquad X_3 = x^2\frac{\partial}{\partial x} + xy\frac{\partial}{\partial y}$$

leaves invariant the equation

$$y'' = y^{-3}, \tag{3.43}$$

which has the general solution

$$Ay^2 = (Ax + B)^2 + 1, \qquad A = \text{const}, \quad B = \text{const}.$$

By using the complete set

$$\left\{y(y_x - y_{\bar{x}}), \quad \frac{1}{y}\left(\frac{h_+}{y_+} + \frac{h_-}{y_-}\right), \quad \frac{1}{y^2}\frac{h_+ h_-}{h_+ + h_-}\right\}$$

of difference invariants, we can write out an invariant scheme as

$$y_{x\bar{x}} = \frac{1}{y^2}\left(\frac{h_+}{h_+ + h_-}\frac{1}{y_+} + \frac{h_-}{h_+ + h_-}\frac{1}{y_-}\right)f\left(\frac{1}{y^2}\frac{h_+ h_-}{h_+ + h_-}\right),$$

$$\frac{1}{y}\left(\frac{h_+}{y_+} + \frac{h_-}{y_-}\right) = \frac{4}{y^2}\frac{h_+ h_-}{h_+ + h_-}g\left(\frac{1}{y^2}\frac{h_+ h_-}{h_+ + h_-}\right).$$

An approximation to Eq. (3.43) is obtained for $f = g = 1$.

12. The operator algebra

$$X_1 = \frac{\partial}{\partial x}, \qquad X_2 = x\frac{\partial}{\partial x} + y\frac{\partial}{\partial y}, \qquad X_3 = (x^2 - y^2)\frac{\partial}{\partial x} + 2xy\frac{\partial}{\partial y}$$

gives the invariant equation

$$yy'' = C(1 + (y')^2)^{3/2} - (1 + (y')^2), \qquad C = \text{const}, \qquad (3.44)$$

with the general solution

$$(Ax - B)^2 + (Ay - C)^2 = 1.$$

The set of difference invariants can be taken in the form

$$I_1 = \frac{h_-^2 + (y - y_-)^2}{yy_-}, \qquad I_2 = \frac{h_+^2 + (y_+ - y)^2}{yy_+},$$

$$I_3 = \frac{2y(h_+ + h_- + h_+ y_x^2 + h_- y_{\bar{x}}^2 + 2y(y_x - y_{\bar{x}}))}{4y^2 - (h_+(1 + y_x^2) + 2yy_x)(h_-(1 + y_{\bar{x}}^2) - 2yy_{\bar{x}})}.$$

Since these expressions are very cumbersome, we write out the invariant difference model in terms of invariants:

$$I_3 = \frac{1}{2}(\sqrt{I_1} + \sqrt{I_2})f(I_1), \qquad I_2 = I_1 g(I_1).$$

A difference approximation to Eq. (3.44) is obtained for $f = C$ and $g = 1$.

13. The operator algebra is

$$X_1 = \frac{\partial}{\partial x} + \frac{\partial}{\partial y}, \qquad X_2 = x\frac{\partial}{\partial x} + y\frac{\partial}{\partial y}, \qquad X_3 = x^2\frac{\partial}{\partial x} + y^2\frac{\partial}{\partial y}.$$

The invariant ODE

$$y'' + \frac{2}{x - y}(y' + y'^2) = \frac{2C}{x - y}y'^{3/2} \qquad (3.45)$$

has the general solution

$$y = \frac{1}{A\left(B + \frac{1}{2}C\right) - Ax} + \frac{2B - C}{2A}, \qquad A \neq 0,$$

and the particular solution

$$y = ax,$$

where the constant a is found from the algebraic equation

$$a - Ca\sqrt{a} + a^2 = 0.$$

The complete set

$$I_1 = \frac{h_+^2 y_x}{(x - y_+)(x_+ - y)}, \qquad I_2 = \frac{h_-^2 y_{\bar{x}}}{(x - y_-)(x_- - y)}, \qquad I_3 = \frac{x_+ - y}{x - y} \frac{h_-}{h_- + h_+}$$

of difference invariants permits writing out the invariant equation and mesh as

$$\frac{I_1}{(1 - I_3)^2} - \frac{I_2}{(I_3)^2} = f(I_3) \left(\frac{I_1}{(1 - I_3)^2} + \frac{I_2}{(I_3)^2} \right)^{3/2}, \qquad I_1 = I_2 g(I_3).$$

For $f = C$ and $g = 1$, we obtain an approximation to the ODE (3.45).

14. The basis set

$$X_1 = (1 + x^2) \frac{\partial}{\partial x} + xy \frac{\partial}{\partial y}, \qquad X_2 = xy \frac{\partial}{\partial x} + (1 + y^2) \frac{\partial}{\partial y}, \qquad X_3 = y \frac{\partial}{\partial x} - x \frac{\partial}{\partial y}$$

of operators of the three-dimensional algebra permits calculating the differential invariants and hence writing out the invariant differential equation

$$y'' = C \left(\frac{1 + y'^2 + (y - xy')^2}{1 + x^2 + y^2} \right)^{3/2}. \tag{3.46}$$

Equation (3.46) can be integrated as follows:

$$\left(Bx - Ay + C\sqrt{1 + x^2 + y^2} \right)^2 = 1 + C^2 - A^2 - B^2.$$

The difference invariants can be calculated as

$$I_1 = \frac{h_+^2 (1 + y_x^2 + (y - xy_x)^2)}{(1 + x^2 + y^2)(1 + x_+^2 + y_+^2)}, \qquad I_2 = \frac{h_-^2 (1 + y_{\bar{x}}^2 + (y - xy_{\bar{x}})^2)}{(1 + x_-^2 + y_-^2)(1 + x^2 + y^2)},$$

$$I_3 = \frac{h_+ h_- (y_x - y_{\bar{x}})}{\sqrt{(1 + x_-^2 + y_-^2)(1 + x^2 + y^2)(1 + x_+^2 + y_+^2)}}.$$

We use them to write out the invariant difference equation and mesh as follows:

$$\frac{h_+ h_- (y_x - y_{\bar{x}})}{\sqrt{(1 + x_-^2 + y_-^2)(1 + x^2 + y^2)(1 + x_+^2 + y_+^2)}}$$
$$= f \left(\frac{h_-^2 (1 + y_{\bar{x}}^2 + (y - xy_{\bar{x}})^2)}{(1 + x_-^2 + y_-^2)(1 + x^2 + y^2)} \right),$$

$$\frac{h_+^2 (1 + y_x^2 + (y - xy_x)^2)}{(1 + x^2 + y^2)(1 + x_+^2 + y_+^2)} = g \left(\frac{h_-^2 (1 + y_{\bar{x}}^2 + (y - xy_{\bar{x}})^2)}{(1 + x_-^2 + y_-^2)(1 + x^2 + y^2)} \right).$$

For the difference approximation to Eq. (3.46) we can take the equation

$$\frac{h_+h_-(y_x - y_{\bar{x}})}{\sqrt{(1 + x_-^2 + y_-^2)(1 + x^2 + y^2)(1 + x_+^2 + y_+^2)}}$$
$$= C\left[\left(\frac{h_+^2(1 + y_x^2 + (y - xy_x)^2)}{(1 + x^2 + y^2)(1 + x_+^2 + y_+^2)}\right)^{\frac{3}{2}} + \left(\frac{h_-^2(1 + y_{\bar{x}}^2 + (y - xy_{\bar{x}})^2)}{(1 + x_-^2 + y_-^2)(1 + x^2 + y^2)}\right)^{\frac{3}{2}}\right]$$

on the difference mesh

$$\frac{h_+^2(1 + y_x^2 + (y - xy_x)^2)}{(1 + x^2 + y^2)(1 + x_+^2 + y_+^2)} = \frac{h_-^2(1 + y_{\bar{x}}^2 + (y - xy_{\bar{x}})^2)}{(1 + x_-^2 + y_-^2)(1 + x^2 + y^2)} = \varepsilon^2,$$

where $\varepsilon = \mathrm{const}$.

15. The operator algebra

$$X_1 = \frac{\partial}{\partial y}, \qquad X_2 = y\frac{\partial}{\partial y}, \qquad X_3 = y^2\frac{\partial}{\partial y}$$

contains three linearly connected elements, and the independent variable x is the only invariant in the continuous space (x, y, y', y''). The only invariant manifold is the first-order equation $y' = 0$.

In the difference case, the situation is similar: there are only three invariants x, x_-, and x_+. The dependent variables y, y_-, y_+ do not participate in the formation of invariants, and hence we cannot write out invariant difference equations.

16. The situation with the last three-dimensional subalgebra

$$X_1 = \frac{\partial}{\partial y}; \qquad X_2 = x\frac{\partial}{\partial y}; \qquad X_3 = \phi(x)\frac{\partial}{\partial y}, \qquad \phi''(x) \neq 0,$$

is similar. In this case as well, there is neither an invariant second-order ODE nor a second-order difference equation invariant under this group.

3.2.3. Difference equations invariant under four-dimensional groups

In the differential case, the maximum point symmetry group of a second-order ODE is eight-dimensional. Any differential equation admitting a group of dimension 4, 5, or 6 is equivalent to the equation $y'' = 0$, which admits the eight-dimensional group.

The situation is different in the case of difference equations.

1. The four-dimensional operator algebra

$$X_1 = \frac{\partial}{\partial x}, \qquad X_2 = \frac{\partial}{\partial y}, \qquad X_3 = x\frac{\partial}{\partial x} + y\frac{\partial}{\partial y}, \qquad X_4 = y\frac{\partial}{\partial x} - x\frac{\partial}{\partial y}$$

in the space (x, y, y', y'') does not have any differential invariants, but the equation $y'' = 0$ is an invariant manifold, which admits four additional operators (see below). In the space of difference variables, there are two invariants

$$I_1 = \frac{y_x - y_{\bar{x}}}{1 + y_x y_{\bar{x}}}, \qquad I_2 = \frac{h_+}{h_-}\left(\frac{1 + y_x^2}{1 + y_{\bar{x}}^2}\right)^{1/2}.$$

Hence the invariant difference model can be written as

$$y_x - y_{\bar{x}} = C_1(1 + y_x y_{\bar{x}}), \qquad h_+ = C_2 h_-\left(\frac{1 + y_{\bar{x}}^2}{1 + y_x^2}\right)^{1/2}, \qquad (3.47)$$

where C_1 and C_2 are arbitrary constants. But system (3.47) has the continuum limit only for $C_1 = 0$. In this case, the system approximates the equation $y'' = 0$.

2. The group determined by the infinitesimal operators

$$X_1 = \frac{\partial}{\partial y}, \qquad X_2 = x\frac{\partial}{\partial y}, \qquad X_3 = y\frac{\partial}{\partial y}, \qquad X_4 = (1 + x^2)\frac{\partial}{\partial x} + xy\frac{\partial}{\partial y}$$

does not have differential invariants in the space (x, y, y', y'') as well, but it does have the invariant manifold $y'' = 0$. In the difference case, we have the invariants

$$I_1 = \frac{h_+}{1 + xx_+}, \qquad I_2 = \frac{h_-}{1 + xx_-}$$

and the invariant difference manifold $y_{x\bar{x}} = 0$. Thus, we have the invariant difference model

$$y_{x\bar{x}} = 0, \qquad h_+ = h_-\frac{1 + xx_+}{1 + xx_-}g\left(\frac{h_-}{1 + xx_-}\right).$$

For $g = 1$, this system approximates the ODE $y'' = 0$.

3. The four-dimensional algebra

$$X_1 = \frac{\partial}{\partial x}, \qquad X_2 = \frac{\partial}{\partial y}, \qquad X_3 = x\frac{\partial}{\partial y}, \qquad X_4 = x\frac{\partial}{\partial x} + ay\frac{\partial}{\partial y}$$

has the only differential invariant y'' if the constant a is 2. For $a \neq 2$, we have the only invariant manifold $y'' = 0$. The expressions

$$I_1 = \frac{h_+}{h_-}, \qquad I_2 = y_{x\bar{x}}h_+^{2-a}$$

give a complete set of difference invariants, which can be used to write out the general form of the invariant models as

$$y_{x\bar{x}} = C_1 h_+^{a-2}, \qquad h_+ = C_2 h_-.$$

For $a > 2$, the continuum limit is $y'' = 0$. For $a < 2$, the continuum limit exists only if $C_1 = 0$.

4. The Lie point transformation group determined by the infinitesimal operators

$$X_1 = \frac{\partial}{\partial x}, \qquad X_2 = \frac{\partial}{\partial y}, \qquad X_3 = x\frac{\partial}{\partial y}, \qquad X_4 = x\frac{\partial}{\partial x} + (2y + x^2)\frac{\partial}{\partial y}$$

has neither differential invariants nor an invariant manifold in the continuous case. In the difference case, there are two invariants

$$I_1 = \frac{h_+}{h_-}, \qquad I_2 = y_{x\bar{x}} - \ln(h_- h_+),$$

which permit writing out the invariant difference model as

$$y_{x\bar{x}} = \ln(h_- h_+) + C_1, \qquad h_+ = C_2 h_-.$$

This difference model has no continuum limit.

5. The operator algebra

$$X_1 = \frac{\partial}{\partial x}, \qquad X_2 = \frac{\partial}{\partial y}, \qquad X_3 = x\frac{\partial}{\partial y}, \qquad X_4 = y\frac{\partial}{\partial y}$$

does not have invariants in the space (x, y, y', y'') but has the invariant manifold $y'' = 0$. In the difference case, the mesh spacings h_+ and h_- form a basis of invariants, and the difference equation $y_{x\bar{x}} = 0$ is an invariant manifold. The invariant model can be written as

$$y_{x\bar{x}} = 0, \qquad h_+ = h_- g(h_-).$$

6. The operator algebra

$$X_1 = \frac{\partial}{\partial x}, \qquad X_2 = \frac{\partial}{\partial y}, \qquad X_3 = x\frac{\partial}{\partial x}, \qquad X_4 = y\frac{\partial}{\partial y}$$

has no differential invariants, but the ODE $y'' = 0$ is invariant. In the difference case, there exist invariants

$$I_1 = \frac{h_+}{h_-}, \qquad I_2 = \frac{y_x}{y_{\bar{x}}}.$$

Needless to say, the difference model

$$y_{x\bar{x}} = C_1 \frac{y_{\bar{x}}}{h_-}, \qquad h_+ = C_2 h_-$$

has a continuum limit only for $C_1 = 0$.

7. For the four-dimensional algebra

$$X_1 = \frac{\partial}{\partial y}, \qquad X_2 = x\frac{\partial}{\partial y}, \qquad X_3 = x\frac{\partial}{\partial x}, \qquad X_4 = y\frac{\partial}{\partial y},$$

there are no differential invariants, but the ODE $y'' = 0$ is invariant. The expressions $\frac{h_+}{x}$ and $\frac{h_-}{x}$ are difference invariants, and the difference equation $y_{x\bar{x}} = 0$ is an invariant manifold. The invariant scheme in general form can be written as

$$y_{x\bar{x}} = 0, \qquad h_+ = h_- g\left(\frac{h_-}{x}\right).$$

8. The four-dimensional operator algebra

$$X_1 = \frac{\partial}{\partial x}, \qquad X_2 = x\frac{\partial}{\partial x}, \qquad X_3 = y\frac{\partial}{\partial y}, \qquad X_4 = x^2\frac{\partial}{\partial x} + xy\frac{\partial}{\partial y}$$

does not have differential invariants, but the ODE $y'' = 0$ is invariant. The expressions

$$I_1 = h_+ h_- \frac{y_{x\bar{x}}}{y}, \qquad I_2 = \frac{y_- h_+}{y_+ h_-}$$

form a complete set of difference invariants, which implies the general form

$$y_{x\bar{x}} = \frac{C_1}{h_+ h_-} y, \qquad h_+ y_- = C_2 h_- y_+$$

of an invariant model. This system has a continuum limit only if $C_1 = 0$.

3.2.4. Difference equations invariant under five-dimensional groups

1. The five-dimensional operator algebra

$$X_1 = \frac{\partial}{\partial x}, \qquad X_2 = \frac{\partial}{\partial y}, \qquad X_3 = x\frac{\partial}{\partial x}, \qquad X_4 = x\frac{\partial}{\partial y}, \qquad X_5 = y\frac{\partial}{\partial y}$$

does not have differential invariants, but the equation $y'' = 0$ admits it. This group has one difference invariant $\frac{h_+}{h_-}$, and the invariant manifold is

$$(x - x_-)(y_+ - y) - (x_+ - x)(y - y_-) = 0.$$

The invariant difference system in general form can be written as

$$y_{x\bar{x}} = 0, \qquad h_+ = Ch_-, \tag{3.48}$$

where C is an arbitrary constant.

2. The operator algebra

$$X_1 = \frac{\partial}{\partial x}, \quad X_2 = \frac{\partial}{\partial y}, \quad X_3 = y\frac{\partial}{\partial x}, \quad X_4 = x\frac{\partial}{\partial y}, \quad X_5 = x\frac{\partial}{\partial x} - y\frac{\partial}{\partial y}$$

does not have differential invariants as well, and the equation $y'' = 0$ admits it. This group does not have difference invariants but has the invariant manifold (3.48).

3.2.5. Six-dimensional group and an invariant difference model

The six-dimensional operator algebra

$$X_1 = \frac{\partial}{\partial x}, \quad X_2 = \frac{\partial}{\partial y}, \quad X_3 = x\frac{\partial}{\partial x}, \quad X_4 = y\frac{\partial}{\partial x}, \quad X_5 = x\frac{\partial}{\partial y}, \quad X_6 = y\frac{\partial}{\partial y}$$

has the only invariant manifold $y'' = 0$, and the invariant manifold in the difference case is the system

$$y_{x\bar{x}} = 0, \qquad h_+ = Ch_-.$$

There are no three-point difference equations and meshes invariant under the seven-dimensional Lie algebras.

3.2.6. Eight-dimensional Lie transformation group

The linear equation

$$y'' = 0 \tag{3.49}$$

(the equation of free motion of a particle), as well as all the ODE equivalent to it up to a point change of variables, admits the algebra

$$X_1 = \frac{\partial}{\partial x}, \quad X_2 = \frac{\partial}{\partial y}, \quad X_3 = x\frac{\partial}{\partial x}, \quad X_4 = y\frac{\partial}{\partial x}, \quad X_5 = x\frac{\partial}{\partial y},$$
$$X_6 = y\frac{\partial}{\partial y}, \quad X_7 = x^2\frac{\partial}{\partial x} + xy\frac{\partial}{\partial y}, \quad X_8 = xy\frac{\partial}{\partial x} + y^2\frac{\partial}{\partial y}. \tag{3.50}$$

The three-point discretization of Eq. (3.49) must have the form

$$y_{x\bar{x}} = 0, \qquad \Omega(x, x_-, x_+, y_{\bar{x}} + y_x) = 0, \tag{3.51}$$

where the equation $\Omega = 0$ determines the difference mesh. In the difference case, the eight-dimensional algebra (3.50) does not have any invariants, and the only invariant manifold is the difference equation

$$y_{x\bar{x}} = 0.$$

But the equation alone is insufficient to determine a difference model; one needs to have another equation for the mesh. However, the use of any equation for the mesh immediately decreases the model symmetry.

Thus, the maximum symmetry of a second-order difference model is 6, and the symmetry may vary depending on the type of the difference mesh. Note, however, that the linear equation (3.51) on a uniform mesh has the same solution set as the original ODE. Thus, although we do not succeed in constructing a three-point scheme admitting all 8 operators for linear ODE, the simplest linear scheme is exact.

Comments

Let us compare the group classification of second-order ordinary differential equations with that of three-point difference equations and meshes.

1. For each ODE invariant under a Lie group G of dimension $1 \leq n \leq 3$, there exists a family of difference models invariant under the same group G. In particular, for $n = 3$ the second-order invariant ODE is specified up to a constant, while the ambiguity in invariant schemes in general involves two arbitrary functions.

2. All ODE invariant under a group of dimension $n = 4, 5$, or 6 can be transformed into the linear equation $y'' = 0$. In the same dimensions, the invariant schemes either have the equation $y'' = 0$ as their continuum limit or this limit does not exist at all.

3. The difference equation $y_{x\bar{x}} = 0$ has significantly different group properties than its continuum limit. This equation is invariant under an at most six-dimensional point transformation group, and in this case the mesh is given by the equation $h_+ = Ch_-$, where $C > 0$. The equation $y_{x\bar{x}} = 0$ is invariant under groups of dimension $1 \leq n \leq 4$ for meshes of more general forms.

Several examples of invariant difference schemes and lattices and their numerical implementations can be found in [17, 18]

Chapter 4

Invariant Difference Models
of Partial Differential Equations

4.1. Symmetry Preserving Difference Schemes
for the Nonlinear Heat Equation with a Source

The aim of this section is to develop the entire set of invariant difference schemes for the heat equation with a source,

$$u_t = (K(u)u_x)_x + Q(u), \tag{4.1}$$

for all special cases of the coefficients $K(u)$ and $Q(u)$ in which the symmetry group admitted by Eq. (4.1) is extended. This set of invariant difference models corresponds to the Lie group classification [28] (see also [58,74]) of Eq. (4.1) with arbitrary $K(u)$ and $Q(u)$. This classification contains the result due to Ovsyannikov [112] for Eq. (4.1) with $Q = 0$ as well as the symmetries for the linear case ($K \equiv 1$, $Q \equiv 0$), which were already known to S. Lie.

A few examples of invariant difference schemes and meshes for the heat equation were constructed in [8, 9] and completed in [42]. In the present section (see [42]) we go through all cases of $K(u)$ and $Q(u)$ identified in the group classification in [28], and construct difference equations and meshes which admit the same Lie point transformation groups as their continuous counterparts. We single out all linear cases of Eq. (4.1) and consider them in the next section.

When developing an invariant difference scheme, we have to choose a difference stencil sufficient for approximating all derivatives that occur in the equation. We shall consider six-point stencils whose have three points on each of the two time layers. Such stencils allow us to write out explicit as well as implicit difference schemes. For different transformation groups, we consider different meshes: an orthogonal mesh that is uniform in space, an orthogonal mesh that is nonuniform in space, and a mesh nonorthogonal in time-space, i.e., a moving mesh. The corresponding stencils are different. Furthermore, the corresponding spaces of discrete variables are of different dimensions, and so they have different numbers of difference invariants $I = (I_1, I_2, \ldots, I_l)$ for the same Lie group G_n.

For example, let us take an orthogonal mesh that is uniform in space. (Later, we shall describe the groups for which this mesh can be used.) The stencil of this mesh is shown in Fig. 4.1. The corresponding subspace of difference variables is

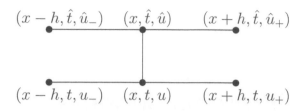

Figure 4.1: The stencil of the orthogonal mesh.

ten-dimensional, $M \sim (t, x, \tau, h, u, u_-, u_+, \hat{u}, \hat{u}_-, \hat{u}_+)$, where $\tau = \hat{t} - t$.

The symmetry operator

$$X = \xi^t \frac{\partial}{\partial t} + \xi^x \frac{\partial}{\partial x} + \eta \frac{\partial}{\partial u}$$

prolonged to the difference stencil variables has the form

$$\widetilde{X} = \xi^t \frac{\partial}{\partial t} + \xi^x \frac{\partial}{\partial x} + (\hat{\xi}^t - \xi^t) \frac{\partial}{\partial \tau} + (\xi^x_+ - \xi^x) \frac{\partial}{\partial h}$$
$$+ \eta \frac{\partial}{\partial u} + \eta_- \frac{\partial}{\partial u_-} + \eta_+ \frac{\partial}{\partial u_+} + \hat{\eta} \frac{\partial}{\partial \hat{u}} + \hat{\eta}_- \frac{\partial}{\partial \hat{u}_-} + \hat{\eta}_+ \frac{\partial}{\partial \hat{u}_+},$$

where we have used the notation $\hat{f} = f(t + \tau, x, u)$, $f_- = f(t, x - h, u)$, and $f_+ = f(t, x + h, u)$ for the time and space shifts.

Once chosen, an invariant mesh serves as a background for the application of the method of finite-difference invariants. Having found finite-difference invariants I_j as solutions of the system of linear equations

$$\widetilde{X_i} \Phi(t, x, \tau, h, u, u_-, u_+, \hat{u}, \hat{u}_-, \hat{u}_+) = 0, \qquad i = 1, \ldots, n,$$

we can use them to approximate the differential invariants,

$$J_j = f_j(I_1, I_2, \ldots, I_l) + O(\tau^\alpha, h^\beta), \qquad j = 1, \ldots, k,$$

where α and β specify some given order of approximation. Note that the approximation error $O(\tau^\alpha, h^\beta)$, together with other terms in the above representation, is invariant. The substitution of the difference invariants I_i instead of the differential invariants J_i into the invariant representation of Eq. (4.1) results in an invariant difference scheme. Practically, we can often omit the representation of the original differential equation in terms of its invariants and just approximate it by difference invariants. The use of difference invariants is the key point in both methods.

Thus, the first step in the invariant approximation is the choice of an invariant mesh. The second step is the choice of an invariant discretization of the original equation on the invariant mesh.

We point out that the invariant approximation is still not unique. For example, by extending the stencil (i.e., by increasing the number of mesh points involved in the approximation) we can find invariant approximations of any higher order.

Now let us start to develop invariant schemes by exhausting all cases of the Lie group classification [28]. Note that the group classification of Eq. (4.1) was obtained in [28] up to the equivalence transformations

$$\bar{t} = at + e, \qquad \bar{x} = bx + f, \qquad \bar{u} = cu + g, \qquad \bar{k} = \frac{b^2}{a}k, \qquad \bar{q} = \frac{c}{a}q,$$

where a, b, c, e, f, and g are arbitrary constants such that $abc \neq 0$. These transformations do not change the differential structure of Eq. (4.1) and transform the group admitted by the equation into a similar point transformation group.

4.1.1. An arbitrary heat conductivity coefficient $K(u)$

1. We start from the general case in which the coefficients $K(u)$ and $Q(u)$ are arbitrary. Then Eq. (4.1) only admits the two-parameter translation group. This group is defined by the infinitesimal operators

$$X_1 = \frac{\partial}{\partial t}, \qquad X_2 = \frac{\partial}{\partial x}, \tag{4.2}$$

which generate the translations of the independent variables. In this case, there are virtually no constraints on the mesh and the difference equation. In particular, we can use an orthogonal mesh regular in both directions in the plane (x, t) as long as the invariant orthogonality and uniformness conditions hold for the operators (4.2).

The group with operators (4.2) in the subspace $(x, t, h, \tau, u, u_-, u_+, \hat{u}, \hat{u}_-, \hat{u}_+)$ corresponding to the stencil shown in Fig. 4.1 has eight invariants

$$\tau, \quad h, \quad u, \quad u_+, \quad u_-, \quad \hat{u}, \quad \hat{u}_-, \quad \hat{u}_+.$$

That is why any difference approximation to Eq. (4.1) using the above invariants can give a difference equation that admits the operators (4.2). For example, the explicit model

$$\frac{\hat{u} - u}{\tau} = \frac{1}{h}\left(K\left(\frac{u_+ + u}{2}\right) u_x - K\left(\frac{u + u_-}{2}\right) u_{\bar{x}} \right) + Q(u), \tag{4.3}$$

where $K(u)$ and $Q(u)$ represent any approximation to the corresponding coefficients by invariants and $u_x = (u_+ - u)/h$ and $u_{\bar{x}} = (u - u_-)/h$ are the right and left difference derivatives, admits the operators (4.2).

2. If $K(u)$ is an arbitrary function and $Q(u) \equiv 0$, then the equation

$$u_t = (K(u)u_x)_x \tag{4.4}$$

admits the three-parameter algebra of operators [112]

$$X_1 = \frac{\partial}{\partial t}, \qquad X_2 = \frac{\partial}{\partial x}, \qquad X_3 = 2t\frac{\partial}{\partial t} + x\frac{\partial}{\partial x}. \tag{4.5}$$

This case is almost similar to the previous one. The operators (4.5) do not violate the conditions of invariant orthogonality and invariant uniformness of the mesh. For example, in this case we could use the orthogonal mesh shown in Fig. 4.1. Any approximation to Eq. (4.4) using the seven invariants

$$\frac{h^2}{\tau}, \quad u, \quad u_+, \quad u_-, \quad \hat{u}, \quad \hat{u}_-, \quad \hat{u}_+$$

gives an invariant model for Eq. (4.4). In particular, the explicit scheme (4.3) with $Q \equiv 0$,

$$\frac{\hat{u} - u}{\tau} = \frac{1}{h}\left(K\left(\frac{u_+ + u}{2}\right)u_x - K\left(\frac{u + u_-}{2}\right)u_{\bar{x}}\right),$$

can be applied.

4.1.2. The exponential heat conductivity coefficient $K = e^u$

In this subsection, we consider three cases of group classification for $K = e^u$ in accordance with [28] and [112].

1. If $Q = 0$, then the equation

$$u_t = (e^u u_x)_x \tag{4.6}$$

admits the four-dimensional algebra of infinitesimal operators

$$X_1 = \frac{\partial}{\partial t}, \qquad X_2 = \frac{\partial}{\partial x}, \qquad X_3 = 2t\frac{\partial}{\partial t} + x\frac{\partial}{\partial x}, \qquad X_4 = t\frac{\partial}{\partial t} - \frac{\partial}{\partial u}. \tag{4.7}$$

As in the cases considered above, the invariant uniformness and invariant orthogonality conditions hold. A difference model for Eq. (4.6) can be constructed by approximation to the differential equation with the help of the difference invariants

$$e^u \frac{\tau}{h^2}, \quad \hat{u} - u, \quad u_+ - u, \quad u - u_-, \quad \hat{u}_+ - \hat{u}, \quad \hat{u} - \hat{u}_-.$$

An example is given by the simple explicit difference model

$$\frac{\hat{u} - u}{\tau} = \frac{1}{h}\left(\exp\left(\frac{u_+ + u}{2}\right)u_x - \exp\left(\frac{u + u_-}{2}\right)u_{\bar{x}}\right), \tag{4.8}$$

but one has still enough freedom to construct invariant schemes using difference invariants.

2. For $Q = \delta = \pm 1$, we can eliminate the constant source from the equation

$$u_t = (e^u u_x)_x + \delta \tag{4.9}$$

by the change of variables

$$\bar{u} = u - \delta t, \qquad \bar{t} = \delta(e^{\delta t} - 1). \tag{4.10}$$

Equation (4.9) is transformed by this change of variables into Eq. (4.6), but the mesh uniformness in the t-direction is destroyed. Equation (4.9) admits the four-dimensional algebra of infinitesimal operators

$$X_1 = \frac{\partial}{\partial t}, \qquad X_2 = \frac{\partial}{\partial x}, \qquad X_3 = e^{-\delta t}\frac{\partial}{\partial t} + \delta e^{-\delta t}\frac{\partial}{\partial u}, \qquad X_4 = x\frac{\partial}{\partial x} + 2\frac{\partial}{\partial u},$$

and we can readily see that the operator X_3 does not preserve the mesh uniformness in the time direction. The difference invariants

$$\frac{e^u(e^{\delta \tau} - 1)}{h^2}, \quad \hat{u} - u - \delta\tau, \quad u_+ - u, \quad u - u_-, \quad \hat{u}_+ - \hat{u}, \quad \hat{u} - \hat{u}_-$$

permit us to construct the following version of the difference model for Eq. (4.9):

$$\frac{\delta(\hat{u} - u) - \tau}{e^{\delta\tau} - 1} = \frac{1}{h}\left(\exp\left(\frac{u_+ + u}{2}\right)\frac{u_x}{h} - \exp\left(\frac{u + u_-}{2}\right)\frac{u_{\bar{x}}}{h}\right). \tag{4.11}$$

Note that the change of variables (4.10) transforms the model (4.11) on the orthogonal mesh given on the time interval $[0, T]$ by the formula

$$t_n - \delta\ln\left(1 + \frac{n}{k}(e^{\delta T} - 1)\right), \qquad n = 0, \ldots, k, \tag{4.12}$$

where k is the number of time steps of the mesh, into the model (4.8) with a uniform time mesh on the time interval $[0, \delta(e^{\delta T} - 1)]$.

3. If $Q = \pm e^{\alpha u}$, $\alpha \neq 0$, then the equation

$$u_t = (e^u u_x)_x \pm e^{\alpha u} \tag{4.13}$$

admits the three infinitesimal operators

$$X_1 = \frac{\partial}{\partial t}, \qquad X_2 = \frac{\partial}{\partial x}, \qquad X_3 = 2\alpha t\frac{\partial}{\partial t} + (\alpha - 1)x\frac{\partial}{\partial x} - 2\frac{\partial}{\partial u}. \tag{4.14}$$

These operators satisfy the conditions of invariant orthogonality and uniformness of the meshes, and we shall consider the stencil in Fig. 4.1. Any approximation to Eq. (4.13) using the difference invariants

$$\frac{\tau^{\frac{\alpha-1}{2\alpha}}}{h}, \quad e^{\alpha u}\tau, \quad \hat{u} - u, \quad u_+ - u, \quad u - u_-, \quad \hat{u}_+ - \hat{u}, \quad \hat{u} - \hat{u}_-$$

gives a version of the difference model for Eq. (4.13) admitting the symmetries (4.14); for example, we obtain the following model:

$$\frac{\hat{u} - u}{\tau} = \frac{1}{h}\left(\exp\left(\frac{u_+ + u}{2}\right)\frac{u_x}{h} - \exp\left(\frac{u + u_-}{2}\right)\frac{u_{\bar{x}}}{h}\right) \pm e^{\alpha u}. \tag{4.15}$$

4. In accordance with the group classification in [28], we shall also consider the case in which $Q = \pm e^u + \delta$, $\delta = \pm 1$. We can eliminate the constant source from the equation

$$u_t = (e^u u_x)_x \pm e^u + \delta \tag{4.16}$$

by the change of variables (4.10). Equation (4.16) is then transformed into (4.13). Equation (4.16) admits the following infinitesimal operators:

$$X_1 = \frac{\partial}{\partial t}, \qquad X_2 = \frac{\partial}{\partial x}, \qquad X_3 = e^{-\delta t}\frac{\partial}{\partial t} + \delta e^{-\delta t}\frac{\partial}{\partial u}. \tag{4.17}$$

The difference invariants

$$e^u(e^{\delta\tau} - 1), \quad h, \quad \hat{u} - u - \delta\tau, \quad u_+ - u, \quad u - u_-, \quad \hat{u}_+ - \hat{u}, \quad \hat{u} - \hat{u}_-$$

of (4.17) permit constructing the following version of the difference model on the invariant mesh (4.12):

$$\frac{\delta(\hat{u} - u) - \tau}{e^{\delta\tau} - 1} = \frac{1}{h}\left(\exp\left(\frac{u_+ + u}{2}\right)u_{x}{}_{h} - \exp\left(\frac{u + u_-}{2}\right)u_{\bar{x}}{}_{h}\right) \pm e^u.$$

The model for the considered equation can be obtained from the model (4.15) with the help of the transformation (4.10).

4.1.3. The power-law heat conductivity coefficient $K = u^\sigma$, $\sigma \neq 0, -4/3$

For $K = u^\sigma$, further classification depends on the source.

1. Let us start from the simplest case $Q \equiv 0$,

$$u_t = (u^\sigma u_x)_x. \tag{4.18}$$

The symmetries of Eq. (4.18) are described by the four-dimensional algebra of infinitesimal operators

$$X_1 = \frac{\partial}{\partial t}, \quad X_2 = \frac{\partial}{\partial x}, \quad X_3 = 2t\frac{\partial}{\partial t} + x\frac{\partial}{\partial x}, \quad X_4 = \sigma x\frac{\partial}{\partial x} + 2u\frac{\partial}{\partial u}. \tag{4.19}$$

For any σ, the operators (4.19) preserve the mesh uniformness and orthogonality. The difference invariants corresponding to the stencil in Fig. 4.1 are

$$u^\sigma\frac{\tau}{h^2}, \quad \frac{\hat{u}}{u}, \quad \frac{u_+}{u}, \quad \frac{u_-}{u}, \quad \frac{\hat{u}_+}{\hat{u}}, \quad \frac{\hat{u}_-}{\hat{u}}.$$

They permit us to write out, for example, the following version of the difference model on the orthogonal uniform mesh in both directions:

$$\frac{\hat{u} - u}{\tau} = \frac{1}{h}\left(\left(\frac{u_+ + u}{2}\right)^\sigma u_{x}{}_{h} - \left(\frac{u + u_-}{2}\right)^\sigma u_{\bar{x}}{}_{h}\right). \tag{4.20}$$

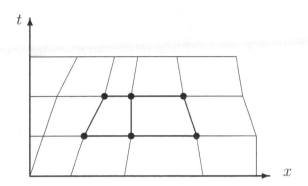

Figure 4.2: A moving mesh with flat time layers

An orthogonal mesh is not the only possible way to discretize the problem. Now let us show how to introduce a *moving mesh* of the form shown in Fig. 4.2.

One can use an adaptive mesh defined by the evolution equation (see also [21])

$$\frac{dx}{dt} = \varphi(t, x, u, u_x). \tag{4.21}$$

In this case, the heat equation acquires the form

$$\frac{du}{dt} = (u^\sigma u_x)_x + \varphi(t, x, u, u_x)u_x.$$

Different requirements can be imposed on the function φ. If we require the invariance of Eq. (4.21) with respect to the whole set of operators (4.19), our freedom to choose φ is limited by the function

$$\varphi = Cu^{\sigma-1}u_x, \qquad C = \text{const.}$$

In what follows, we show how to introduce the Lagrangian type of evolution $\frac{dx}{dt}$. Note that Eq. (4.18) has the form of a conservation law that presents the conservation of heat. Hence we can seek a moving mesh of Lagrangian type which evolves in accordance with heat diffusion. We should find an evolution $\frac{dx}{dt}$ that satisfies the equation

$$\frac{d}{dt} \int_{x1(t)}^{x2(t)} u\,dx = 0.$$

Since

$$\frac{d}{dt} \int_{x1}^{x2} u\,dx = \int_{x1}^{x2} \frac{\partial u}{\partial t}dx + \left[u\frac{dx}{dt}\right]_{x1}^{x2} = \left[u^\sigma u_x + u\frac{dx}{dt}\right]_{x1}^{x2}$$

we obtain the evolution $dx/dt = -u^{\sigma-1}u_x$. Note that this evolution is invariant with respect to the operators (4.19). Our original differential equation (4.18) can now be represented in the form of the system

$$\frac{dx}{dt} = -u^{\sigma-1}u_x, \qquad \frac{du}{dt} = u^\sigma u_{xx} + (\sigma - 1)u^{\sigma-1}u_x^2. \qquad (4.22)$$

Note that Eq. (4.18) has two conservation laws

$$u_t = (u^\sigma u_x)_x, \qquad (xu)_t = \left(xu^\sigma u_x - \frac{u^{\sigma+1}}{\sigma + 1} \right)_x.$$

For the evolution system (4.22) it is convenient to represent the conservation laws in the integral form

$$\frac{d}{dt} \int_{x1(t)}^{x2(t)} u\,dx = 0, \qquad \frac{d}{dt} \int_{x1(t)}^{x2(t)} xu\,dx = -\frac{u^{\sigma+1}}{\sigma + 1}\bigg|_{x_1}^{x_2}. \qquad (4.23)$$

For finite-difference modeling of system (4.22), we can take the stencil shown in Fig. 4.3.

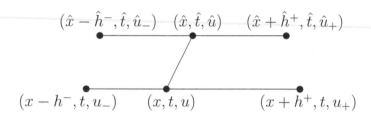

Figure 4.3: The stencil of the evolution mesh

In the space of the variables $(t, x, \tau, h^+, h^-, \hat{h}^+, \hat{h}^-, \Delta x, u, u_+, u_-, \hat{u}, \hat{u}_+, \hat{u}_-)$ corresponding to this stencil, there are ten finite-difference invariants

$$u^\sigma \frac{\tau}{h^{+2}}, \quad \frac{\hat{u}}{u}, \quad \frac{u_+}{u}, \quad \frac{u_-}{u}, \quad \frac{\hat{u}_+}{\hat{u}}, \quad \frac{\hat{u}_-}{\hat{u}}, \quad \frac{h^-}{h_+}, \quad \frac{\hat{h}^-}{h^+}, \quad \frac{\hat{h}^+}{h^+}, \quad \frac{\Delta x}{h^+}.$$

Approximating system (4.22) with the use of these invariants, we can obtain, say, the system of two equations

$$\frac{\Delta x}{\tau} = -\frac{1}{2\sigma}\left(\frac{u_+^\sigma - u^\sigma}{h^+} + \frac{u^\sigma - u_-^\sigma}{h^-} \right), \qquad \frac{\hat{u} + \hat{u}_+}{2}\hat{h}_+ = \frac{u + u_+}{2}h_+, \qquad (4.24)$$

where we have approximated the heat conservation law to obtain the equation for the solution u.

The first equation in system (4.22) shows that the evolution of x depends on the solution. System (4.24) may be inconvenient for computations, because the step length will be changed automatically and the nature of this process is not clear. To avoid this uncertainty, we introduce a new space variable whose values characterize the evolution trajectories of x. Consider the variable s defined by the system

$$s_t = u^\sigma u_x, \qquad s_x = u.$$

One can readily see that each trajectory of x is determined by a fixed value of s, since

$$\frac{ds}{dt} = s_t + s_x \frac{dx}{dt} = u^\sigma u_x - \frac{u}{\sigma}(u^\sigma)_x = 0.$$

In the new coordinate system with independent variables (t, s), Eq. (4.18) has the form

$$\left(\frac{1}{u}\right)_t = -(u^\sigma u_s)_s, \tag{4.25}$$

and the former space variable x satisfies

$$x_t = -u^\sigma u_s, \qquad x_s = \frac{1}{u}. \tag{4.26}$$

For the discrete modeling of Eq. (4.18), one can use Eq. (4.25) in the new independent variables (t, s) to describe the diffusion process and the first equation in system (4.26) to trace the evolution of the coordinate x. Equation (4.25), considered together with system (4.26), admits the symmetries

$$X_1 = \frac{\partial}{\partial t}, \quad X_2 = \frac{\partial}{\partial x}, \quad X_3 = \frac{\partial}{\partial s}, \quad X_4 = 2t\frac{\partial}{\partial t} + s\frac{\partial}{\partial s} + x\frac{\partial}{\partial x},$$

$$X_5 = (\sigma + 2)s\frac{\partial}{\partial s} + \sigma x\frac{\partial}{\partial x} + 2u\frac{\partial}{\partial u}.$$

In the new variables (t, s), the stencil becomes orthogonal, so that there is no need to consider a nonuniform mesh in the variable s. We have the following invariants for this set of operators in the space

$$(t, \tau, s, h_s, x, h_x^+, h_x^-, \hat{h}_x^+, \hat{h}_x^-, \Delta x, u, u_+, u_-, \hat{u}, \hat{u}_+, \hat{u}_-)$$

corresponding to the orthogonal stencil in (t, s) extended by the additional dependent variable x:

$$u^\sigma \frac{\tau}{h_x^{+2}}, \quad \frac{\hat{u}}{u}, \quad \frac{u_+}{u}, \quad \frac{u_-}{u}, \quad \frac{\hat{u}_+}{\hat{u}}, \quad \frac{\hat{u}_-}{\hat{u}}, \quad \frac{h_x^-}{h_x^+}, \quad \frac{\hat{h}_x^-}{h_x^+}, \quad \frac{\hat{h}_x^+}{h_x^+}, \quad \frac{\Delta x}{h_x^+}, \quad \frac{h_s}{h_x^+}.$$

By means of these invariants, we obtain an approximation to (4.25) that has the form of the conservation law

$$\frac{1}{\tau}\left(\frac{1}{\hat{u}} - \frac{1}{u}\right) = -\frac{\alpha}{\sigma + 1}\left(\frac{u_+^{\sigma+1} - 2u^{\sigma+1} + u_-^{\sigma+1}}{h_s^2}\right)$$

$$-\frac{1 - \alpha}{\sigma + 1}\left(\frac{\hat{u}_+^{\sigma+1} - 2\hat{u}^{\sigma+1} + \hat{u}_-^{\sigma+1}}{h_s^2}\right), \tag{4.27}$$

where $0 \leq \alpha \leq 1$. Note that the variable x is introduced in the coordinates (t, s) by the system (4.26) as some sort of potential for Eq. (4.25). In a similar way, we can introduce x as a discrete potential with the help of the system

$$\frac{\Delta x}{\tau} = -\frac{\alpha}{\sigma + 1}\left(\frac{u_+^{\sigma+1} - u_-^{\sigma+1}}{2h_s}\right) - \frac{1-\alpha}{\sigma+1}\left(\frac{\hat{u}_+^{\sigma+1} - \hat{u}_-^{\sigma+1}}{2h_s}\right),$$

$$\frac{h_x^+}{h_s} = \frac{1}{2}\left(\frac{1}{u} + \frac{1}{u_+}\right). \tag{4.28}$$

In computations, only Eq. (4.27) and the first equation in (4.28) are needed. The second equation in system (4.28) is needed only to establish the relationship between the solutions $u(x)$ and $u(s)$ for a fixed time. For given initial data $u(0, x) = u_0(x)$, we choose an appropriate step length h_s for the Lagrangian coordinate s. Then we can introduce the mesh points x_i in the original coordinates satisfying

$$\frac{x_{i+1} - x_i}{h_s} = \frac{1}{2}\left(\frac{1}{u_0(x_i)} + \frac{1}{u_0(x_{i+1})}\right);$$

i.e., we use this equation to establish a difference relation between the original space coordinate x and the Lagrangian substantive coordinate s. Computing the solution with the help of the numerical scheme (4.27) and the first equation in (4.28), we preserve the relation

$$\frac{x_{i+1} - x_i}{h_s} = \frac{1}{2}\left(\frac{1}{u_i} + \frac{1}{u_{i+1}}\right).$$

Introducing the material type variable s, we can rewrite the conservation laws (4.23) as

$$\frac{\partial}{\partial t}\int_{s1}^{s2} ds = 0, \qquad \frac{\partial}{\partial t}\int_{s1}^{s2} x\,ds = -\frac{u^{\sigma+1}}{\sigma+1}\bigg|_{s_1}^{s_2}.$$

The proposed discrete model possesses the difference analogs

$$\sum_{i=1}^{N-1} h_s = \text{const},$$

$$\sum_{i=1}^{N-1}\frac{\hat{x}_i + \hat{x}_{i+1}}{2}h_s - \sum_{i=1}^{N-1}\frac{x_i + x_{i+1}}{2}h_s = -\frac{\alpha}{\sigma+1}\left(\frac{u_{N+1}^{\sigma+1} + u_N^{\sigma+1}}{2}\right)$$

$$-\frac{1-\alpha}{\sigma+1}\left(\frac{\hat{u}_{N+1}^{\sigma+1} + \hat{u}_N^{\sigma+1}}{2}\right) + \frac{\alpha}{\sigma+1}\left(\frac{u_{-1}^{\sigma+1} + u_0^{\sigma+1}}{2}\right) + \frac{1-\alpha}{\sigma+1}\left(\frac{\hat{u}_{-1}^{\sigma+1} + \hat{u}_0^{\sigma+1}}{2}\right)$$

of these conservation laws.

2. $Q = \delta u$, $\delta = \pm 1$. In this case, the symmetry of the equation

$$u_t = (u^\sigma u_x)_x + \delta u \qquad (4.29)$$

is described by the infinitesimal operators

$$X_1 = \frac{\partial}{\partial t}, \quad X_2 = \frac{\partial}{\partial x}, \quad X_3 = \sigma x \frac{\partial}{\partial x} + 2u \frac{\partial}{\partial u}, \quad X_4 = e^{-\delta\sigma t} \frac{\partial}{\partial t} + \delta e^{-\delta\sigma t} u \frac{\partial}{\partial u}.$$

The change of variables

$$\bar{u} = u e^{-\delta t}, \qquad \bar{t} = \frac{\delta}{\sigma}(e^{\delta\sigma t} - 1) \qquad (4.30)$$

transforms Eq. (4.29) into Eq. (4.18). The finite-difference invariants

$$\frac{u^\sigma(e^{\delta\sigma\tau} - 1)}{h^2}, \quad \delta \ln \frac{\hat{u}}{u} - \tau, \quad \frac{u_+}{u}, \quad \frac{u_-}{u}, \quad \frac{\hat{u}_+}{\hat{u}}, \quad \frac{\hat{u}_-}{\hat{u}}$$

give the following possibility for an explicit difference model:

$$\frac{\sigma u}{e^{\delta\sigma\tau} - 1}\left(\delta \ln \frac{\hat{u}}{u} - \tau\right) = \frac{1}{h}\left(\left(\frac{u_+ + u}{2}\right)^\sigma u_x \frac{}{h} - \left(\frac{u + u_-}{2}\right)^\sigma u_{\bar{x}} \frac{}{h}\right).$$

Note that the change of variables (4.30) transforms this equation considered on the orthogonal mesh with time layers

$$t_n = \frac{\delta}{\sigma} \ln\left(1 + \frac{n}{k}(e^{\delta\sigma T} - 1)\right), \qquad n = 0, \dots, k, \qquad (4.31)$$

on the time interval $[0, T]$ into Eq. (4.20) on the uniform mesh on the time interval $[0, \delta\sigma^{-1}(e^{\delta\sigma T} - 1)]$.

3. $Q = \pm u^{\sigma+1} + \delta u^n$, $\delta = \pm 1$. The equation

$$u_t = (u^\sigma u_x)_x \pm u^n, \qquad \sigma, n = \text{const}, \qquad (4.32)$$

admits a three-parameter symmetry group. One possible representation of this group is given by the following infinitesimal operators:

$$X_1 = \frac{\partial}{\partial t}, \quad X_2 = \frac{\partial}{\partial x}, \quad X_3 = 2(n-1)t\frac{\partial}{\partial t} + (n-\sigma-1)x\frac{\partial}{\partial x} - 2u\frac{\partial}{\partial u}. \quad (4.33)$$

The set (4.33) satisfies all invariant orthogonality and regularity conditions. Thus, one can use an orthogonal mesh uniform in both t- and x-directions. By considering the set of operators (4.33) in the space $(t, \hat{t}, x, h^+, h^-, u, u_+, u_-, \hat{u}, \hat{u}_+, \hat{u}_-)$ corresponding to the stencil shown in Fig. 4.1, we find seven difference invariants of the Lie algebra:

$$\frac{\tau^{\frac{n-\sigma-1}{2(n-1)}}}{h}, \quad \tau u^{n-1}, \quad \frac{\hat{u}}{u}, \quad \frac{u_+}{u}, \quad \frac{u_-}{u}, \quad \frac{\hat{u}_+}{\hat{u}}, \quad \frac{\hat{u}_-}{\hat{u}}.$$

There are few symmetry operators and hence many difference invariants. Thus we are left with freedom in the invariant difference modeling of Eq. (4.32). Using the difference invariants we, for example, obtain the following explicit invariant scheme:

$$\frac{\hat{u} - u}{\tau} = \frac{1}{h}\left(\left(\frac{u_+ + u}{2}\right)^\sigma u_x - \left(\frac{u + u_-}{2}\right)^\sigma u_{\bar{x}}\right) \pm u^n, \qquad (4.34)$$

where

$$u_x = \frac{u_+ - u}{h}, \qquad u_{\bar{x}} = \frac{u - u_-}{h}.$$

4. $Q = \pm u^{\sigma+1} + \delta u$, $\delta = \pm 1$. The equation

$$u_t = (u^\sigma u_x)_x \pm u^{\sigma+1} + \delta u$$

is related to Eq. (4.32) by the transformation (4.30). This equation admits the infinitesimal operators

$$X_1 = \frac{\partial}{\partial t}, \qquad X_2 = \frac{\partial}{\partial x}, \qquad X_3 = e^{-\delta\sigma t}\frac{\partial}{\partial t} + \delta e^{-\delta\sigma t}u\frac{\partial}{\partial u}. \qquad (4.35)$$

Using the invariants

$$u^\sigma(e^{\delta\sigma\tau} - 1), \quad h, \quad \delta\ln\frac{\hat{u}}{u} - \tau, \quad \frac{u_+}{u}, \quad \frac{u_-}{u}, \quad \frac{\hat{u}_+}{\hat{u}}, \quad \frac{\hat{u}_-}{\hat{u}}$$

of the operators (4.35), we obtain the following example of an explicit difference model:

$$\frac{\sigma u}{e^{\delta\sigma\tau} - 1}\left(\delta\ln\frac{\hat{u}}{u} - \tau\right) = \frac{1}{h}\left(\left(\frac{u_+ + u}{2}\right)^\sigma u_x - \left(\frac{u + u_-}{2}\right)^\sigma u_{\bar{x}}\right) \pm u^{\sigma+1}.$$

This equation on the mesh (4.31) is transformed by the change of variables (4.30) into Eq. (4.34) on a mesh with regular time spacing.

4.1.4. The special case $K = u^{-4/3}$ of a power-law heat conductivity coefficient

1. If $Q \equiv 0$, then the symmetry of the equation

$$u_t = (u^{-4/3}u_x)_x \qquad (4.36)$$

is described by the five-dimensional algebra with infinitesimal operators (see [112])

$$X_1 = \frac{\partial}{\partial t}, \qquad X_2 = \frac{\partial}{\partial x}, \qquad X_3 = 2t\frac{\partial}{\partial t} + x\frac{\partial}{\partial x},$$

$$X_4 = 2x\frac{\partial}{\partial x} - 3u\frac{\partial}{\partial u}, \qquad X_5 = x^2\frac{\partial}{\partial x} - 3xu\frac{\partial}{\partial u}. \qquad (4.37)$$

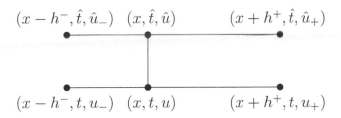

$$(x - h^-, \hat{t}, \hat{u}_-) \quad (x, \hat{t}, \hat{u}) \qquad (x + h^+, \hat{t}, \hat{u}_+)$$

$$(x - h^-, t, u_-) \quad (x, t, u) \qquad (x + h^+, t, u_+)$$

Figure 4.4: The stencil of a nonuniform mesh

These operators preserve the mesh orthogonality and uniformness in the time direction. The operator X_5 preserves the mesh uniformness in the t-direction but does not preserve the mesh uniformness in the x-direction; however, orthogonality is not disturbed. We shall consider the stencil shown in Fig. 4.4.

The finite-difference invariants

$$\frac{\hat{u}}{u}, \quad \frac{\hat{u}_+}{u_+}, \quad \frac{\hat{u}_-}{u_-}, \quad u_+^{1/3} u^{1/3} \frac{h^+}{\sqrt{\tau}}, \quad u_-^{1/3} u^{1/3} \frac{h^-}{\sqrt{\tau}}, \quad \frac{u^{2/3}}{\sqrt{\tau}} \left(\frac{h^+ h^-}{h^+ + h^-} \right) \qquad (4.38)$$

corresponding to this stencil give, among others, the explicit difference equation

$$\frac{\hat{u} - u}{\tau} = -\frac{h^+ + h^-}{6 h^+ h^-} \left(\frac{u_+^{-1/3} - u^{-1/3}}{h^+} - \frac{u^{-1/3} - u_-^{-1/3}}{h^-} \right). \qquad (4.39)$$

Note that one cannot apply the spatial mesh $h_+ = f(x, h_-)$, which is preserved under all transformations of the group (4.37). Using the difference invariants (4.38), one can apply, for example, the following invariant mesh:

$$u_+^{1/3} h^+ = u_-^{1/3} h^-,$$

which has the obvious integral

$$u_+^{1/3} u^{1/3} h^+ = \varepsilon, \qquad \varepsilon = \text{const.}$$

2. If $Q = \delta u$ and $\delta = \pm 1$, then the equation

$$u_t = (u^{-4/3} u_x)_x + \delta u \qquad (4.40)$$

admits the operators

$$X_1 = \frac{\partial}{\partial t}, \qquad X_2 = \frac{\partial}{\partial x}, \qquad X_3 = 2x \frac{\partial}{\partial x} - 3u \frac{\partial}{\partial u},$$

$$X_4 = e^{4\delta t/3} \frac{\partial}{\partial t} + \delta e^{4\delta t/3} u \frac{\partial}{\partial u}, \qquad X_5 = x^2 \frac{\partial}{\partial x} - 3xu \frac{\partial}{\partial u}. \qquad (4.41)$$

The change of variables (4.30) transforms this equation into Eq. (4.36). Let us write out the difference invariants for the set of operators (4.41):

$$\delta \ln \frac{\hat{u}}{u} - \tau, \qquad u^{2/3}\left(\frac{h^+ h^-}{h^+ + h^-}\right)\frac{1}{\sqrt{(e^{\delta\sigma\tau} - 1)}},$$

$$\frac{u_+^{1/3} u^{1/3} h^+}{\sqrt{(e^{\delta\sigma\tau} - 1)}}, \qquad \frac{u_-^{1/3} u^{1/3} h^-}{\sqrt{(e^{\delta\sigma\tau} - 1)}}, \qquad \frac{\hat{u}_+^{1/3} \hat{u}^{1/3} h^+}{\sqrt{(e^{\delta\sigma\tau} - 1)}}, \qquad \frac{\hat{u}_-^{1/3} \hat{u}^{1/3} h^-}{\sqrt{(e^{\delta\sigma\tau} - 1)}}.$$

These invariants can be used to construct a difference model for Eq. (4.40). Let us present the explicit version

$$\frac{\sigma u}{e^{\delta\sigma\tau} - 1}\left(\delta \ln \frac{\hat{u}}{u} - \tau\right) = -\frac{h^+ + h^-}{6h^+ h^-}\left(\frac{u_+^{-1/3} - u^{-1/3}}{h^+} - \frac{u^{-1/3} - u_-^{-1/3}}{h^-}\right) \quad (4.42)$$

of the difference model and the example

$$u_+^{1/3} h^+ = u_-^{1/3} h^-$$

of an invariant mesh.

3. $Q = \pm u^n$, $n \neq -1/3$. The equation

$$u_t = (u^{-4/3} u_x)_x \pm u^n \tag{4.43}$$

admits the infinitesimal operators

$$X_1 = \frac{\partial}{\partial t}, \qquad X_2 = \frac{\partial}{\partial x}, \qquad X_3 = 2(n-1)t\frac{\partial}{\partial t} + \left(n + \frac{1}{3}\right)x\frac{\partial}{\partial x} - 2u\frac{\partial}{\partial u}.$$

Although this equation is specified in the group classification (see [28]), it is a special case of Eq. (4.32): there is no extension of the admitted group. That is why we can use the model (4.34) with parameter $\sigma = -4/3$ corresponding to the given equation as an invariant difference model for Eq. (4.43).

4. If $Q = \alpha u^{-1/3}$ and $\alpha = \pm 1$, then the version of the difference model for the equation

$$u_t = (u^{-4/3} u_x)_x \pm u^{-\frac{1}{3}} \tag{4.44}$$

depends on the sign of the coefficient α. Equation (4.44) admits the five-dimensional algebra of infinitesimal operators

$$X_1 = \frac{\partial}{\partial t}, \qquad X_2 = \frac{\partial}{\partial x}, \qquad X_3 = \frac{4}{3}t\frac{\partial}{\partial t} + 2u\frac{\partial}{\partial u},$$

$$X_4 = e^{2\sqrt{\alpha/3}x}\frac{\partial}{\partial x} - \sqrt{3\alpha}e^{2\sqrt{\alpha/3}x}u\frac{\partial}{\partial u}, \quad X_5 = e^{-2\sqrt{\alpha/3}x}\frac{\partial}{\partial x} + \sqrt{3\alpha}e^{-2\sqrt{\alpha/3}x}u\frac{\partial}{\partial u}.$$

(a) The case of $\alpha = 1$. The change of variables

$$\bar{u} = u \cosh^3 \frac{x}{\sqrt{3}}, \qquad \bar{x} = \sqrt{3} \tanh \frac{x}{\sqrt{3}} \qquad (4.45)$$

transforms the considered equation into Eq. (4.36) (see [74]). Using the difference invariants

$$\frac{\hat{u}}{u}, \qquad \frac{\hat{u}_+}{u_+}, \qquad \frac{\hat{u}_-}{u_-}, \qquad \sqrt{\tau} u^{-2/3} \left(\frac{1}{\tanh(h^+/\sqrt{3})} + \frac{1}{\tanh(h^-/\sqrt{3})} \right),$$

$$\frac{u^{1/3} u_+^{1/3}}{\sqrt{\tau}} \sinh \frac{h^+}{\sqrt{3}}, \qquad \frac{u^{1/3} u_-^{1/3}}{\sqrt{\tau}} \sinh \frac{h^-}{\sqrt{3}},$$

one can construct a difference model. Let us write out one possible version of the difference model, namely, the explicit equation

$$\frac{\hat{u} - u}{\tau} = -\frac{1}{18} \left(\frac{1}{\tanh(h^+/\sqrt{3})} + \frac{1}{\tanh(h^-/\sqrt{3})} \right)$$

$$\times \left(\frac{u_+^{-1/3} - u^{-1/3} \cosh(h^+/\sqrt{3})}{\sinh(h^+/\sqrt{3})} - \frac{u^{-1/3} \cosh(h^-/\sqrt{3}) - u_-^{-1/3}}{\sinh(h^-/\sqrt{3})} \right), \qquad (4.46)$$

and the example

$$u_+^{1/3} \sinh \left(\frac{h^+}{\sqrt{3}} \right) = u_-^{1/3} \sinh \left(\frac{h^-}{\sqrt{3}} \right)$$

of an invariant mesh.

(b) The case of $\alpha = -1$. The change of variables

$$\bar{u} = u \cos^3 \frac{x}{\sqrt{3}}, \qquad \bar{x} = \sqrt{3} \tan \frac{x}{\sqrt{3}} \qquad (4.47)$$

transforms the given equation into Eq. (4.36). The set of difference invariants

$$\frac{\hat{u}}{u}, \qquad \frac{\hat{u}_+}{u_+}, \qquad \frac{\hat{u}_-}{u_-}, \qquad \sqrt{\tau} u^{-2/3} \left(\frac{1}{\tan(h^+/\sqrt{3})} + \frac{1}{\tan(h^-/\sqrt{3})} \right),$$

$$\frac{u^{1/3} u_+^{1/3}}{\sqrt{\tau}} \sin \frac{h^+}{\sqrt{3}}, \qquad \frac{u^{1/3} u_-^{1/3}}{\sqrt{\tau}} \sin \frac{h^-}{\sqrt{3}}$$

permits us to construct an invariant difference scheme. For example, one can use the explicit difference equation

$$\frac{\hat{u} - u}{\tau} = -\frac{1}{18} \left(\frac{1}{\tan(h^+/\sqrt{3})} + \frac{1}{\tan(h^-/\sqrt{3})} \right)$$

$$\times \left(\frac{u_+^{-1/3} - u^{-1/3} \cos(h^+/\sqrt{3})}{\sin(h^+/\sqrt{3})} - \frac{u^{-1/3} \cos(h^-/) - u_-^{-1/3}}{\sin(h^-/\sqrt{3})} \right), \qquad (4.48)$$

and an example of an invariant mesh is given by

$$u_+^{1/3} \sin \frac{h^+}{\sqrt{3}} = u_-^{1/3} \sin \frac{h^-}{\sqrt{3}}.$$

We point out that the obtained difference models (4.46) and (4.48) are related to the difference model (4.39) for Eq. (4.36) by the changes of variables (4.45) and (4.47), respectively, as is the case for the original differential equations.

5. $Q = \alpha u^{-\frac{1}{3}} + \delta u$, $|\alpha| = |\delta| = 1$. As in the preceding item, two cases of the parameter α in the equation

$$u_t = (u^\sigma u_x)_x \pm u^{\sigma+1} + \delta u \qquad (4.49)$$

should be considered separately, and two difference models should be constructed. Let us write out the infinitesimal operators admitted by Eq. (4.49):

$$X_1 = \frac{\partial}{\partial t}, \qquad X_2 = \frac{\partial}{\partial x}, \qquad X_3 = e^{4\delta t/3}\frac{\partial}{\partial t} + \delta e^{4\delta t/3} u \frac{\partial}{\partial u},$$

$$X_4 = e^{2\sqrt{\alpha/3}x}\frac{\partial}{\partial x} - \sqrt{3\alpha}e^{2\sqrt{\alpha/3}x} u \frac{\partial}{\partial u}, \quad X_5 = e^{-2\sqrt{\alpha/3}x}\frac{\partial}{\partial x} + \sqrt{3\alpha}e^{-2\sqrt{\alpha/3}x} u \frac{\partial}{\partial u}.$$

(a) The case of $\alpha = 1$. The change of variables (4.30) transforms the considered equation into Eq. (4.44), and the change (4.45) transforms it into Eq. (4.36).

Let us write out the set of finite-difference invariants for Eq. (4.49) with $\alpha = 1$:

$$\delta \ln \frac{\hat{u}}{u} - \tau, \qquad \sqrt{(e^{\delta\sigma\tau} - 1)}u^{-2/3}\left(\frac{1}{\tanh(h^+/\sqrt{3})} + \frac{1}{\tanh(h^-/\sqrt{3})}\right),$$

$$\frac{u^{1/3}u_+^{1/3}}{\sqrt{(e^{\delta\sigma\tau} - 1)}}\sinh\frac{h^+}{\sqrt{3}}, \qquad \frac{u^{1/3}u_-^{1/3}}{\sqrt{(e^{\delta\sigma\tau} - 1)}}\sinh\frac{h^-}{\sqrt{3}},$$

$$\frac{\hat{u}^{1/3}\hat{u}_+^{1/3}}{\sqrt{(e^{\delta\sigma\tau} - 1)}}\sinh\frac{h^+}{\sqrt{3}}, \qquad \frac{\hat{u}^{1/3}\hat{u}_-^{1/3}}{\sqrt{(e^{\delta\sigma\tau} - 1)}}\sinh\frac{h^-}{\sqrt{3}}.$$

The explicit version of the difference model for Eq. (4.49) on the time mesh (4.31) has the form

$$\frac{\sigma u}{e^{\delta\sigma\tau} - 1}\left(\delta \ln \frac{\hat{u}}{u} - \tau\right) = -\frac{1}{18}\left(\frac{1}{\tanh(h^+/\sqrt{3})} + \frac{1}{\tanh(h^-/\sqrt{3})}\right)$$

$$\times \left(\frac{u_+^{-1/3} - u^{-1/3}\cosh(h^+/\sqrt{3})}{\sinh(h^+/\sqrt{3})} - \frac{u^{-1/3}\cosh(h^-/\sqrt{3}) - u_-^{-1/3}}{\sinh(h^-/\sqrt{3})}\right), \qquad (4.50)$$

and an example of an invariant mesh is given by

$$u_+^{1/3} \sinh \frac{h^+}{\sqrt{3}} = u_-^{1/3} \sinh \frac{h^-}{\sqrt{3}}.$$

(b) The case of $\alpha = -1$. The change of variables (4.47) transforms this equation into Eq. (4.40), and the change of variables (4.30) transforms it into Eq. (4.44). A difference model for Eq. (4.49) can be obtained with the help of the invariants

$$\delta \ln \frac{\hat{u}}{u} - \tau, \qquad \sqrt{(e^{\delta\sigma\tau} - 1)} u^{-2/3} \left(\frac{1}{\tan(h^+/\sqrt{3})} + \frac{1}{\tan(h^-/\sqrt{3})} \right),$$

$$\frac{u^{1/3} u_+^{1/3}}{\sqrt{(e^{\delta\sigma\tau} - 1)}} \sin \frac{h^+}{\sqrt{3}}, \qquad \frac{u^{1/3} u_-^{1/3}}{\sqrt{(e^{\delta\sigma\tau} - 1)}} \sin \frac{h^-}{\sqrt{3}},$$

$$\frac{\hat{u}^{1/3} \hat{u}_+^{1/3}}{\sqrt{(e^{\delta\sigma\tau} - 1)}} \sin \frac{h^+}{\sqrt{3}}, \qquad \frac{\hat{u}^{1/3} \hat{u}_-^{1/3}}{\sqrt{(e^{\delta\sigma\tau} - 1)}} \sin \frac{h^-}{\sqrt{3}}.$$

One possible difference model for Eq. (4.49) is

$$\frac{\sigma u}{e^{\delta\sigma\tau} - 1} \left(\delta \ln \frac{\hat{u}}{u} - \tau \right) = -\frac{1}{18} \left(\frac{1}{\tan(h^+/\sqrt{3})} + \frac{1}{\tan(h^-/\sqrt{3})} \right)$$

$$\times \left(\frac{u_+^{-1/3} - u^{-1/3} \cos(h^+/\sqrt{3})}{\sin(h^+/\sqrt{3})} - \frac{u^{-1/3} \cos(h^-/\sqrt{3}) - u_-^{-1/3}}{\sin(h^-/\sqrt{3})} \right), \qquad (4.51)$$

and an example of an invariant mesh is given by

$$u_+^{1/3} \sin \frac{h^+}{\sqrt{3}} = u_-^{1/3} \sin \frac{h^-}{\sqrt{3}}.$$

The difference models (4.50) and (4.51) are related to the model (4.42) by the changes of variables (4.45) and (4.47), respectively. The change (4.30) transforms the difference models obtained in this item into the model of item 4 for the corresponding values of the parameter α. This example shows that in invariant difference modeling it is possible to obtain consistent models related to each other by the same point transformations as their original differential counterparts.

4.1.5. Linear heat conduction with a nonlinear source

In this section, we consider the semilinear heat equation

$$u_t = u_{xx} + Q(u)$$

with various types of source.

1. With $Q = \pm e^u$, the equation becomes

$$u_t = u_{xx} \pm e^u.$$

It admits the three-dimensional algebra of infinitesimal operators

$$X_1 = \frac{\partial}{\partial t}, \qquad X_2 = \frac{\partial}{\partial x}, \qquad X_3 = 2t \frac{\partial}{\partial t} + x \frac{\partial}{\partial x} - 2 \frac{\partial}{\partial u}.$$

One can readily verify that the conditions for the preservation of the mesh orthogonality and regularity hold for these operators. An approximation to the equation with the use of the invariants

$$\frac{h^2}{\tau}, \quad \tau e^u, \quad \hat{u} - u, \quad u_+ - u, \quad u - u_-, \quad \hat{u}_+ - \hat{u}, \quad \hat{u} - \hat{u}_-$$

gives various types of difference equations. An explicit equation can be as follows:

$$\frac{\hat{u} - u}{\tau} = \frac{1}{h}(u_{\underset{h}{x}} - u_{\underset{h}{\bar{x}}}) \pm e^u.$$

2. $Q = \pm u^n$. The equation

$$u_t = u_{xx} \pm u^n$$

admits the infinitesimal operators

$$X_1 = \frac{\partial}{\partial t}, \qquad X_2 = \frac{\partial}{\partial x}, \qquad X_3 = 2(n-1)t\frac{\partial}{\partial t} + (n-1)x\frac{\partial}{\partial x} - 2u\frac{\partial}{\partial u}.$$

These operators satisfy the orthogonality and regularity invariance conditions, and the difference invariants

$$\frac{h^2}{\tau}, \quad \tau u^{n-1}, \quad \frac{\hat{u}}{u}, \quad \frac{u_+}{u}, \quad \frac{u_-}{u}, \quad \frac{\hat{u}_+}{\hat{u}}, \quad \frac{\hat{u}_-}{\hat{u}}$$

permit us to construct, for example, the following difference scheme:

$$\frac{\hat{u} - u}{\tau} = \frac{1}{h}(u_{\underset{h}{x}} - u_{\underset{h}{\bar{x}}}) \pm u^n.$$

3. $Q = \delta u \ln u, \delta = \pm 1$. The semilinear heat equation

$$u_t = u_{xx} + \delta u \ln u, \qquad \delta = \pm 1, \tag{4.52}$$

admits the four-parameter Lie symmetry point transformation group corresponding to the following set of infinitesimal operators:

$$X_1 = \frac{\partial}{\partial t}, \quad X_2 = \frac{\partial}{\partial x}, \quad X_3 = 2e^{\delta t}\frac{\partial}{\partial x} - \delta e^{\delta t}xu\frac{\partial}{\partial u}, \quad X_4 = e^{\delta t}u\frac{\partial}{\partial u}. \tag{4.53}$$

Before constructing a difference equation and a mesh that approximate (4.52) and inherit the whole Lie algebra (4.53), we should first verify the orthogonality invariance condition. The operators X_1, X_2, and X_4 preserve orthogonality, while X_3 does not. Consequently, an orthogonal mesh cannot be used for the invariant modeling of (4.52). Condition of invariant flatness of the time layer is true for the complete set of operators, so that it is possible to use a nonorthogonal mesh with flat time layers, and we shall use the mesh shown in Fig. 4.2.

A possible reformulation of Eq. (4.52) with the use of the four differential invariants

$$J_1 = dt, \quad J_2 = \left(\frac{u_x}{u}\right)^2 - \frac{u_{xx}}{u}, \quad J_3 = 2\frac{u_x}{u} + \frac{dx}{dt}, \quad J_4 = \frac{du}{udt} - \delta \ln u + \frac{1}{4}\left(\frac{dx}{dt}\right)^2$$

in the subspace $(t, x, u, u_x, u_{xx}, dt, dx, du)$ is given by the system

$$J_3 = 0, \qquad J_4 = J_2, \tag{4.54}$$

that is,

$$\frac{dx}{dt} = -2\frac{u_x}{u}, \qquad \frac{du}{dt} = u_{xx} + \delta u \ln u - 2\frac{u_x^2}{u}. \tag{4.55}$$

Thus, the structure of the admitted group suggests approximating two evolution equations.

As the next step, we shall find difference invariants for the operators $X_1, \ldots, X_4$ of the group (4.53). We shall use the six-point difference stencil in Fig. 4.3, on which we shall approximate system (4.55). The stencil defining the subspace of the variables $(t, \hat{t}, x, \hat{x}, h^+, h^-, \hat{h}^+, \hat{h}^-, u, u_+, u_-, \hat{u}, \hat{u}_+, \hat{u}_-)$ and the operators (4.53) has the following difference invariants:

$$I_1 = \tau, \qquad I_2 = h^+, \qquad I_3 = h^-, \qquad I_4 = \hat{h}^+, \qquad I_5 = \hat{h}^-,$$

$$I_6 = (\ln u)_x - (\ln u)_{\bar{x}}, \qquad I_7 = (\ln \hat{u})_x - (\ln \hat{u})_{\bar{x}},$$

$$I_8 = \delta \Delta x + 2(e^{\delta\tau} - 1)\left(\frac{h^-}{h^+ + h^-}(\ln u)_x + \frac{h^+}{h^+ + h^-}(\ln u)_{\bar{x}}\right),$$

$$I_9 = \delta \Delta x + 2(1 - e^{-\delta\tau})\left(\frac{\hat{h}^-}{\hat{h}^+ + \hat{h}^-}(\ln \hat{u})_x + \frac{\hat{h}^+}{\hat{h}^+ + \hat{h}^-}(\ln \hat{u})_{\bar{x}}\right),$$

$$I_{10} = \delta(\Delta x)^2 + 4(1 - e^{-\delta\tau})(\ln \hat{u} - e^{\delta\tau} \ln u),$$

where

$$\Delta x = \hat{x} - x, \qquad (\ln u)_x = \frac{\ln u_+ - \ln u}{h^+}, \qquad (\ln u)_{\bar{x}} = \frac{\ln u - \ln u_-}{h^-}.$$

An explicit model can be chosen as follows:

$$I_8 = 0, \qquad I_{10} = \frac{8}{\delta}\frac{(e^{\delta I_1} - 1)^2}{I_2 + I_3}I_6,$$

i.e.,

$$\delta \Delta x + 2(e^{\delta\tau} - 1)\left(\frac{h^-}{h^+ + h^-}(\ln u)_x + \frac{h^+}{h^+ + h^-}(\ln u)_{\bar{x}}\right) = 0,$$

$$\delta(\Delta x)^2 + 4(1 - e^{-\delta\tau})(\ln \hat{u} - e^{\delta\tau} \ln u) = \frac{8}{\delta}\frac{(e^{\delta\tau} - 1)^2}{h^+ + h^-}[(\ln u)_x - (\ln u)_{\bar{x}}]. \tag{4.56}$$

Consider a symmetry reduction and an appropriate family of exact invariant solutions of this scheme [8, 9].

Consider solutions of Eq. (4.56) invariant with respect to the operator

$$2\alpha X_2 + X_3, \qquad \alpha = \text{const.}$$

This operator has the three difference invariants

$$J_1 = t, \qquad J_2 = u \exp\left(\frac{\delta e^{\delta t}}{\alpha + e^{\delta t}} \frac{x^2}{4}\right), \qquad J_3 = \left(\frac{\Delta x}{e^{\delta t}(e^{\delta \tau} - 1)} - \frac{x}{\alpha + e^{\delta t}}\right),$$

and hence we seek an exact solution in the form

$$u(x,t) = \exp\left(-\frac{\delta e^{\delta t}}{\alpha + e^{\delta t}} \frac{x^2}{4}\right) e^{f(t)}, \qquad \frac{\Delta x}{e^{\delta t}(e^{\delta \tau} - 1)} = \frac{x}{\alpha + e^{\delta t}} + g(t).$$

The substitution of this solution into system (4.56) yields the system of ordinary difference equations

$$\frac{f(t+\tau) - e^{\delta \tau} f(t)}{e^{\delta \tau}(e^{\delta \tau} - 1)} = -\frac{1}{2}\frac{e^{\delta t}}{\alpha + e^{\delta t}}, \qquad g(t) = 0,$$

for *two* unknown functions $f(t)$ and $g(t)$. The solution of the last system is the expression

$$u(x,t) = \exp\left(e^{\delta t}\left(f(0) - \frac{e^{\delta \tau} - 1}{2}\sum_{j=1}^{n-1}\frac{e^{-\delta t_j}}{1 + \alpha e^{-\delta t_j}}\right) - \frac{\delta e^{\delta t}}{\alpha + e^{\delta t}}\frac{x^2}{4}\right)$$

for u and the expression

$$x = x^0 \frac{e^{\delta t} + \alpha}{1 + \alpha}$$

for the mesh, where $x = x_i^j = x_i(t_j)$ and $t = t_j$. The mesh can be arbitrary at the initial time $t = 0$. If it is originally regular, then it will be regular on any further time layer.

The obtained exact solution yields the solution of the Cauchy problem with the initial conditions

$$u(x,0) = \exp\left(f(0) - \frac{\delta e^{\delta t}}{\alpha + e^{\delta t}}\frac{x^2}{4}\right).$$

4.2. Symmetry Preserving Difference Schemes for the Linear Heat Equation

In this section, we complete the set of invariant difference schemes for the heat transfer equation with a source in accordance with the Lie group classification [28] (see also [74]). Namely, we consider a linear heat equation without a source and with a linear source [8, 9, 42].

4.2.1. Linear heat equation without a source

The linear heat equation

$$u_t = u_{xx} \tag{4.57}$$

admits the six-parameter point transformation group with infinitesimal operators

$$X_1 = \frac{\partial}{\partial t}, \quad X_2 = \frac{\partial}{\partial x}, \quad X_3 = 2t\frac{\partial}{\partial x} - xu\frac{\partial}{\partial u}, \quad X_4 = 2t\frac{\partial}{\partial t} + x\frac{\partial}{\partial x},$$

$$X_5 = 4t^2\frac{\partial}{\partial t} + 4tx\frac{\partial}{\partial x} - (x^2 + 2t)u\frac{\partial}{\partial u}, \quad X_6 = u\frac{\partial}{\partial u} \tag{4.58}$$

and the infinite-dimensional symmetry

$$X^* = a(x, t)\frac{\partial}{\partial u},$$

where $a(t, x)$ is an arbitrary solution of Eq. (4.57). The symmetry X^* represents the linearity of Eq. (4.57).

Probably the simplest approximation to the linear equation is given by the explicit scheme

$$\frac{\hat{u} - u}{\tau} = \frac{u_+ - 2u + u_-}{h^2} \tag{4.59}$$

considered on a uniform orthogonal mesh. As was shown in the Introduction, this equation is invariant with respect to the operators X_1, X_2, X_4, and X_6 in the set (4.58). Since the equation is linear, it obeys the superposition principle, which is reflected in the invariance with respect to the operator

$$X_h^* = a_h(x, t)\frac{\partial}{\partial u},$$

where $a_h(x, t)$ is an arbitrary solution of Eq. (4.59). It was shown in [8, 9, 42] how to construct a discrete model which admits the six-dimensional group (4.58). To preserve the Galilei operator X_3 and the projective operator X_5, one has to introduce a moving mesh.

Heat equation as a system of equations

With the help of the differential invariants

$$J_1 = \frac{dx + 2\frac{u_x}{u}dt}{dt^{1/2}}, \quad J_2 = \frac{du}{u} + \frac{1}{4}\frac{dx^2}{dt} + \left(-\frac{u_{xx}}{u} + \frac{u_x^2}{u^2}\right)dt$$

of the operators (4.58) in the space $(t, x, u, u_x, u_{xx}, dt, dx, du)$, we can represent the heat equation (4.57) as the system

$$J_1 = 0, \quad J_2 = 0,$$

that is,

$$\frac{dx}{dt} = -2\frac{u_x}{u}, \qquad \frac{du}{dt} = u_{xx} - 2\frac{u_x^2}{u}. \tag{4.60}$$

This system is invariant with respect to the six-dimensional group generated by the operators (4.58), because it was constructed by means of differential invariants.

Invariant schemes on moving meshes

For the difference modeling of system (4.60), we need the whole set of difference invariants of the symmetry group (4.58) in the difference space corresponding to the chosen stencil $(t, \hat{t}, x, \hat{x}, h^+, h^-, \hat{h}^+, \hat{h}^-, u, \hat{u}, u_+, u_-, \hat{u}_+, \hat{u}_-)$:

$$I_1 = \frac{h^+}{h^-}, \qquad I_2 = \frac{\hat{h}^+}{\hat{h}^-}, \qquad I_3 = \frac{\hat{h}^+ h^+}{\tau}, \qquad I_4 = \frac{\tau^{1/2}}{h^+}\frac{\hat{u}}{u}\exp\left(\frac{1}{4}\frac{(\Delta x)^2}{\tau}\right),$$

$$I_5 = \frac{1}{4}\frac{h^{+2}}{\tau} - \frac{h^{+2}}{h^+ + h^-}\left(\frac{1}{h^+}\ln\frac{u_+}{u} + \frac{1}{h^-}\ln\frac{u_-}{u}\right),$$

$$I_6 = \frac{1}{4}\frac{\hat{h}^{+2}}{\tau} + \frac{\hat{h}^{+2}}{\hat{h}^+ + \hat{h}^-}\left(\frac{1}{\hat{h}^+}\ln\frac{\hat{u}_+}{\hat{u}} + \frac{1}{\hat{h}^-}\ln\frac{\hat{u}_-}{\hat{u}}\right),$$

$$I_7 = \frac{\Delta x h^+}{\tau} + \frac{2h^+}{h^+ + h^-}\left(\frac{h^-}{h^+}\ln\frac{u_+}{u} - \frac{h^+}{h^-}\ln\frac{u_-}{u}\right),$$

$$I_8 = \frac{\Delta x \hat{h}^+}{\tau} + \frac{2\hat{h}^+}{\hat{h}^+ + \hat{h}^-}\left(\frac{\hat{h}^-}{\hat{h}^+}\ln\frac{\hat{u}_+}{\hat{u}} - \frac{\hat{h}^+}{\hat{h}^-}\ln\frac{\hat{u}_-}{\hat{u}}\right).$$

Approximating system (4.60) by invariants, we obtain a system of difference evolution equations. By way of example, here we present the invariant difference model

$$\Delta x = \frac{2\tau}{h^+ + h^-}\left(-\frac{h^-}{h^+}\ln\frac{u_+}{u} + \frac{h^+}{h^-}\ln\frac{u_-}{u}\right),$$

$$\left(\frac{u}{\hat{u}}\right)^2 \exp\left(-\frac{1}{2}\frac{(\Delta x)^2}{\tau}\right) = 1 - \frac{4\tau}{h^+ + h^-}\left(\frac{1}{h^+}\ln\frac{u_+}{u} + \frac{1}{h^-}\ln\frac{u_-}{u}\right), \tag{4.61}$$

which has explicit equations for the solution u and the mesh trajectory. We also can write out the implicit model

$$\Delta x = \frac{2\tau}{\hat{h}^+ + \hat{h}^-}\left(-\frac{\hat{h}^-}{\hat{h}^+}\ln\frac{\hat{u}_+}{\hat{u}} + \frac{\hat{h}^+}{\hat{h}^-}\ln\frac{\hat{u}_-}{\hat{u}}\right),$$

$$\left(\frac{\hat{u}}{u}\right)^2 \exp\left(\frac{1}{2}\frac{\Delta x^2}{\tau}\right) = 1 + \frac{4\tau}{\hat{h}^+ + \hat{h}^-}\left(\frac{1}{\hat{h}^+}\ln\frac{\hat{u}_-}{\hat{u}} + \frac{1}{\hat{h}^-}\ln\frac{\hat{u}_-}{\hat{u}}\right).$$

It is also possible to combine an explicit equation for the mesh and an implicit approximation to the partial differential equation, or vice versa. Other ways to approximate system (4.60) by difference invariants are also possible.

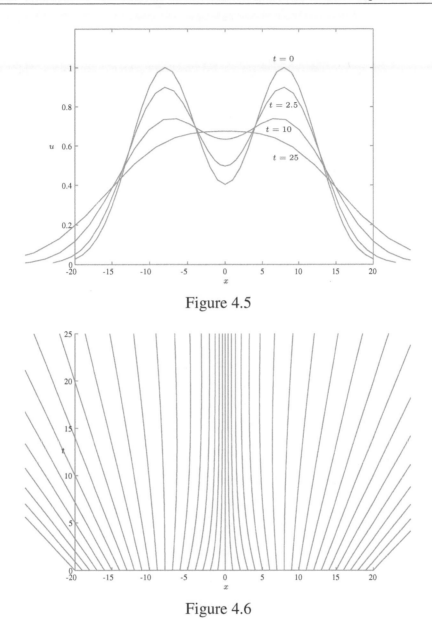

Figure 4.5

Figure 4.6

It should be noticed that an invariant mesh adapts to a given solution regardless of whether the solution itself is invariant or not. Examples of two evolutions of meshes and appropriate *noninvariant* solutions are shown in Figs. 4.5–4.8. The invariant scheme (4.61) was used in the numerical implementation.

Optimal system of subalgebras and reduced systems

Among all invariant solutions, there is a minimal set of such solutions, called a *optimal system of invariant solutions* [111]. Any invariant solution can be obtained from this set of invariant solutions by an appropriate group transformation.

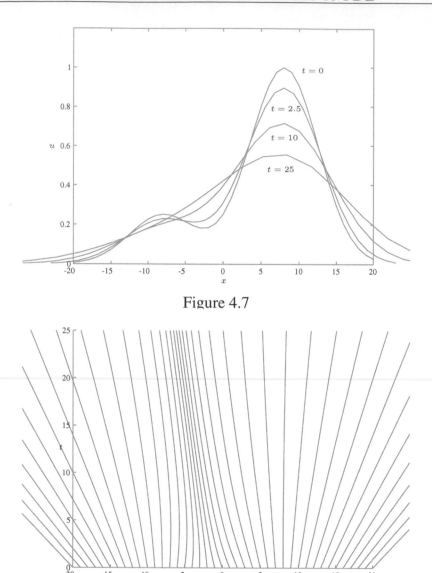

Figure 4.7

Figure 4.8

The difference model (4.61) is a system of two evolution equations. To find its invariant solutions, we need to provide a time mesh that is invariant with respect to the considered operator. An invariant time mesh giving flat time layers can be represented by the equation

$$\tau_i = g(t_i), \qquad i = 0, 1, 2, \ldots . \qquad (4.62)$$

We require this equation to be invariant with respect to the considered symmetry. Since the coefficients ξ^t for the operators (4.58) do not depend on x and u, we can propose an invariant time mesh for any symmetry. In the case of $\xi^t = 0$, the function g can be taken arbitrarily. For example, we can choose the uniform mesh

$t_j = j\tau$, $\tau = $ const. Thus, different invariant solutions can have different time meshes.

The action of admitted Lie group transforms an invariant solution into another one [111]. In our case, it also transforms the time mesh equation (4.62). Thus, the group action gives a new invariant solution with the corresponding invariant mesh.

Using the difference model (4.61) as an example, we shall construct an optimal system of solutions invariant with respect to one-parameter groups. The optimal system of one-dimensional subalgebras of the symmetry algebra for the linear heat equation consists of the algebras corresponding to the operators (see [111])

$$Y_1 = X_2 = \frac{\partial}{\partial x}, \qquad Y_2 = X_6 = u\frac{\partial}{\partial u}, \qquad Y_3 = X_1 + cX_6 = \frac{\partial}{\partial t} + cu\frac{\partial}{\partial u},$$

$$Y_4 = X_1 - X_3 = \frac{\partial}{\partial t} - 2t\frac{\partial}{\partial x} + xu\frac{\partial}{\partial u},$$

$$Y_5 = X_4 + 2cX_6 = 2t\frac{\partial}{\partial t} + x\frac{\partial}{\partial x} + 2cu\frac{\partial}{\partial u},$$

$$Y_6 = X_1 + X_5 + cX_6 = (4t^2 + 1)\frac{\partial}{\partial t} + 4tx\frac{\partial}{\partial x} + (c - x^2 - 2t)u\frac{\partial}{\partial u}.$$

Let us find the invariant solutions corresponding to these one-dimensional subalgebras.

1. The subalgebra corresponding to the operator Y_1 has only the constant solutions $u = C$, $C = $ const, considered on the orthogonal mesh $\Delta x = 0$.

2. The subalgebra corresponding to the operator Y_2 does not have invariant solutions. (The necessary condition for the existence of invariant solutions does not hold [21].)

3. The operator Y_3 has the following invariants: $u\exp(-ct)$, τ, and Δx. The time step τ is invariant, and so we can consider a uniform time mesh. We seek a solution of the difference model in the form

$$u = \exp(ct)f(x).$$

By substituting this invariant form of the solution into system (4.61), we obtain

$$\Delta x = \frac{2\tau}{h^+ + h^-}\left(-\frac{h^+}{h^-}\ln\left(\frac{f(x + h^+)}{f(x)}\right) + \frac{h^+}{h^-}\ln\left(\frac{f(x - h^-)}{f(x)}\right)\right),$$

$$\left(\frac{f(x)}{f(x + \Delta x)}\right)^2 \exp\left(-2c\tau - \frac{1}{2}\frac{\Delta x^2}{\tau}\right) \tag{4.63}$$

$$= 1 - \frac{4\tau}{h^+ + h^-}\left(\frac{1}{h^+}\ln\left(\frac{f(x + h^+)}{f(x)}\right) + \frac{1}{h^-}\ln\left(\frac{f(x - h^-)}{f(x)}\right)\right).$$

System (4.63) becomes a system of two ordinary difference equations if we project it onto the space of invariants. To project the system, we have to impose the condition

$$\Delta x = -h^-, \ 0, \ \text{or} \ h^+. \tag{4.64}$$

The solution of system (4.63) with one of conditions (4.64) provides the solution of system (4.61) invariant with respect to the operator Y_3.

4. The operator Y_4 has the invariants

$$u \exp\left(-xt - \frac{2}{3}t^3\right), \quad x + t^2, \quad \tau, \quad \frac{\Delta x}{2\tau} - t.$$

Let us seek the solution of the difference model (4.61) in the form

$$u = \exp\left(tx + \frac{2}{3}t^3\right) f(x + t^2).$$

By the change of variables

$$y = x + t^2, \qquad y - h_y^- = x - h^- + t^2,$$
$$y + h_y^+ = x + h^+ + t^2, \qquad y + \Delta y = x + \Delta x + (t + \tau)^2,$$

we obtain the following system for the invariant solution of system (4.61):

$$\Delta y - \tau^2 = \frac{2\tau}{h_y^+ + h_y^-}\left(-\frac{h_y^-}{h_y^+}\ln\left(\frac{f(y + h_y^+)}{f(y)}\right) + \frac{h_y^+}{h_y^-}\ln\left(\frac{f(y - h_y^-)}{f(y)}\right)\right),$$

$$\left(\frac{f(y)}{f(y + \Delta y)}\right)^2 \exp\left(-\frac{1}{2\tau}\Delta y^2 - \tau(2y + \Delta y) + \frac{1}{6}\tau^3\right)$$

$$= 1 - \frac{4\tau}{h_y^+ + h_y^-}\left(\frac{1}{h_y^+}\ln\left(\frac{f(y + h_y^+)}{f(y)}\right) + \frac{1}{h_y^-}\ln\left(f\left(\frac{y - h_y^-}{f(y)}\right)\right)\right),$$

where Δy can have one of the following values:

$$\Delta y = -h_y^-, \ 0, \ \text{or} \ h_y^+. \tag{4.65}$$

The solution of the above system with one of conditions (4.65) allows us to find the solution invariant under the operator Y_4.

5. The expressions

$$\frac{x}{\sqrt{t}}, \quad t^{-c}u, \quad \frac{\tau}{t}, \quad \frac{\Delta x}{x}$$

are invariants of the operator Y_5. Let us seek the solution of the difference model in the form

$$u = t^c f\left(\frac{x}{\sqrt{t}}\right).$$

In the variables

$$y = \frac{x}{\sqrt{t}}, \qquad y - h_y^- = \frac{x - h^-}{\sqrt{t}}, \qquad y + h_y^+ = \frac{x + h^+}{\sqrt{t}}, \qquad y + \Delta y = \frac{x + \Delta x}{\sqrt{t + \tau}},$$

we obtain the following system of equations:

$$\sqrt{1+a}(y + \Delta y) - y$$

$$= \frac{2a}{h_y^+ + h_y^-} \left(-\frac{h_y^-}{h_y^+} \ln \left(\frac{f(y + h_y^+)}{f(y)} \right) + \frac{h_y^+}{h_y^-} \ln \left(\frac{f(y - h_y^-)}{f(y)} \right) \right),$$

$$(1 + a)^{-2c} \left(\frac{f(y)}{f(y + \Delta y)} \right)^2 \exp \left(-\frac{1}{2} \left((y + \Delta y) \sqrt{\frac{1+a}{a}} - y \frac{1}{\sqrt{a}} \right)^2 \right)$$

$$= 1 - \frac{4a}{h_y^+ + h_y^-} \left(\frac{1}{h_y^+} \ln \left(\frac{f(y + h_y^+)}{f(y)} \right) + \frac{1}{h_y^-} \ln \left(\frac{f(y - h_y^-)}{f(y)} \right) \right).$$

Here Δy can have one of the values determined by conditions (4.65) and a is the constant in the condition $a = \frac{\tau}{t}$, which determines an invariant time spacing. This condition can be found if we look for a time spacing $\tau = g(t)$ invariant with respect to the operator Y_5.

6. For the operator Y_6, we have the following invariants:

$$\frac{x}{\sqrt{4t^2 + 1}}, \qquad (4t^2 + 1)^{1/4} u \exp \left(\frac{tx^2}{4t^2 + 1} + \frac{c}{2} \arctan(2t) \right),$$

$$\frac{4t^2 + 1}{\tau} + 4t, \qquad \frac{\Delta x}{x} \frac{4t^2 + 1}{\tau} - 4t.$$

We seek the solution of the difference model in the form

$$u = (4t^2 + 1)^{-1/4} \exp \left(-\frac{tx^2}{4t^2 + 1} - \frac{c}{2} \arctan(2t) \right) f \left(\frac{x}{\sqrt{4t^2 + 1}} \right).$$

In the new variables

$$y = \frac{x}{\sqrt{4t^2 + 1}}, \qquad y - h_y^- = \frac{x - h^-}{\sqrt{4t^2 + 1}},$$

$$y + h_y^+ = \frac{x + h^+}{\sqrt{4t^2 + 1}}, \qquad y + \Delta y = \frac{x + \Delta x}{\sqrt{4(t + \tau)^2 + 1}},$$

system (4.61) can be represented in the form

$$
\sqrt{b^2+1}(y+\Delta y)-by
$$
$$
=\frac{1}{h_y^+ + h_y^-}\left(-\frac{h_y^-}{h_y^+}\ln\left(\frac{f(y+h_y^+)}{f(y)}\right)+\frac{h_y^+}{h_y^-}\ln\left(\frac{f(y-h_y^-)}{f(y)}\right)\right),
$$

$$
\sqrt{b^2+1}\left(\frac{f(y)}{f(y+\Delta y)}\right)^2
$$
$$
\times\exp\left(c\arctan\left(\frac{1}{b}\right)-b(y^2+(y+\Delta y)^2)+2\sqrt{b^2+1}\,y(y+\Delta y)\right)
$$
$$
=b-\frac{2}{h_y^+ + h_y^-}\left(\frac{1}{h_y^+}\ln\left(\frac{f(y+h_y^+)}{f(y)}\right)+\frac{1}{h_y^-}\ln\left(\frac{f(y-h_y^-)}{f(y)}\right)\right),
$$

where Δy has one of the values (4.65) and b is the constant in the necessary condition

$$
2b=4t+\frac{4t^2+1}{\tau}
$$

for the existence of an invariant mesh.

Therefore, the obtained reduced systems of equations determine the optimal system of invariant solutions for the difference model of the liner heat equation. It means that each invariant solution can be found by a transformation of a solution from the optimal system with the help of the corresponding element of the group. As was mentioned before, the invariant time mesh for the new solution is obtained from the time mesh of the solution from the optimal system with the help of the same group transformation. For example, the transformation corresponding to the operator X_1 with parameter value $-t_0$ gives the time shift $\hat{t}=t-t_0$. Since the appropriate operator becomes

$$
X_*=Y_5+2t_0X_1,
$$

it follows that the action of this transformation takes the invariant solution with respect to the operator Y_5 into the solution invariant with respect to the operator X_*. By this transformation, the spacing $\frac{\tau}{t}=a$ is transformed into the spacing $\frac{\tau}{t+t_0}=a$.

EXAMPLE (of an exact solution). Among all group invariant solutions for the difference model (4.61), there is an interesting solution that can be integrated exactly [9]. This is the solution invariant with respect to the operator

$$
2t_0X_2+X_3, \qquad t_0=\text{const},
$$

namely, the solution

$$
u(x,t)=C\left(\frac{t_0}{t+t_0}\right)^{1/2}\exp\left(-\frac{x^2}{4(t+t_0)}\right) \tag{4.66}
$$

considered on the mesh

$$x_i = x_i^0 \left(\frac{t + t_0}{t_0} \right),$$
(4.67)

where the x_i^0 are the space mesh points at $t = t_0$. In the case of $t_0 = 0$, we obtain the well-known fundamental solution of the linear heat equation. Note that it has a "singular" mesh.

Let us show how this solution can be obtained from the optimal system of the invariant solutions. We consider the operator

$$X_* - X_2 + 2\varepsilon X_3,$$

and see that the solution (4.66) can be obtained from the solution invariant with respect to operator Y_1 by the transformation corresponding to X_* with $\varepsilon = 0.25t_0^{-1}$. If we take the original solution on the orthogonal mesh that is uniform in space and has the special time spacing

$$u_i^j = C, \qquad x_i = ih, \quad i = 0, \pm 1, \pm 2, \dots, \qquad t_j = \frac{j\tau t_0}{t_0 + j\tau}, \quad j = 0, 1, 2, \dots,$$

on the interval $[0, t_0]$, then the proposed transformation provides the solution (4.66) on the uniform space mesh (4.67) and the uniform time mesh $t_j = j\tau$.

Thus, we see that the difference model (4.61) inherits both the group admitted by the original differential equation and the integrability on a subgroup.

A way to stop a moving mesh

The obtained difference models have self-adaptive nonorthogonal moving meshes. We can find a way to stop the moving mesh, i.e., a change of variables that orthogonalizes the mesh. The differentiation operator d/dt of Lagrangian type can be represented in the form

$$\frac{d}{dt} = D_t - 2\frac{u_x}{u}D_x,$$

where

$$D_t = \frac{\partial}{\partial t} + u_t \frac{\partial}{\partial u} + \cdots, \qquad D_x = \frac{\partial}{\partial x} + u_x \frac{\partial}{\partial u} + \cdots.$$

The operator d/dt, in contrast to the operators D_t and D_x, does not commute with the operators of total differentiation with respect to t and x:

$$\left[\frac{d}{dt}, D_t \right] = 2 \left(\frac{u_{xt}}{u} - \frac{u_x u_t}{u^2} \right) D_x, \qquad \left[\frac{d}{dt}, D_x \right] = 2 \left(\frac{u_{xx}}{u} - \frac{u_x^2}{u^2} \right) D_x.$$

We need to find an operator of total differentiation with respect to a new space variable s such that

$$\left[\frac{d}{dt}, D_s \right] = 0.$$
(4.68)

The last commutativity property is possible if we involve a new dependent variable $\rho > 0$ (density). The operator $D_s = \rho^{-1} D_x$ satisfies (4.68) if ρ satisfies the equation

$$\rho_t - 2\rho \left(\frac{u_{xx}}{u} - \frac{u_x^2}{u^2} \right) - 2\frac{u_x}{u}\rho_x = 0.$$

The new space variable s is introduced with the help of the equations

$$s_t = 2\rho \frac{u_x}{u}, \qquad s_x = \rho.$$

For convenience, we can take the initial data $\rho(0, x) \equiv 1$. Then $s = x$ for $t = 0$.

In the variables (t, s), the heat equation becomes the system

$$u_t = \rho^2 \left(u_{ss} - 2\frac{u_s^2}{u} \right) + \rho\rho_s u_s, \quad \rho_t = 2\rho^3 \left(\frac{u_{ss}}{u} - \frac{u_s^2}{u^2} \right) + 2\rho^2 \rho_s \frac{u_s}{u}, \qquad (4.69)$$

which can be rewritten in the form of the conservation laws

$$\left(\frac{1}{\rho} \right)_t = \left(-2\rho \frac{u_s}{u} \right)_s, \qquad \left(\frac{u}{\rho} \right)_t = (-\rho u_s)_s.$$

The space coordinate x is defined by the system of equations

$$x_t = -2\rho \frac{u_s}{u}, \qquad x_s = \frac{1}{\rho}. \qquad (4.70)$$

System (4.69) in the space of independent variables (t, s) and the extended set of dependent variables (u, ρ, x) admits the point transformation group determined by the infinitesimal operators

$$X_1 = \frac{\partial}{\partial t}, \quad X_2 = \frac{\partial}{\partial x}, \quad X_3 = 2t\frac{\partial}{\partial x} - xu\frac{\partial}{\partial u}, \quad X_4 = 2t\frac{\partial}{\partial t} + x\frac{\partial}{\partial x} + s\frac{\partial}{\partial s},$$

$$X_5 = 4t^2 \frac{\partial}{\partial t} + 4tx\frac{\partial}{\partial x} - (x^2 + 2t)u\frac{\partial}{\partial u} - 4t\rho\frac{\partial}{\partial \rho}, \qquad X_6 = u\frac{\partial}{\partial u},$$

$$X^* = f(s)\frac{\partial}{\partial s} + \rho f'(s)\frac{\partial}{\partial \rho},$$

$$(4.71)$$

where $f(s)$ is an arbitrary function of s.

The condition of mesh orthogonality and the condition of spatial mesh uniformness are satisfied, and it gives us the opportunity to construct a difference model that is invariant with respect to the operators $X_1, \ldots, X_6$ on the orthogonal mesh.

Let us rewrite system (4.69),(4.70) in the form of differential invariants. In the space of the variables $(t, x, s, u, \rho, dt, dx, ds, du, d\rho, u_s, \rho_s, x_s, u_{ss})$, there are five invariants

$$J_1 = x_s \rho, \qquad J_2 = \frac{\rho}{ds}\left(dx + 2\rho\frac{u_s}{u}dt\right), \qquad J_3 = \frac{(ds)^2}{\rho^2 dt},$$

$$J_4 = \frac{(ds)^2}{\rho^3}\left(\frac{d\rho}{dt} - \frac{\rho_s ds}{dt} - 2\rho^3\left(\frac{u_{ss}}{u} - \frac{u_s^2}{u^2}\right) - 2\rho^2\rho_s\frac{u_s}{u}\right),$$

$$J_5 = \left(\frac{ds}{\rho}\right)^2\left(-\frac{2}{u}\frac{du}{dt} - \frac{1}{2}\left(\frac{dx}{dt}\right)^2 + 2\rho^2\left(\frac{u_{ss}}{u} - \frac{u_s^2}{u^2}\right) + 2\rho\rho_s\frac{u_s}{u}\right).$$

With the help of these invariants, we rewrite system (4.69), (4.70) as

$$u_t = \rho^2\left(u_{ss} - \frac{u_s^2}{u}\right) + \rho\rho_s u_s, \qquad \rho_t = 2\rho^3\left(\frac{u_{ss}}{u} - \frac{u_s^2}{u^2}\right) + 2\rho^2\rho_s\frac{u_s}{u},$$

$$x_t = -2\rho\frac{u_s}{u} \tag{4.72}$$

with the constraint equation $x_s = \rho^{-1}$.

Now we can find a system of equations that approximates (4.72) and is invariant with respect to the set of operators (4.71). We can use a six-point stencil which corresponds to the space

$$(t, \hat{t}, s, h_s^+, h_s^-, x, \hat{x}, h_x^+, h_x^-, \hat{h}_x^+, \hat{h}_x^-, u, \hat{u}, u_+, u_-, \hat{u}_+, \hat{u}_-, \rho, \hat{\rho}, \rho_+, \rho_-, \hat{\rho}_+, \hat{\rho}_-).$$

For the set of operators (4.71), where the operator X^* is replaced by its difference analog

$$X_h^* = f(s)\frac{\partial}{\partial s} + \rho \underset{+s}{D}(f(s))\frac{\partial}{\partial \rho}$$

we have the following set of invariants

$$I_1 = \frac{h_x^+}{h_x^-}, \quad I_2 = \frac{\hat{h}_x^+}{\hat{h}_x^-}, \quad I_3 = \frac{h_x^+ \hat{h}_x^+}{\tau}, \quad I_4 = \frac{\tau^{1/2}}{h_x^+}\frac{\hat{u}}{u}\exp\left(\frac{1}{4}\frac{\Delta x^2}{\tau}\right),$$

$$I_5 = \frac{1}{4}\frac{h_x^{+2}}{\tau} - \frac{h_x^{+2}}{h^+ + h^-}\left(\frac{1}{h_x^+}\ln\frac{u_+}{u} + \frac{1}{h_x^-}\ln\frac{u_-}{u}\right),$$

$$I_6 = \frac{1}{4}\frac{\hat{h}_x^{+2}}{\tau} + \frac{\hat{h}_x^{+2}}{\hat{h}_x^+ + \hat{h}_x^-}\left(\frac{1}{\hat{h}_x^+}\ln\frac{\hat{u}_+}{\hat{u}} + \frac{1}{\hat{h}_x^-}\ln\frac{\hat{u}_-}{\hat{u}}\right),$$

$$I_7 = \frac{\Delta x h_x^+}{\tau} + \frac{2h_x^+}{h_x^+ + h_x^-}\left(\frac{h_x^-}{h_x^+}\ln\frac{u_+}{u} - \frac{h_x^+}{h_x^-}\ln\frac{u_-}{u}\right),$$

$$I_8 = \frac{\Delta x \hat{h}_x^+}{\tau} + \frac{2\hat{h}_x^+}{\hat{h}_x^+ + \hat{h}_x^-}\left(\frac{\hat{h}_x^-}{\hat{h}_x^+}\ln\frac{\hat{u}_+}{\hat{u}} - \frac{\hat{h}_x^+}{\hat{h}_x^-}\ln\frac{\hat{u}_-}{\hat{u}}\right),$$

$$I_9 = \frac{\hat{\rho}_-}{\rho_-}, \quad I_{10} = \frac{\hat{\rho}}{\rho}, \quad I_{11} = \frac{\hat{\rho}_+}{\rho_+}, \quad I_{12} = \frac{h_s^+}{\rho h_x^+}, \quad I_{13} = \frac{h_s^-}{\rho_- h_x^-}.$$

With the help of these invariants, we can write out the difference model in the form of the following system of evolution difference equations (here we present only one invariant difference model, which corresponds to system (4.61) in the variables (t, x)):

$$\Delta x = 2\tau \frac{-\dfrac{h_s^-}{h_s^+}\dfrac{\rho}{\rho_-}\ln\dfrac{u_+}{u} + \dfrac{h_s^+}{h_s^-}\dfrac{\rho_-}{\rho}\ln\dfrac{u_-}{u}}{\dfrac{h_s^+}{\rho} + \dfrac{h_s^-}{\rho_-}}, \qquad \hat{\rho} = \rho\frac{h_x^+}{\hat{h}_x^+},$$

$$\left(\frac{u}{\hat{u}}\right)^2 \exp\left(-\frac{1}{2}\frac{\Delta x^2}{\tau}\right) = 1 - 4\tau\frac{\dfrac{\rho}{h_s^+}\ln\dfrac{u_+}{u} + \dfrac{\rho_-}{h_s^-}\ln\dfrac{u_-}{u}}{\dfrac{h_s^+}{\rho} + \dfrac{h_s^-}{\rho_-}}.$$

In the case of a uniform mesh ($h_s^+ = h_s^- = h_s$), this model can be simplified as follows:

$$\Delta x = 2\tau\frac{-\rho^2\ln\frac{u_+}{u} + \rho_-^2\ln\frac{u_-}{u}}{h_s(\rho + \rho_-)}, \qquad \hat{\rho} = \rho\frac{h_x^+}{\hat{h}_x^+},$$

$$\left(\frac{u}{\hat{u}}\right)^2\exp\left(-\frac{1}{2}\frac{\Delta x^2}{\tau}\right) = 1 - \frac{4\tau\rho\rho_-}{h_s^2(\rho + \rho_-)}\left(\rho\ln\frac{u_+}{u} + \rho_-\ln\frac{u_-}{u}\right).$$

System (4.69) has only two dependent variables u and ρ and can be approximated without involvement of the space variable x. However, the Galilei symmetry X_3 and the projective symmetry X_5 are nonlocal in the coordinate system (t, s), and we need to consider the dependent variable x to have these symmetries. When constructing a difference model invariant with respect to the set of operators (4.71), we inevitably include x in the difference equations.

It is important to note that moving meshes can be stopped in all cases by using a Lagrange type coordinate system. (For the introduction of Lagrange type coordinate systems, e.g., see [115].) Note that most of the obtained schemes are quite different from the "traditional" schemes [127].

4.2.2. Linear heat equation with a linear source

1. If $Q = \delta u$ and $\delta = \pm 1$, then the equation

$$u_t = u_{xx} + \delta u \tag{4.73}$$

can be transformed into Eq. (4.57) by the change of variables

$$\bar{u} = ue^{-\delta t}.$$

Reversing this transformation, one can get an invariant model for equation (4.73) from an invariant model for the heat equation without a source.

2. $Q = \delta = $ const. The equation has the form

$$u_t = u_{xx} + \delta. \tag{4.74}$$

The constant source can be eliminated by the obvious transformation

$$\bar{u} = u - \delta t.$$

It means that we can obtain a difference model for (4.74) from the model for (4.57).

4.3. Invariant Difference Models for the Burgers Equation

The Burgers equation

$$v_t + v v_x = v_{xx} \tag{4.75}$$

admits the transformation group determined by the infinitesimal operators (e.g., see [74])

$$X_1 = \frac{\partial}{\partial t}, \qquad X_2 = \frac{\partial}{\partial x}, \qquad X_3 = t\frac{\partial}{\partial x} + \frac{\partial}{\partial v},$$

$$X_4 = 2t\frac{\partial}{\partial t} + x\frac{\partial}{\partial x} - v\frac{\partial}{\partial v}, \qquad X_5 = t^2\frac{\partial}{\partial t} + tx\frac{\partial}{\partial x} + (x - tv)\frac{\partial}{\partial v}, \tag{4.76}$$

$$X^* = \left(a\exp\left(\frac{w}{2}\right)\right)_x \frac{\partial}{\partial v},$$

where $a = a(x, t)$ is an arbitrary solution of the heat equation $a_t = a_{xx}$. The function $w(t, x)$ is the *potential* of the Burgers equation and is introduced by the system

$$w_x = v, \qquad w_t = v_x - \frac{v^2}{2}.$$

The function $w(t, x)$ satisfies the equation

$$w_{tx} + w_x w_{xx} = w_{xxx},$$

which, after the integration, implies

$$w_t + \frac{w_x^2}{2} = w_{xx}.$$

This equation is often referred to as the potential Burgers equation.

Note that the first five operators describe the point symmetry of Eq. (4.75), while X^* is the nonlocal symmetry operator.

One can readily verify that the set of operators admitted by Eq. (4.75) preserves the uniformness of the spatial mesh structure but does not preserve the mesh uniformness in time. One can see that the transformation group of Eq. (4.75) does not preserve the mesh orthogonality.

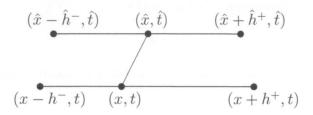

Figure 4.9

Thus, for the invariant difference modeling of Eq. (4.75) it is necessary to use meshes that are not orthogonal in (x, t) or find a "straightening" change of variables (which, of course, also transforms the equation itself).

One can see that all operators (4.76) preserve the flatness of the time layer. Thus, a preliminary analysis of the set of operators (4.76) permits concluding that, in the difference modeling of Eq. (4.75) which preserves the symmetry of the original model, it is possible to use a nonorthogonal mesh with parallel structure of the time layers (see Fig. 4.9).

For the Burgers equation (4.75), we construct several discrete models inheriting the transformation group (4.76) (see [8, 9]).

Let us rewrite the Burgers equation in terms of differential invariants. In the space of the variables $(t, x, v, v_x, v_{xx}, dt, dx, dv)$, there are three such invariants

$$J^1 = (dx - vdt)v_{xx}^{1/3}, \qquad J^2 = v_{xx}^{2/3}dt, \qquad J^3 = \frac{dv - v_x(dx - vdt)}{v_{xx}^{1/3}}.$$

This permits rewriting the Burgers equation as the system

$$J^1 = 0, \qquad J^2 = J^3,$$

or

$$\frac{dx}{dt} = v, \qquad \frac{dv}{dt} = v_{xx}. \tag{4.77}$$

Note that the structure of the group (4.76) forces us to use two evolution equations.

As the next step, we find the difference invariants of the set of point operators $X_1, \ldots, X_5$ of the group (4.76), which are required to approximate system (4.77). We consider the six-point stencil of the evolution difference mesh in Fig. 4.9.

This stencil, on which we approximate system (4.77), determines the space $(t, \hat{t}, x, \hat{x}, h^+, h^-, \hat{h}^+, \hat{h}^-, v, v_+, v_-, \hat{v}, \hat{v}_+, \hat{v}_-)$, where the group (4.76) has the following finite-difference invariants:

$$I^1 = \frac{\hat{h}^|}{h^-}, \qquad I^2 = \frac{\hat{h}^|}{\hat{h}^-}, \qquad I^3 = \frac{\hat{h}^| h^|}{\tau},$$

$$I^4 = h^- h^+ (v_x - v_{\bar{x}}), \qquad I^5 = \hat{h}^- \hat{h}^+ (\hat{v}_x - \hat{v}_{\bar{x}}), \qquad I^6 = h^+ \left(\frac{\Delta x}{\tau} - v \right),$$

$$I^7 = \hat{h}^+ \left(\frac{\Delta x}{\tau} - \hat{v} \right), \qquad I^8 = h^{+2} \left(\frac{1}{\tau} + v_x \right), \qquad I^9 = \hat{h}^{+2} \left(\frac{1}{\tau} + \hat{v}_x \right),$$

where

$$\Delta x = \hat{x} - x, \qquad v_x = \frac{v_+ - v}{h^+}, \qquad v_{\bar{x}} = \frac{v - v_-}{h^-}.$$

These invariants permit writing out the following invariant difference models.

A. A model with explicit approximation of the difference mesh evolution

1. Explicit scheme for the Burgers equation:

$$\Delta x = \tau v, \qquad \frac{\hat{v} - v}{\tau} \frac{\hat{h}^+ \hat{h}^-}{h^+ h^-} = v_{x\bar{x}} = \frac{2}{h^+ + h^-}(v_x - v_{\bar{x}}). \qquad (4.78)$$

2. Implicit scheme for the Burgers equation:

$$\Delta x = \tau v, \qquad \frac{\hat{v} - v}{\tau} \frac{h^+}{\hat{h}^+} = \hat{v}_{x\bar{x}} \equiv \frac{2}{\hat{h}^+ + \hat{h}^-}(\hat{v}_x - \hat{v}_{\bar{x}}).$$

3. Implicit scheme for the Burgers equation:

$$\Delta x = \tau v, \qquad \frac{\hat{v} - v}{\tau} = \frac{2}{h^+ + h^-}(\hat{v}_x - \hat{v}_{\bar{x}}).$$

B. A model with implicit approximation of the difference mesh evolution

1. Explicit scheme for the Burgers equation:

$$\Delta x = \tau \hat{v}, \qquad \frac{\hat{v} - v}{\tau} \frac{\hat{h}^+}{h^+} = v_{x\bar{x}}.$$

2. Implicit scheme for the Burgers equation:

$$\Delta x = \tau \hat{v}, \qquad \frac{\hat{v} - v}{\tau} \frac{h^+ h^-}{\hat{h}^+ \hat{h}^-} = \hat{v}_{x\bar{x}}.$$

3. Implicit scheme for the Burgers equation:

$$\Delta x = \tau \hat{v}, \qquad \frac{\hat{v} - v}{\tau} = \frac{2}{\hat{h}^+ + \hat{h}^-}(v_x - v_{\hat{x}}).$$

All above difference schemes have a moving nonorthogonal difference mesh. Let us find a change of variables rectifying the difference mesh.

In system (4.77), we have used the Lagrange type differentiation operator d/dt, which can be represented as

$$\frac{d}{dt} = D_t + vD_x,$$

where

$$D_t = \frac{\partial}{\partial t} + v_t \frac{\partial}{\partial v} + \cdots, \qquad D_x = \frac{\partial}{\partial x} + v_x \frac{\partial}{\partial v} + \cdots.$$

In contrast to D_t and D_x, the operator d/dt does not commute with the total differentiation operators in t and x,

$$\left[\frac{d}{dt}, D_x\right] = -v_x D_x, \qquad \left[\frac{d}{dt}, D_t\right] = -v_t D_x.$$

To "rectify" the coordinate system, it is necessary to find the spatial differentiation operator with respect to the variable s such that

$$\left[\frac{d}{dt}, D_s\right] = 0.$$

This property holds for the operator $D_s = \frac{1}{\rho}D_x$ with a new dependent variable ρ satisfying the equation

$$\rho_t + v\rho_x + \rho v_x = 0$$

and the condition $\rho > 0$ in the entire domain under study.

The function $\rho = \rho(t, x)$ is called the "initial data density." The new variable s is introduced by the system of equations

$$s_x = \rho, \qquad s_t = -\rho v.$$

In the variables (t, s), the Burgers equation can be rewritten in the form of the system

$$\frac{dv}{dt} = \rho^2 v_{ss} + \rho \rho_s v_s, \qquad \frac{d\rho}{dt} = -\rho^2 v_s. \qquad (4.79)$$

Now this system can be rewritten in the divergence form

$$\frac{d}{dt}\left(\frac{1}{\rho}\right) = D_s(v), \qquad \frac{d}{dt}\left(\frac{v}{\rho}\right) = D_s\left(\rho v_s + \frac{v^2}{2}\right).$$

Let us construct a difference analog of system (4.79) (also see [6]).

We supplement system (4.79) with the dependent variable x determined by the system of equations

$$x_t = v, \qquad x_s = \frac{1}{\rho}. \tag{4.80}$$

System (4.79), (4.80) admits the transformation group determined by the infinitesimal operators

$$X_1 = \frac{\partial}{\partial t}, \qquad X_2 = \frac{\partial}{\partial x}, \qquad X_3 = t\frac{\partial}{\partial x} + \frac{\partial}{\partial v},$$

$$X_4 = 2t\frac{\partial}{\partial t} + s\frac{\partial}{\partial s} + x\frac{\partial}{\partial x} - v\frac{\partial}{\partial v},$$

$$X_5 = t^2\frac{\partial}{\partial t} + tx\frac{\partial}{\partial x} + (x - tv)\frac{\partial}{\partial v} - t\rho\frac{\partial}{\partial \rho}, \tag{4.81}$$

$$X^* = f(s)\frac{\partial}{\partial s} + \rho f'(s)\frac{\partial}{\partial s},$$

where $f(s)$ is an arbitrary function of s.

In the new variables (t, s), the difference mesh orthogonality condition and the mesh uniformness condition in the spatial variable are satisfied for the operators $X_1, \ldots, X_5$, which are the operators of Lie algebra factorized by the operator X^*. In the variables (t, s), this permits constructing a difference model that is invariant under the operators $X_1, \ldots, X_5$ on an *orthogonal uniform* mesh.

We write out system (4.79), (4.80) in differential invariants, which have the following form in the space of dependent and independent variables, differentials, and spatial derivatives $(t, x, s, v, \rho, dt, dx, ds, dv, d\rho, v_s, \rho_s, x_s, v_{ss})$:

$$J^1 = x_s\rho, \quad J^2 = \frac{\rho}{ds}(dx - vdt), \quad J^3 = \left(\frac{ds}{\rho}\right)^3(\rho^2 v_{ss} + \rho\rho_s v_s), \quad J^4 = \frac{(ds)^2}{\rho^2 dt},$$

$$J^5 = \frac{(ds)^2}{\rho^3}\left(\frac{d\rho}{dt} - \frac{\rho_s ds}{dt} + \rho^2 v_s\right), \quad J^6 = \frac{ds}{\rho}(dv - \rho v_s(dx - vdt)).$$

This permits rewriting system (4.79), (4.80) as

$$\frac{dv}{dt} = \rho^2 v_{ss} + \rho\rho_s v_s, \qquad \frac{d\rho}{dt} = -\rho^2 v_s, \qquad \frac{dx}{dt} = v, \qquad \frac{\partial x}{\partial s} = \frac{1}{\rho}. \tag{4.82}$$

Let us find the difference system of four equations approximating (4.82) and invariant under the set of operators (4.81). We consider a six-point orthogonal mesh stencil with three points on two neighboring time layers (see Fig. 4.10).

Note that the new operator X^* in (4.81) is not related to the symmetry of the Burgers equation but is related to the extension of the space of dependent variables

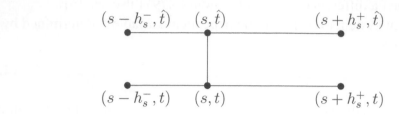

Figure 4.10

and the introduction of a new coordinate system. In the difference version, we replace it by the operator

$$X^* = f(s)\frac{\partial}{\partial s} + \rho D_s f(s)\frac{\partial}{\partial \rho},$$

which also determines an infinite symmetry but has different coefficients. In the space

$$(t, \hat{t}, s, h_s^+, h_s^-, x, \Delta x, h_x^+, h_x^-, \hat{h}_x^+, \hat{h}_x^-, v, v_+, v_-, \hat{v}, \hat{v}_+, \hat{v}_-, \rho, \rho_+, \rho_-, \hat{\rho}, \hat{\rho}_+, \hat{\rho}_-)$$

corresponding to the chosen stencil, we have the following 14 difference invariants:

$$I^1 = \frac{h_x^+}{h_x^-}, \quad I^2 = \frac{\hat{h}_x^+}{\hat{h}_x^-}, \quad I^3 = h_x^- h_x^+(v_x - v_{\bar{x}}), \quad I^4 = \hat{h}_x^- \hat{h}_x^+(\hat{v}_x - \hat{v}_{\bar{x}}),$$

$$I^5 = h_x^+\left(\frac{\Delta x}{\tau} - v\right), \quad I^6 = \hat{h}_x^+\left(\frac{\Delta x}{\tau} - \hat{v}\right), \quad I^7 = h_x^{+2}\left(\frac{1}{\tau} + v_x\right),$$

$$I^8 = \hat{h}_x^{+2}\left(\frac{1}{\tau} + \hat{v}_x\right), \quad I^9 = \frac{\hat{h}_x^+ h_x^+}{\tau}, \quad I^{10} = \frac{\hat{\rho}_-}{\rho_-},$$

$$I^{11} = \frac{\hat{\rho}}{\rho}, \quad I^{12} = \frac{\hat{\rho}_+}{\rho_+}, \quad I^{13} = \frac{h_s^+}{\rho h_x^+}, \quad I^{14} = \frac{h_s^-}{\rho_- h_x^-}.$$

These invariants permit writing out a difference model as the following system of difference evolution equations (we present one possible version of the difference scheme, namely, the explicit scheme; in the variables (t, x), this scheme is associated with the scheme (4.78)):

$$\Delta x = \tau v, \quad \frac{\hat{v} - v}{\tau} = \frac{2\hat{\rho}\hat{\rho}_-}{h_s^+ \hat{\rho}_- + h_s^- \hat{\rho}}(\rho v_s - \rho_- v_{\bar{s}}), \quad \hat{\rho}\hat{h}_x^+ = \rho h_x^+ = h_s^+.$$

If the mesh is uniform in s, $h_s^+ = h_s^- = h_s$, then this scheme acquires the form

$$\Delta x = \tau v, \quad \frac{\hat{v} - v}{\tau} = \frac{2\hat{\rho}\hat{\rho}_-}{\hat{\rho}_- + \hat{\rho}}(\rho v_s)_{\bar{s}}, \quad \hat{\rho}\hat{h}_x^+ = \rho h_x^+ = h_s.$$

The above families of difference equations and meshes, which are invariant under the complete set of operators admitted by the Burgers equation, allow us to consider solutions that are invariant under a certain one-parameter subgroup. In this case, the "scheme–mesh" system reduces to ordinary difference equations. We again point out that, in contrast to the original continuous case, *two* equations are reduced in our case, namely, the difference scheme and the difference equation for the mesh.

By way of example, consider the invariant solution of the difference model of the Burgers equation. For the subgroup we take the case corresponding to

$$\alpha X_2 + X_3 = (t + \alpha)\frac{\partial}{\partial x} + \frac{\partial}{\partial v}, \qquad \alpha = \text{const.} \tag{4.83}$$

Consider the simplest explicit scheme (4.78). In the subspace $(t, x, \Delta x, v)$, we have three invariants of the operators (4.83):

$$J_1 = t, \qquad J_2 = v - \frac{x}{t + \alpha}, \qquad J_3 = \frac{\Delta x}{\tau} - \frac{x}{t + \alpha}.$$

According to this, we seek the invariant solution in the form

$$v(x, t) = f(t) + \frac{x}{t + \alpha}, \qquad \frac{\Delta x}{\tau} = g(t) + \frac{x}{t + \alpha}.$$

By substituting such a solution into the scheme (4.78), we obtain a system of difference equations for $f(t)$ and $g(t)$,

$$f(t + \tau) = \frac{t + \tau}{t + \tau + \alpha} f(t), \qquad g(t) = f(t).$$

By integrating this system, we obtain the solution

$$v(x, t) = \frac{x}{t + \alpha} + \frac{f(0)\alpha}{t + \alpha}$$

of the difference equation for v and the solution

$$x = x^0 \left(\frac{t + \alpha}{\alpha}\right) + f(0)t$$

of the equation for the difference mesh. Here $x = x_i^j = x_i(t_j)$ and $t = t_j$. The mesh can be arbitrary at the initial time moment.

The solution thus obtained is a solution of the Cauchy problem with invariant initial conditions,

$$v(x, 0) = f(0) + \frac{x}{\alpha}.$$

Difference scheme for the potential Burgers equation

In addition to the Burgers equation (4.75), the following potential Burgers equation is also often considered:

$$w_t + \frac{1}{2}w_x^2 = w_{xx}. \tag{4.84}$$

The well-known Hopf transformation

$$v = -2\frac{u_x}{u}$$

relates the solutions of the Burgers equation (4.75) to the solutions of the heat equation

$$u_t = u_{xx}, \tag{4.85}$$

which was considered above. For Eqs. (4.84) and (4.85), this is a point relation,

$$w = -2\ln u. \tag{4.86}$$

The potential Burgers equation admits the point transformation group determined by the operators

$$X_1 = \frac{\partial}{\partial t}, \qquad X_2 = \frac{\partial}{\partial x}, \qquad X_3 = t\frac{\partial}{\partial x} + x\frac{\partial}{\partial w}, \qquad X_4 = 2t\frac{\partial}{\partial t} + x\frac{\partial}{\partial x},$$

$$X_5 = t^2\frac{\partial}{\partial t} + tx\frac{\partial}{\partial x} + \left(\frac{1}{2}x^2 + t\right)\frac{\partial}{\partial w}, \quad X_6 = \frac{\partial}{\partial w}, \quad X^* = a\exp\left(\frac{w}{2}\right)\frac{\partial}{\partial w},$$

where X^* is now a point symmetry operator.

It is of interest to note that the invariant difference model of the linear heat equation is related *by precisely the same Hopf transformations* to the invariant difference model of the Burgers equation. For example, the change of variables (4.86) reduces the explicit scheme

$$\Delta x = \frac{2\tau}{h^+ + h^-}\left(-\frac{h^-}{h^+}\ln\frac{u_+}{u} + \frac{h^+}{h^-}\ln\frac{u_-}{u}\right),$$

$$\left(\frac{u}{\hat{u}}\right)^2\exp\left(-\frac{1}{2}\frac{(\Delta x)^2}{\tau}\right) = 1 - \frac{4\tau}{h^+ + h^-}\left(\frac{1}{h^+}\ln\frac{u_+}{u} + \frac{1}{h^-}\ln\frac{u_-}{u}\right)$$

obtained above for the heat equation into the explicit moving mesh scheme

$$\Delta x = \frac{\tau}{h^+ + h^-}\left(\frac{h^-}{h^+}(w_+ - w) + \frac{h^+}{h^-}(w - w_-)\right),$$

$$e^{(\hat{w}-w)}\exp\left(-\frac{1}{2}\frac{(\Delta x)^2}{\tau}\right) = 1 + \frac{2\tau}{h^+ + h^-}\left(\frac{w_+ - w}{h^+} + \frac{w - w_-}{h^-}\right)$$

for the potential Burgers equation.

The implicit invariant schemes for the potential Burgers equation can be obtained precisely in the same way. Another approach to the discretization of the Burgers equation was developed in [67].

4.4. Invariant Difference Model of the Heat Equation with Heat Flux Relaxation

In addition to the above-considered parabolic models of heat transfer, let us consider the hyperbolic heat equation with heat flux relaxation taken into account ("hyperbolic heat transfer"):

$$\tau(u_x)u_{tt} + u_t = k(u_x)u_{xx}. \tag{4.87}$$

In particular, the group classification of this equation (see [74, p. 163]) contains the case

$$\tau_0 u_{tt} + u_t = k_0 u_x u_{xx},$$

where k_0 and τ_0 are positive constants.

In this case, the transformation group admitted by the equation is wider than that in the general case (4.87). This group is determined by the operators

$$X_1 = \frac{\partial}{\partial t}, \qquad X_2 = \frac{\partial}{\partial x}, \qquad X_3 = \frac{\partial}{\partial u},$$
$$X_4 = e^{\frac{t}{\tau_0}}\frac{\partial}{\partial u}, \qquad X_5 = x\frac{\partial}{\partial x} + 3u\frac{\partial}{\partial u}. \tag{4.88}$$

We use the method of difference invariants to construct a second-order explicit scheme for the approximation in the special case of Eq. (4.87). One can readily see that all five operators satisfy the uniformness invariance condition and preserve the mesh orthogonality, and hence we can use the rectangular mesh $\underset{h}{\omega}$, which is uniform in each direction.

In the space $(t, x, u, u_t, u_x, u_{tt}, u_{xx})$, the complete set of invariants of the five-parameter group (4.88) consists of two invariants, which, for example, can be chosen in the form

$$J_1 = \frac{\tau_0 u_{tt} + u_t}{u_x^{3/2}}, \qquad J_2 = \frac{u_{xx}}{u_x^{1/2}}. \tag{4.89}$$

The invariants (4.89) permit representing Eq. (4.87) in the invariant form

$$J_1 = k_0 J_2, \tag{4.90}$$

or

$$\frac{\tau_0 u_{tt} + u_t}{u_x^{3/2}} = k_0 \frac{u_{xx}}{u_x^{1/2}}.$$

We prolong the operators (4.88) to the difference derivatives $u_t, u_{\bar{t}}, u_x, u_{\bar{x}}, u_{t\bar{t}},$

and $\underset{h}{u_{x\bar{x}}}$:

$$X_1 = \frac{\partial}{\partial t}, \qquad X_2 = \frac{\partial}{\partial x}, \qquad X_3 = \frac{\partial}{\partial u},$$

$$X_4 = e^{-t/\tau_0}\frac{\partial}{\partial u} + \frac{1}{\tau}e^{-t/\tau_0}(e^{-\tau/\tau_0}-1)\frac{\partial}{\partial \underset{\tau}{u_t}} + \frac{1}{\tau}e^{-t/\tau_0}(1-e^{-\tau/\tau_0})\frac{\partial}{\partial \underset{\tau}{u_{\bar{t}}}}$$

$$+ \frac{1}{\tau^2}e^{-t/\tau_0}(e^{-\tau/\tau_0}-2+e^{\tau/\tau_0})\frac{\partial}{\partial \underset{\tau}{u_{t\bar{t}}}}, \tag{4.91}$$

$$X_5 = x\frac{\partial}{\partial x} + 3u\frac{\partial}{\partial u} + 2\underset{h}{u_x}\frac{\partial}{\partial \underset{h}{u_x}} + 2\underset{h}{u_{\bar{x}}}\frac{\partial}{\partial \underset{h}{u_{\bar{x}}}} + 3\underset{\tau}{u_t}\frac{\partial}{\partial \underset{\tau}{u_t}} + 3\underset{\tau}{u_{\bar{t}}}\frac{\partial}{\partial \underset{\tau}{u_{\bar{t}}}}$$

$$+ \underset{h}{u_x}\frac{\partial}{\partial \underset{h}{u_x}} + 3\underset{\tau}{u_{t\bar{t}}}\frac{\partial}{\partial \underset{\tau}{u_{t\bar{t}}}} + h\frac{\partial}{\partial h}.$$

In the space $(x, t, u, \underset{\tau}{u_t}, \underset{\tau}{u_{\bar{t}}}, \underset{h}{u_x}, \underset{h}{u_{\bar{x}}}, \underset{\tau}{u_{t\bar{t}}}, \underset{h}{u_{x\bar{x}}}, \tau, h)$, the group (4.88) has a complete set of six invariants. But it suffices for us to choose only two finite-difference invariants approximating (4.90) up to $O(\tau^2 + h^2)$.

One can show that the second-order finite-difference forms

$$I_1 = \left(\frac{1}{2}\underset{h}{u_x} + \frac{1}{2}\underset{h}{u_{\bar{x}}}\right)^{-3/2}\left\{\tau_0\underset{\tau}{u_{t\bar{t}}} + \frac{\tau_0}{2\tau}\left[\underset{\tau}{u_t}\left(e^{\tau/\tau_0}-1\right) + \underset{\tau}{u_{\bar{t}}}\left(1-e^{-\tau/\tau_0}\right)\right]\right\}$$

$$I_2 = \underset{h}{u_{x\bar{x}}}\left(\frac{1}{2}\underset{h}{u_x} + \frac{1}{2}\underset{h}{u_{\bar{x}}}\right)^{-1/2}$$

are invariants of all operators (4.91) and approximate J_1 and J_2 up to $O(\tau^2 + h^2)$:

$$I_1 = J_1 + O(\tau^2 + h^2), \qquad I_2 = J_2 + O(h^2).$$

Substituting them into the invariant representation (4.90) for J_1 and J_2, we obtain the invariant equation

$$I_1 = k_0 I_2,$$

which is equivalent to the difference scheme

$$\tau_0\underset{\tau}{u_{t\bar{t}}} + \frac{1}{2}\frac{\tau_0}{\tau}\left[\underset{\tau}{u_{\bar{t}}}\left(1-e^{-\frac{\tau}{\tau_0}}\right) + \underset{\tau}{u_t}\left(e^{\frac{\tau}{\tau_0}}-1\right)\right] = \frac{k_0}{2}\underset{h}{u_{x\bar{x}}}(\underset{h}{u_x} + \underset{h}{u_{\bar{x}}}). \tag{4.92}$$

The finite-difference equation (4.92) admits all five operators (4.91) and approximates the differential equation (4.87) up to $O(\tau^2 + h^2)$ on the uniform rectangular mesh $\underset{h}{\omega}$.

4.5. Invariant Difference Model of the Korteweg–de Vries Equation

It is well known [79] that the KdV equation

$$u_t = uu_x + u_{xxx} \qquad (4.93)$$

admits the four-parameter Lie point transformation group with infinitesimal operators[1]

$$X_1 = \frac{\partial}{\partial t}, \quad X_2 = \frac{\partial}{\partial x}, \quad X_3 = t\frac{\partial}{\partial x} - \frac{\partial}{\partial u}, \quad X_4 = x\frac{\partial}{\partial x} + 3t\frac{\partial}{\partial t} - 2u\frac{\partial}{\partial u}. \quad (4.94)$$

Before constructing a difference equation and a mesh that approximate (4.93) with given order of approximation and inherit the whole Lie algebra (4.94), we should verify conditions for the invariance of the mesh geometry structure. One can readily confirm that the operators X_1, X_2, and X_4 preserve the orthogonality, but X_3 gives

$$D_{+\tau}(t) \neq -D_{+h}(0),$$

where τ is the step in the t-direction and h is the step in the x-direction. Consequently, *an orthogonal mesh cannot be used for the invariant modeling of* (4.93). Thus, we should seek an invariant moving mesh scheme.

The next question arising in this situation is the possibility of using a nonorthogonal moving mesh with flat time layers. The condition of invariance of the time layer flatness is satisfied for the complete set of operators (4.94),

$$D_{+h}D_{+\tau}(\xi_\alpha^t) = 0, \qquad \alpha = 1, 2, 3, 4.$$

To approximate (4.93) on such a mesh, we have to use at least four points in the x-direction and two points in the other direction. We shall use the minimum difference stencil of explicit type, as shown in Fig. 4.11.

In accordance with this stencil, we have a subspace of difference variables where Eq. (4.93) can be represented. In the subspace $(x, \hat{x}, t, \tau, h^-, h^+, h^{++}, u, \hat{u}, u^+, u^{++}, u^-)$, we have eight difference invariants, the set of which can be the following:

$$J_1 = \frac{\hat{x} - x + \tau u}{h^+}, \qquad J_2 = (\hat{u} - u)(h^+)^2, \qquad J_3 = \tau \underset{h}{u_x} \equiv \tau \frac{u^+ - u}{h^+},$$

$$J_4 = \tau \underset{h}{u_x^-} \equiv \tau \frac{u - u^-}{h^-}, \qquad J_5 = \tau \underset{h}{u_x^+} \equiv \tau \frac{u^{++} - u^+}{h^{++}}, \qquad (4.95)$$

$$J_6 = \frac{(h^+)^3}{\tau}, \qquad J_7 = \frac{h^-}{h^+}, \qquad J_8 = \frac{h^{++}}{h^+}.$$

---

[1]Equation (4.93) also admits well-known series of higher-order and nonlocal symmetries, which we do not consider here.

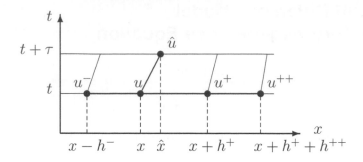

Figure 4.11

To generate an invariant difference mesh, we should use the set (4.95), which has the only invariant J_1 involving the coordinate $\hat{x}$ of the point x on the next time layer. The mesh generating equation in general has the form

$$J_1 = \Phi(J_2, J_3, J_4, J_5, J_6, J_7, J_8),$$

where Φ is an arbitrary function.

We shall use the simplest form

$$J_1 = 0, \quad \text{or} \quad \hat{x} = x - \tau u,$$

for describing this evolution mesh. Using the set (4.95), we can construct the following invariant difference equation:

$$J_2 = 2\frac{J_5 - J_3}{J_8 + 1} - 2\frac{J_3 - J_4}{J_7 + 1},$$

The meaning of the last equation will shortly become clear.

Thus, we have two evolution equations,

$$\frac{\Delta x}{\tau} = -u, \quad \text{where } \Delta x \equiv \hat{x} - x \tag{4.96}$$

and

$$\frac{\hat{u} - u}{\tau} = \frac{1}{h^+}\left\{\left(\frac{u^{++} - u^+}{h^{++}} - \frac{u^+ - u}{h^+}\right)\frac{2}{h^{++} + h^+} \right.$$
$$\left. - \left(\frac{u^+ - u}{h^+} - \frac{u - u^-}{h^-}\right)\frac{2}{h^+ + h^-}\right\}. \tag{4.97}$$

Let us return to the continuous space so as to understand more clearly what the discrete model (4.96)–(4.97) is. In the continuous limit $h \to 0$, $\tau \to 0$, from (4.96), (4.97) we have

$$\frac{dx}{dt} = -u, \qquad \frac{du}{dt} = u_{xxx}. \tag{4.98}$$

Now let us introduce the following new operator of differentiation with respect to time:

$$\frac{d}{dt} = D_t - uD_x, \tag{4.99}$$

where D_t and D_x are the operators of total differentiation with respect to t and x, respectively,

$$D_t = \frac{\partial}{\partial t} + u_t \frac{\partial}{\partial u} + u_{xt} \frac{\partial}{\partial u_x} + u_{tt} \frac{\partial}{\partial u_t} + \cdots,$$

$$D_x = \frac{\partial}{\partial x} + u_x \frac{\partial}{\partial u} + u_{tx} \frac{\partial}{\partial u_t} + u_{xx} \frac{\partial}{\partial u_x} + \cdots.$$

The operator (4.99) can be viewed as the Lagrangian operator of differentiation with respect to time. By applying (4.99) to x and u, we obtain

$$\frac{dx}{dt} = -u, \qquad \frac{du}{dt} = u_t - uu_x. \tag{4.100}$$

Consequently, system (4.98) is equivalent to the KdV equation (4.93). These equations provide the relation between Eq. (4.93) in the Cartesian coordinate system and system (4.98) in the Lagrange coordinate system. The transformation (4.100) means that we now have the pair of operators $(\frac{d}{dt}, D_x)$ instead of (D_t, D_x).

Let us slightly transform the difference model (4.96), (4.97) so as to increase the order of approximation.

By using the full set of difference invariants (4.95), we can add some terms to Eqs. (4.96), (4.97) without destroying their invariance. We would like to symmetrize the scheme as follows:

$$\hat{x} = x + \frac{h^+}{2} - \tau \frac{u + u^+}{2}, \qquad \hat{u} = \frac{u + u^+}{2} + \tau \underset{h}{u_{x\bar{x}x}}, \tag{4.101}$$

where

$$\underset{h}{u_{x\bar{x}x}} \equiv \frac{1}{h^+} \left(\frac{u^{++} - u^+}{h^{++}} - \frac{u^+ - u}{h^+} \right) \frac{2}{(h^{++} + h^+)}$$

$$- \frac{1}{h^+} \left(\frac{u^+ - u}{h^+} - \frac{u - u^-}{h^-} \right) \frac{2}{(h^+ + h^-)}.$$

The scheme (4.101) means that we use the stencil presented in Fig. 4.12.

One can readily estimate the order of approximation of the schemes (4.101) and (4.96), (4.97) on a uniform mesh with $h^{++} = h^+ = h^- = h$. For the scheme (4.96), (4.97), the order is $O(\tau + h)$, and the scheme (4.101) is of order $O(\tau + h^3)$. To estimate the order of approximation on a nonuniform mesh, one needs to use appropriate norms in a space of difference variables (see [122, 125]).

From the algebraic point of view, the schemes (4.96), (4.97) and (4.101) are similar, both being invariant under the operators (4.94). We point out that although

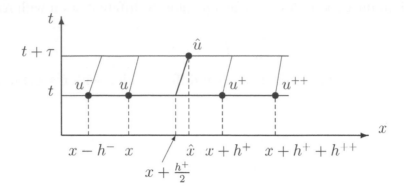

Figure 4.12

the invariant mesh and scheme for the KdV system (4.98) are not unique, all other invariant meshes and difference equations can be constructed using difference invariants.

Consider the symmetry reduction and the invariant solution of the difference equations (4.96), (4.97).

The operator $X_3 = t\partial/\partial x - \partial/\partial u$ destroys the mesh orthogonality, and so this symmetry cannot be preserved in a discrete model on an orthogonal mesh in the original coordinate system. It is thus of interest to verify whether the Galilei symmetry is respected by (4.96), (4.97) (or (4.101)).

The operator X_3 has two invariants, $J_1 = t$ and $J_2 = u + x/t$, in the subspace (x, t, u), and so we shall seek invariant solutions in the form

$$u(x,t) = v(t) - \frac{x}{t} \tag{4.102}$$

Reducing both Eqs.(4.96) and (4.97) using the invariant form (4.102), we obtain

$$\frac{\hat{x} - x}{\tau} = \frac{x}{t} - v(t), \qquad \frac{\hat{x}}{t + \tau} - \frac{x}{t} = v(t + \tau) - v(t). \tag{4.103}$$

The solution of system (4.103) is $x = at + c$, $v(t) = c/t$ (see the appropriate mesh on Fig. 4.13), and it gives the solution

$$x = at + c, \qquad u(x,t) = \frac{c - x}{t}$$

of the original system (4.96), (4.97). This solution coincides with the solution of the differential KdV system (4.98).

Thus, (4.96) corresponds to a family of meshes that are *self-adapted to the symmetries of subgroups*.

System (4.98) does indeed correspond to the KdV equation, but it seems impossible to establish conservation laws for (4.98) in this Lagrangian coordinate system.

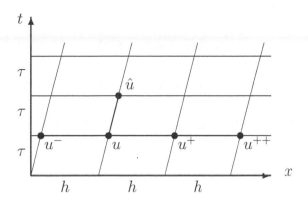

Figure 4.13

It is well known (see [74] and references therein) that, as a consequence of the symmetry group (4.94), the KdV equation possesses the following set of conservation laws:

$$D_t(u) - D_x\left(\frac{u^2}{2} + u_{xx}\right) = 0, \qquad D_t(u^2) + D_x\left(u_x^2 - 2uu_{xx} - \frac{2}{3}u^3\right) = 0,$$

$$D_t\left(t\frac{u^2}{2} + xu\right) + D_x\left[t\left(\frac{u_x^2}{2} - uu_{xx} - \frac{u^3}{3}\right) + u_x - xu_{xx} - x\frac{u^2}{3}\right] = 0,$$

$$D_t(u^3 - 3u_x^2) + D_x\left(6u_tu_x - 3u_{xx}^2 - 3u^2u_{xx} - \frac{3}{4}u^4\right) = 0.$$

To construct the corresponding set of conservation laws for (4.98), we need a divergence-like operator in the Lagrangian coordinate system.

Now we have the following commutators for d/dt, D_x, and D_t:

$$\left[\frac{d}{dt}, D_t\right] = u_t D_x, \qquad \left[\frac{d}{dt}, D_x\right] = u_x D_x,$$

and we would like to change our coordinate system so as to have a pair of differential operators that commute,

$$\left[\frac{d}{dt}, D_s\right] = 0,$$

where $s = s(t,x)$ is a new independent coordinate that obeys the condition $s_t - us_x = 0$.

Introducing the operator

$$D_s = \frac{1}{\rho}D_x, \qquad D_s = \frac{\partial}{\partial s} + u_s\frac{\partial}{\partial u} + \rho_s\frac{\partial}{\partial \rho} + \cdots,$$

we find that the new dependent variable ρ satisfies the equation

$$\rho_t - u_x\rho - u\rho_x = 0$$

and the condition $\rho > 0$, which means that we have no "vacuum gap" between x and s.

Thus, we change the coordinate system $\{t, x, u(t, x)\}$ to $\{t, s, u(t, s), \rho(t, s)\}$ by the transformation

$$ds = \rho\, dx + \rho u\, dt, \qquad u(t, x) = u(t, s), \qquad \frac{dx}{dt} = -u, \qquad \frac{\partial x}{\partial s} = \frac{1}{\rho}.$$

In the new coordinate system, we have the differential equations

$$\frac{du}{dt} = \rho^3 u_{sss} + 3\rho^2\rho_s u_{ss} + (\rho\rho_s^2 + \rho^2\rho_{ss})u_{ss}, \qquad \frac{d\rho}{dt} = \rho^2 u_s. \qquad (4.104)$$

System (4.104) has the conservation laws

$$\frac{d}{dt}\left(\frac{1}{\rho}\right) + D_s(u) = 0, \qquad \frac{d}{dt}\left(\frac{u}{\rho}\right) + D_s\left(\frac{u^2}{2} - \rho^2 u_{ss} + \rho\rho_s u_s\right) = 0. \quad (4.105)$$

Note that we can add the variable x to this coordinate system as some type of potential and simultaneously add the following equations to (4.104):

$$\frac{dx}{dt} = -u, \qquad \frac{\partial x}{\partial s} = \frac{1}{\rho}. \qquad (4.106)$$

Finally, for system (4.104), (4.106) we have the conservation laws (4.105) and the additional law

$$\frac{d}{dt}\left(\frac{tu^2/2 + xu}{\rho}\right)$$
$$+ D_s\left(t(\frac{\rho^2 u_s^2}{2} - \rho u(\rho u_s)_s + \frac{u^3}{6}) + \rho u_s - x\rho(\rho u_s)_s + \frac{xu^2}{2}\right) = 0, \quad (4.107)$$

which corresponds to the Galilei symmetry.

It is not difficult to construct an invariant scheme and mesh for system (4.104) in the same way as we did for the linear heat equation. But constructing an invariant difference scheme for system (4.104), (4.106) preserving the whole set of conservation laws (4.105), (4.107) seems to be a rather complicated task.

In the paper [50] there were constructed invariant difference schemes for Korteweg–de Vries equations with *variable coefficients*.

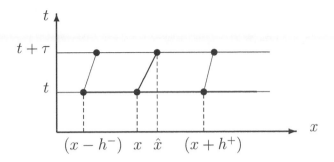

Figure 4.14

4.6. Invariant Difference Model of the Nonlinear Schrödinger Equation

The Schrödinger equation

$$iE_t + E_{xx} + E|E|^2 = 0 \tag{4.108}$$

with cubic nonlinearity admits a four-parameter point transformation group (e.g., see [74]).[2]

First, we perform a change of variables to pass to two real dependent variables:

$$E = A(t,x)e^{i\Phi(t,x)}.$$

The system of equations

$$A_t + 2A_x\Phi_x + A\Phi_{xx} = 0, \qquad A\Phi_t + 2A\Phi_x^2 - A_{xx} - A^3 = 0, \tag{4.109}$$

equivalent to (4.108), admits the operators

$$X_1 = \frac{\partial}{\partial t}, \qquad X_2 = \frac{\partial}{\partial x}, \qquad X_3 = \frac{\partial}{\partial \Phi}, \qquad X_4 = 2t\frac{\partial}{\partial x} + x\frac{\partial}{\partial \Phi},$$

$$X_5 = 2t\frac{\partial}{\partial t} + x\frac{\partial}{\partial x} - A\frac{\partial}{\partial A}.$$

Just as in a majority of the evolution equations considered above, the symmetry of system (4.109) does not permit using an orthogonal mesh; namely, the operator X_4 violates the mesh orthogonality but preserves the flatness of the time layer. Thus, we again should use a mesh evolving in time (see Fig. 4.14).

Let us construct explicit invariant difference equations for calculating the mesh and the solution A, Φ (see [31, 34]). In the subspace

$$(x, \hat{x}, t, \tau, h^+, h^-, \Phi, \Phi^+, \Phi^-, \hat{\Phi}, A, A^+, A^-, \hat{A})$$

---

[2]Equation (4.108) also admits an infinite series of higher-order and nonlocal symmetries (e.g., see [74]), which we do not consider here.

corresponding to the explicit stencil, there are $14 - 5 = 9$ difference invariants

$$\frac{\Delta x - 2\tau \Phi_x}{h^+}, \quad \frac{\Delta x - 2\tau \Phi_{\bar x}}{h^-}, \quad \frac{h^+}{h^-}, \quad \frac{\tau}{(h^+)^2}, \quad \frac{A}{A^+}, \quad \frac{A}{A^3}; \quad \frac{A}{\hat A},$$

$$\tau a^2, \quad \frac{(\Delta x)^2 - 4\tau(\hat \Phi - \Phi)}{(h^+)^2}, \tag{4.110}$$

where $\Delta x = \hat x - x$.

We choose the evolution mesh in the following invariant representation:

$$\Delta x = 2\tau \left(\frac{h^-}{h^+ + h^-} \Phi_x + \frac{h^+}{h^+ + h^-} \Phi_{\bar x} \right). \tag{4.111}$$

The continual limit implies the equation

$$\frac{dx}{dt} = 2\Phi_x. \tag{4.112}$$

Equation (4.112) forces us to write out system (4.109) as

$$\frac{dA}{dt} = -A\Phi_{xx}, \quad \frac{d\Phi}{dt} = \Phi_x^2 + \frac{A_{xx}}{A} + A^2, \tag{4.113}$$

where the Lagrangian differentiation operator

$$\frac{d}{dt} = D_t + 2\Phi_x D_x$$

has been introduced.

System (4.113) can be approximated with the use of the invariants (4.110) on the mesh (4.111) by the following explicit scheme:

$$\frac{\hat A - A}{\tau} + \frac{2A}{h^+ + h^-}(\Phi_x - \Phi_{\bar x}) = 0,$$

$$\left(\frac{\hat \Phi - \Phi}{\tau} - \left(\frac{h^-}{h^+ + h^-} \Phi_x + \frac{h^+}{h^+ + h^-} \Phi_{\bar x} \right)^2 \right) A - 2\frac{A_x - A_{\bar x}}{h^+ + h^-} - A^3 = 0. \tag{4.114}$$

We can show that the invariant difference model (4.111), (4.114) approximates system (4.112)–(4.113) up to $O(\tau + h^2)$ on a spatially uniform mesh. In a similar way, we can construct implicit and explicit-implicit invariant schemes and their versions in the Lagrangian type coordinate system.

Chapter 5

Combined Mathematical Models and Some Generalizations

In the last decades, the classical Lie group analysis has been considerably extended to such mathematical models as integro-differential equations, stochastic differential equations, functional differential equations, etc. This chapter deals with applications of Lie transformation groups to equations that contain difference and differential variables.

5.1. Second-Order Ordinary Delay Differential Equations

Delay ordinary differential equations are similar to ordinary differential equations, but they contain values of solutions as well as of derivatives at earlier instants of time. Many mathematical models based on delay ordinary differential equations have wide applications in biology, physics, engineering (see [51, 78, 103]), etc.

Here we consider a second-order delay ordinary differential equation of the type [117]

$$u'' = F(x, u, u', u_-, u'_-), \tag{5.1}$$

where x is the independent variable, u is a dependent variable, $u_- = u(x - \tau)$, $u'_- = u'(x - \tau)$, $x_- = x - \tau$, and τ is a parameter.

Thus, we actually have a differential-difference equation in the subspace of the variables $(x, u, u', x_-, u_-, u'_-)$. Equation (5.1) is of second order in differential variables and of first order as a difference equation. All notation is clearly seen from Fig. 5.1.

According to [117], the delay parameter τ is considered to be constant. Following our point of view, we supplement (5.1) with the simple lattice equation

$$\tau = x - x_- = \text{const} \tag{5.2}$$

to complete the delay model.

The symmetry generator admitted by the delay ordinary differential equation model (5.1),(5.2) has the form

$$X = \xi(x, u)\frac{\partial}{\partial x} + \eta(x, u)\frac{\partial}{\partial u},$$

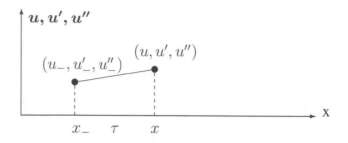

Figure 5.1

which should be prolonged for the derivatives and delay variables. As to the difference equations, the coordinates of the point transformation operator can be simply shifted to the left stencil point as follows:

$$\tilde{X} = \xi(x, u)\frac{\partial}{\partial x} + \eta(x, u)\frac{\partial}{\partial u} + \zeta_1\frac{\partial}{\partial u'} + \zeta_2\frac{\partial}{\partial u''} + \eta^-(x_-, u_-)\frac{\partial}{\partial u_-}$$
$$+ \zeta_1^-\frac{\partial}{\partial u'_-} + (\xi(x, u) - \xi^-(x_-, u_-))\frac{\partial}{\partial \tau}, \quad (5.3)$$

where

$$\zeta_1 = D(\eta) - u'D(\xi) = \eta_x(x, u) + u'\eta_u(x, u) - u'\xi_x(x, u) - (u')^2\xi_u(x, u),$$
$$\zeta_2 = D(\zeta_1) - u''D(\xi) = D^2(\eta) - 2u''D(\xi) - u'D^2(\xi),$$
$$\zeta_1^- = \underset{-h}{S}(\zeta_1) = \eta_x^-(x_-, u_-) + u'_-\eta_u^-(x_-, u_-) - u'_-\xi_x^-(x_-, u_-)$$
$$- (u'_-)^2\xi_u^-(x_-, u_-),$$
$$D = \frac{\partial}{\partial x} + u'\frac{\partial}{\partial u} + u''\frac{\partial}{\partial u'} + u'''\frac{\partial}{\partial u''} + \cdots,$$

and $\underset{-h}{S}$ is the left shift operator. Note that the prolongation for the variable τ is the same as for the lattice step in the discrete case.

To derive the determining equations, we apply the operator (5.3) to Eq. (5.1):

$$D^2(\eta) - 2u''D(\xi) - u'D^2(\xi) = F_x\xi + F_u\eta + F_{u_-}\eta^- + F_{u'}\zeta_1 + F_{u'_-}\zeta_1^-,$$

or, in particular,

$$\eta_{xx} + 2\eta_{xu}u' + \eta_{uu}(u')^2 + \eta_u u'' - 2u''(\xi_x + \xi_u u') - u'(\xi_{xx} + 2\xi_{xu}u' + \xi_{uu}(u')^2 + \xi_u u'')$$
$$= F_x\xi + F_u\eta + F_{u_-}\eta^- + F_{u'}\left(\eta_x(x, u) + u'\eta_u(x, u) - u'\xi_x(x, u) - (u')^2\xi_u(x, u)\right)$$
$$+ F_{u'_-}\left(\eta_x^-(x_-, u_-) + u'_-\eta_u^-(x_-, u_-) - u'_-\xi_x^-(x_-, u_-) - (u'_-)^2\xi_u^-(x_-, u_-)\right).$$
$$(5.4)$$

Now we should substitute u'' from (5.1) into (5.4):

$$\eta_{xx} + 2\eta_{xu}u' + \eta_{uu}(u')^2 + \eta_u F - 2F(\xi_x + \xi_u u') - u'(\xi_{xx} + 2\xi_{xu}u' + \xi_{uu}(u')^2 + \xi_u F)$$
$$= F_x\xi + F_u\eta + F_{u_-}\eta^- + F_{u'}\left(\eta_x(x,u) + u'\eta_u(x,u) - u'\xi_x(x,u) - (u')^2\xi_u(x,u)\right)$$
$$+ F_{u'_-}\left(\eta_x^-(x_-, u_-) + u'_-\eta_u^-(x_-, u_-) - u'_-\xi_x^-(x_-, u_-) - (u'_-)^2\xi_u^-(x_-, u_-)\right). \tag{5.5}$$

To complete the determining equations, we apply the operator (5.3) to Eq. (5.2):

$$\xi(x, u) - \xi^-(x_-, u_-) = 0. \tag{5.6}$$

Thus, Eqs. (5.5) and (5.6) are the determining equations for the delay model (5.1), (5.2). Regardless of the function F in (5.1), one can conclude from (5.6) that $\xi(x, u)$ is independent of u. Indeed, the differentiation of (5.6) with respect to u and u_- yields

$$\xi_u = \xi_{u_-}^- = 0.$$

Consequently, $\xi(x)$ is a periodic function of x,

$$\xi(x) = \xi(x - \tau).$$

In the paper [117] (see also [101, 133]), the complete Lie group classification of Eq. (5.1) was presented. The classification was done in a way similar to that used by Lie himself for second-order ordinary differential equations. Starting from the list of all Lie algebras over the real plane [63], there were singled out 40 distinct classes of delay second-order ordinary differential equations. Note that the classification list in [117] is much longer than that for second-order ordinary differential equations. A similar situation was shown to hold for second-order ordinary difference equations [47].

To illustrate the results, below we reproduce an excerpt from [117].

EXAMPLE. The three-dimensional Lie algebra spanned by the operators

$$X_1 = \frac{\partial}{\partial x}, \qquad X_2 = 2x\frac{\partial}{\partial x} + u\frac{\partial}{\partial u}, \qquad X_3 = x^2\frac{\partial}{\partial x} + xu\frac{\partial}{\partial u}$$

has the following complete set of invariants in the subspace $(x, u, u', u_-, u'_-, u'')$:

$$\left\{ u''u^3, \quad \frac{u_-}{u}, \quad u'u_-\left(\frac{u'_-}{u'} - \frac{u_-}{u}\right) \right\}.$$

Consequently, the general form of an invariant delay equation is

$$u'' = u^{-3} f\left(\frac{u_-}{u}, u'u_-\left(\frac{u'_-}{u'} - \frac{u_-}{u}\right)\right), \tag{5.7}$$

where f is an arbitrary smooth function. In the special case $f = 1$, Eq. (5.7) degenerates into

$$u'' = u^{-3},$$

which is nothing else than an ordinary differential equation from the Lie list of invariant equations.

Remark. Apparently, the list of invariant delay ordinary differential equations can be extended if, instead of (5.2), one uses the more general lattice equation

$$\tau = G(x, x_-, u, u', u_-, u'_-, u'', u''_-),$$

which means that $\tau = x - x_-$ is included in the set $(x, x_-, u, u_-, u', u'_-, u'', u''_-, \tau)$ of variables on which the transformation group acts.

5.2. Partial Delay Differential Equations

In this section, we deal with delay partial differential equations. Namely, we consider Lie group properties of the semi-linear delay heat equation with a source (the reaction-diffusion equation) [102]:

$$u_t = u_{xx} + g(u, \check{u}), \tag{5.8}$$

where $u = u(t, x)$, $\check{u} = u(t - \tau, x)$, and $\tau = \text{const}$ is a delay parameter.

The theory of existence of solutions of (5.8) can be found in [138]. The complete group classification with an arbitrary element g was done in [102].

The delay parameter τ is considered to be a constant [102], and x is not changed for the time $t - \tau$, which corresponds to the lattice equations

$$\tau = t - \check{t} = \text{const}, \qquad x = \check{x}. \tag{5.9}$$

Consider a Lie transformation group acting in the subspace of the variables $(x, t, t - \tau, u, \check{u}, u_x, u_t, \check{u}_x, u_{xx}, \check{u}_{xx})$, where u and $\check{u}$ are two dependent variables. Since Eq. (5.8) is a partial differential equation of second order and a first-order difference equation, it follows that the symmetry generator

$$X = \xi^t(t, x, u)\frac{\partial}{\partial t} + \xi^x(t, x, u)\frac{\partial}{\partial x} + \eta(t, x, u)\frac{\partial}{\partial u} \tag{5.10}$$

should be prolonged for the derivatives and the delay variables occurring in (5.8),

$$\widetilde{X} = \xi^t\frac{\partial}{\partial t} + \xi^x\frac{\partial}{\partial x} + (\xi^t - \check{\xi}^t)\frac{\partial}{\partial \tau} + \eta\frac{\partial}{\partial u} + \check{\eta}\frac{\partial}{\partial \check{u}} + \zeta_x\frac{\partial}{\partial u_x} + \zeta_t\frac{\partial}{\partial u_t} + \zeta_{xx}\frac{\partial}{\partial u_{xx}},$$

where

$$\zeta_x = D_x(\eta) - u_x D_x(\xi^x) - u_t D_x(\xi^t) \qquad \zeta_t = D_t(\eta) - u_x D_t(\xi^x) - u_t D_t(\xi^t),$$
$$\zeta_{xx} = D_x(\zeta_x) - u_{xx} D_x(\xi^x) - u_{xt} D_x(\xi^t)$$

are the standard prolongation formulas,

$$\ddot{\xi}^t = \xi^t(x, t - \tau, \ddot{u}), \qquad \dot{\xi}^x = \xi^x(x, t - \tau, \ddot{u}),$$

are shifted to the left stencil point coordinates ξ^t, ξ^x, and

$$D_x = \frac{\partial}{\partial x} + u_x \frac{\partial}{\partial u} + u_{xx} \frac{\partial}{\partial u_x} + \cdots, \qquad D_t = \frac{\partial}{\partial t} + u_t \frac{\partial}{\partial u} + u_{xt} \frac{\partial}{\partial u_x} + \cdots.$$

By applying the generator (5.10) to (5.8), we obtain

$$\left. \left(\zeta_t - \zeta_{xx} - \xi^x g_x - \eta g_u - \ddot{\eta} g_{\ddot{u}} \right) \right|_{(5.8)} = 0. \tag{5.11}$$

It is supposed in (5.11) that u_t is substituted from (5.8).

To complete the determining system, we should act on (5.9) by the generator (5.10),

$$\xi^t = \ddot{\xi}^t \quad \xi^x = \ddot{\xi}^x. \tag{5.12}$$

Again, the differentiation of (5.12) with respect to u and $\ddot{u}$ yields

$$\xi_u^t = \xi_{\ddot{u}}^t = 0, \qquad \xi_u^x = \xi_{\ddot{u}}^x = 0;$$

consequently, $\xi^t, \ddot{\xi}^t, \xi^x,$ and $\ddot{\xi}^x$ are independent of u and $\ddot{u}$ and are periodic functions of t,

$$\xi^t(x, t) = \xi^t(x, t - \tau), \qquad \xi^x(x, t) = \xi^x(x, t - \tau).$$

The last relations substantially simplify solving the remaining equation (5.11), canceling many terms. The determining equation (5.11) can then be split with respect to u_{xx}, u_x, u, $\ddot{u}$ into several equations. In [102], the group classification was developed and all special cases of the function g were singled out. The core admitted Lie algebra is the two-dimensional algebra

$$X_1 = \frac{\partial}{\partial t}, \qquad X_2 = \frac{\partial}{\partial x}.$$

For four special cases of the function g, there exist additional symmetries. To illustrate the results, below we reproduce an excerpt from [102].

EXAMPLE. Equation (5.8) with

$$g(u, \ddot{u}) = k_1 \ddot{u} + k_2 u + k,$$

where k, k_1, and k_2 are constants ($k_1 \neq 0$), admits the three-dimensional Lie algebra spanned by the operators

$$X_1 = \frac{\partial}{\partial t}, \qquad X_2 = \frac{\partial}{\partial x}, \qquad X_3 = e^{-k_0 t} \frac{\partial}{\partial u}.$$

Consider a special solution invariant with respect to X_3. One can seek the solution in the form

$$u = \beta x e^{-k_0 t} + \phi(t).$$

By substituting the above solution into Eq. (5.8), we obtain the reduced delay ordinary differential equation

$$\phi'(t) = k_1 \phi(t - \tau) + k_2 \phi(t) + k$$

for the unknown function ϕ.

5.3. Symmetry of Differential-Difference Equations

Historically, the Lie group symmetry approach was applied primarily by several authors to differential-difference equations on a fixed regular lattice [62, 86–88, 99, 118, 119, 121]. One motivation of that is that such equations arise as primary mathematical models in physics and mechanics. As typical examples of a differential-difference equation, we consider the Toda lattice [86] and the discrete Volterra equation [85].

EXAMPLE 5.1 (the Toda lattice). Consider the differential-difference equation

$$u_{tt} = e^{u_- - u} - e^{u - u_+}, \tag{5.13}$$

where $u = u(t, x)$, $u_- = u(t, x - h_-)$, and $u_+ = u(t, x + h_+)$ are defined on a fixed regular lattice

$$h_- = h_+ = \text{const}, \qquad t_- = t_+ = t. \tag{5.14}$$

The last relations mean that time is one and the same at the points x, $x + h_+$ and $x - h_-$ (i.e. the orthogonality conditions).

We add the continuous variables $u_{tt}^- = u_{tt}(t, x - h_-)$ and $u_{tt}^+ = u(t, x + h_+)$ using our notation (clear from Fig. 5.2), which is different from that in [86].

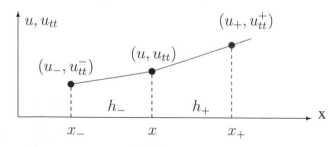

Figure 5.2

Equation (5.13) is a partial differential equation of second order in t and a second-order difference equation in x. Consider a Lie transformation group acting in the subspace $(x, t, x + h_+, x - h_-, u, u_-, u_+, u_{tt}, u_{tt}^-, u_{tt}^+)$. The symmetry generator

$$X = \xi^t(t, x, u)\frac{\partial}{\partial t} + \xi^x(t, x, u)\frac{\partial}{\partial x} + \eta(t, x, u)\frac{\partial}{\partial u}$$

should be prolonged for the derivative u_{tt} and the difference variables occurring in (5.13) and (5.14),

$$\tilde{X} = \xi^t\frac{\partial}{\partial t} + \xi^x\frac{\partial}{\partial x} + \eta\frac{\partial}{\partial u} + \zeta_{tt}\frac{\partial}{\partial u_{tt}} + (\xi_+^x - \xi^x)\frac{\partial}{\partial h_+}$$

$$+ (\xi^x - \xi_-^x)\frac{\partial}{\partial h_-} + \eta_+\frac{\partial}{\partial u_+} + \eta_-\frac{\partial}{\partial u_-}, \quad (5.15)$$

where

$$\zeta_{tt} = D_t(\zeta_t) - u_{tt}D_t(\xi^t) - u_{xt}D_t(\xi^x), \qquad \zeta_t = D_t(\eta) - u_x D_t(\xi^x) - u_t D_t(\xi^t)$$

are the standard prolongation formulas,

$$\xi_+^x = \xi^x(t, x + h_+, u_+), \qquad \xi_-^x = \xi^x(t, x - h_-, u_-),$$
$$\eta_+ = \eta(t, x + h_+, u_+), \qquad \eta_- = \eta(t, x - h_-, u_-)$$

are the coordinates η, ξ^x, shifted to the left and right stencil points, and

$$D_t = \frac{\partial}{\partial t} + u_t\frac{\partial}{\partial u} + u_{xt}\frac{\partial}{\partial u_x} + \cdots .$$

By applying the generator (5.15) to (5.13), we obtain

$$\left(\zeta_{tt} - e^{u_- - u}(\eta_- - \eta) - e^{u - u_+}(\eta - \eta_+)\right)\big|_{(5.13)} = 0. \quad (5.16)$$

It is supposed in (5.16) that u_{tt} is taken from (5.13).

To complete the determining system we should act on (5.14) by the generator (5.15),

$$\xi^x = \xi_-^x = \xi_+^x, \qquad \xi^t = \xi_-^t = \xi_+^t. \quad (5.17)$$

Without using (5.16), one can conclude from (5.17) that $\xi^x(x, t, u)$ and $\xi^t(x, t, u)$ are independent of u. Indeed, the differentiation of (5.17) with respect to u, u_+, u_-, x, x_-, and x_+ yields

$$\xi_u^t = 0, \qquad \xi_u^x = 0, \qquad \xi^t = \xi^t(t).$$

Consequently, $\xi^x(x)$ is at most a periodic function of x,

$$\xi^x(x) = \xi^x(x + h_+) = \xi^x(x - h_-), \qquad h_+ = h_-.$$

The last relations substantially simplify solving the remaining equation (5.16), canceling many terms. The determining equation (5.16) can then be split with respect to u_+ and u_- into several easier-to-solve equations.

The solution of the determining system is given by the following Lie algebra operators [86]:

$$X_1 = \frac{\partial}{\partial t}, \qquad X_2 = \frac{\partial}{\partial u}, \qquad X_3 = t\frac{\partial}{\partial u}.$$

In the special case of the lattice $h_- = h_+ = 1$, one has the additional operator [86]

$$X_4 = t\frac{\partial}{\partial t} + 2x\frac{\partial}{\partial u}.$$

Remark. Let us weaken the condition for the Toda lattice, replacing it by the regular mesh condition. We additionally plug in finite-difference derivatives instead of differences of u. Thus, instead of (5.13) and (5.14) we consider the following system of differential-difference equations:

$$u_{tt} = e^{-(u-u_-)/h_-} - e^{-(u_+-u)/h_+}, \qquad h_- = h_+, \qquad t_- = t_+ = t. \qquad (5.18)$$

The appropriate determining system is

$$\left[\zeta_{tt} - e^{-(u-u_-)/h_-}\left(\frac{\eta_- - \eta}{h_-} + (\xi^x - \xi_-^x)\frac{u - u_-}{h_-^2} \right) \right.$$

$$\left. - e^{-(u_+-u)/h_+}\left(\frac{\eta - \eta_+}{h_+} + (\xi_+^x - \xi^x)\frac{u_+ - u}{h_+^2} \right) \right]\Bigg|_{(5.18)} = 0, \qquad (5.19)$$

$$\xi_+^x - 2\xi^x = \xi_-^x = 0, \qquad \xi^t = \xi_-^t = \xi_+^t. \qquad (5.20)$$

The solution of the determining system (5.19), (5.20) yields the following Lie algebra of generators:

$$X_1 = \frac{\partial}{\partial t}, \qquad X_2 = \frac{\partial}{\partial u}, \qquad X_3 = t\frac{\partial}{\partial u},$$

$$X_4 = \frac{\partial}{\partial x}, \qquad X_5 = 2x\frac{\partial}{\partial x} + t\frac{\partial}{\partial t} + 2u\frac{\partial}{\partial u}.$$

EXAMPLE 5.2 (the discrete Volterra equation). The discrete Volterra equation [85] is the differential-difference equation

$$u_t + u\frac{u_+ - u_-}{x_+ - x_-} = u_t + u\frac{u_x - u_{\bar{x}}}{2} = 0, \qquad (5.21)$$

where $u = u(t, x)$, $u_- = u(t, x - h_-)$, and $u_+ = u(t, x + h_+)$ are defined on the fixed regular lattice

$$h_- = h_+, \qquad t_- = t_+ = t. \qquad (5.22)$$

Equation (5.21) is a partial differential equation of the first order in t and a second-order difference equation in x.

Consider the Lie transformation group acting in the subspace $(x, t, x + h_+, x - h_-, u, u_-, u_+, u_t)$. The symmetry generator

$$X = \xi^t(t, x, u)\frac{\partial}{\partial t} + \xi^x(t, x, u)\frac{\partial}{\partial x} + \eta(t, x, u)\frac{\partial}{\partial u}$$

should be prolonged for the derivative u_t and the difference variables occurring in (5.21) and (5.22),

$$\tilde{X} = \xi^t\frac{\partial}{\partial t} + \xi^x\frac{\partial}{\partial x} + \eta\frac{\partial}{\partial u} + \zeta_t\frac{\partial}{\partial u_t} + (\xi_+^x - \xi^x)\frac{\partial}{\partial h_+}$$
$$+ (\xi^x - \xi_-^x)\frac{\partial}{\partial h_-} + \eta_+\frac{\partial}{\partial u_+} + \eta_-\frac{\partial}{\partial u_-}, \quad (5.23)$$

where $\zeta_t = D_t(\eta) - u_x D_t(\xi^x) - u_t D_t(\xi^t)$ and

$$\xi_+^x = \xi^x(t, x + h_+, u_+), \qquad \xi_-^x = \xi^x(t, x - h_-, u_-),$$
$$\eta_+ = \eta(t, x + h_+, u_+), \qquad \eta_- = \eta(t, x - h_-, u_-)$$

are the coordinates of ξ^x and η shifted to the left and right stencil points.

By applying the generator (5.23) to (5.21), we obtain

$$\left[\zeta_t + \eta\frac{u_+ - u_-}{x_+ - x_-} + u\frac{\eta_+ - \eta_-}{x_+ - x_-} - u\frac{u_+ - u_-}{(x_+ - x_-)^2}(\xi_+^x - \xi_-^x)\right]\Bigg|_{(5.21)} = 0. \quad (5.24)$$

It is supposed in (5.24) that u_t is substituted from (5.21). To complete the determining system, we should act on (5.22) by the generator (5.23),

$$\xi_+^x - 2\xi^x + \xi_-^x = 0, \qquad \xi^t = \xi_-^t = \xi_+^t. \quad (5.25)$$

Then one can conclude from (5.25) that ξ^x and ξ^t are independent of u: $\xi_u^t = 0$, $\xi_u^x = 0$, $\xi^t = \xi^t(t)$, and $\xi^x = \xi^x(x)$. The last relations substantially simplify solving the remaining equation (5.24), canceling many terms. The determining equation (5.24) can then be split with respect to u_+ and u_- into several easier-to-solve equations.

The solution of the determining system is the Lie algebra spanned by the operators [85]

$$X_1 = \frac{\partial}{\partial t}, \quad X_2 = \frac{\partial}{\partial x}, \quad X_3 = t\frac{\partial}{\partial t} - u\frac{\partial}{\partial u}, \quad X_4 = x\frac{\partial}{\partial x} + u\frac{\partial}{\partial u}. \quad (5.26)$$

In the continuous limit, Eq. (5.21) becomes the simplest transport equation

$$u_t + uu_x = 0, \quad (5.27)$$

which is linearizable and admits an infinite-dimensional symmetry group (e.g., see [74], p.178).

Equation (5.21) possesses the four-dimensional Lie algebra (5.26) only and cannot be considered as a good numerical approximation to (5.27).

Many other examples of differential-difference equations and their symmetries can be found in [65]. A number of examples of differential-difference equations that possess noninvariant solutions can be found in [59].

Chapter 6

Lagrangian Formalism
for Difference Equations

The well-known Noether theorem [104] states the relationship between the invariance of a variational functional and the conservativeness of the corresponding Euler differential equations, i.e., the fact that the conservation laws are satisfied on their solutions. In the present chapter, we give a difference analog of this construction (see [29, 30, 36, 39, 48, 49]). We also find necessary and sufficient conditions for the invariance of a difference functional defined on a mesh. We show that the invariance of a finite-difference functional does not automatically imply the invariance of the corresponding Euler equations. We obtain a condition for the difference Euler equation to be invariant. We derive a *new difference equation* (which, in general, does not coincide with the difference Euler equation) such that the functional is stationary under the group transformations on its solutions. This equation, which is said to be *quasi-extremal*, depends on the group operator coordinates and has the corresponding conservation law if the functional is invariant. We study the properties of quasi-extremal equations. If the functional admits more than one symmetry, then it makes sense to consider the set of intersections of solutions of quasi-extremal equations. For the intersection of quasi-extremals of an invariant functional, we state a theorem quite similar to the Noether theorem. Note that the proposed difference construction becomes the classical Noether theorem in the continuum limit.

Note that the preservation of difference analogs of conservation laws in numerical schemes is of great importance (see [64, 122]).

6.1. Discrete Representation of Euler's Operator

Let us find out how the Euler equation of a difference functional can be written on various difference meshes.

1. First, consider the simpler case of one single independent variable x and one or several dependent variables $(u^1, u^2, u^3, \ldots, u^m)$.

It is well known (see [72, 73]) that the Euler operator in the "continuous" space $\tilde{Z}$ that contains u and all the derivatives $(u, u_1, u_2, \ldots)$ can be written as

$$\frac{\delta}{\delta u} = \frac{\partial}{\partial u} + \sum_{s=1}^{\infty} (-1)^s D^s \left(\frac{\partial}{\partial u_s} \right). \tag{6.1}$$

We first represent the Euler operator in $\underset{h}{Z}$ under the assumption that the mesh $\underset{h}{\omega}$ is *uniform* (or *regular*). We assume that it is applied to functions $\mathcal{L}(x, u, \underset{h}{u_x}) \in \underset{h}{A}$, $\underset{h}{u_x} = (u_+ - u)/h$, defined on the mesh $\underset{h}{\omega}$.

It is remarkable that the operator series that plays the key role in variational calculus, i.e., the second term on the right-hand side in (6.1), can be "rolled" into a compact form with the use of the discrete differentiations $\underset{+h}{D}$ and $\underset{-h}{D}$. Indeed, we note that

$$\frac{\partial}{\partial u_s} = \frac{\partial \underset{h}{u_x}}{\partial u_s} \frac{\partial}{\partial \underset{h}{u_x}} = \frac{h^{s-1}}{s!} \frac{\partial}{\partial \underset{h}{u_x}}$$

and obtain

$$\frac{\delta}{\delta u} = \frac{\partial}{\partial u} - \sum_{s=1}^{\infty} (-h)^{s-1} D^s \left(\frac{\partial}{\partial \underset{h}{u_x}} \right) = \frac{\partial}{\partial u} - \underset{-h}{D} \left(\frac{\partial}{\partial \underset{h}{u_x}} \right), \tag{6.2}$$

where

$$\underset{-h}{D} = \sum_{s \geq 1} \frac{(-h)^{s-1}}{s!} D^s$$

is the left discrete differentiation operator.

Note that in formula (6.2) the "continuous" partial differentiation with respect to the first right difference derivative $\underset{h}{u_x}$ is first applied, and then the "discrete" left differentiation is used.

The finite-difference equation

$$\frac{\delta \mathcal{L}}{\delta u} = \frac{\partial \mathcal{L}}{\partial u} - \underset{-h}{D} \left(\frac{\partial \mathcal{L}}{\partial \underset{h}{u_x}} \right) = 0 \tag{6.3}$$

will be called the *difference Euler equation on a uniform mesh*, the function $\mathcal{L} = \mathcal{L}(x, u, \underset{h}{u_x})$ is called a *mesh* (or *discrete*, or *finite-difference*) *Lagrangian function*, and any solution of Eq. (6.3) is called an *extremal*.

EXAMPLE. Let $\mathcal{L} = \frac{1}{2} \underset{h}{u_x^2} + e^u$; then the Euler equation (6.3) acquires the form

$$\underset{h}{u_{x\bar{x}}} - e^u = 0.$$

2. Consider how the Euler operator (6.1) can be written on a one-dimensional *nonuniform* mesh $\underset{h}{\omega}$. If the mesh is nonuniform, then the shift and discrete differentiation operators are of "local" character, i.e., are related to the local mesh spacings h^- and h^+ at a given point x, and hence

$$\frac{\partial}{\partial u_s} = \frac{\partial \underset{h}{u_x}}{\partial u_s} \frac{\partial}{\partial \underset{h}{u_x}} = \frac{(h^+)^{s-1}}{s!} \frac{\partial}{\partial \underset{h}{u_x}}.$$

Therefore, the Euler operator at the point x acquires the form

$$\frac{\delta}{\delta u} = \frac{\partial}{\partial u} - \sum_{s \geq 1} \frac{(-h^+)^{s-1}}{s!} D^s \left(\frac{\partial}{\partial \underset{h}{u_x}} \right),$$

where the second term is the left discrete differentiation but with the right step h^+ rather than the left step h^-, because the left discrete differentiation is written as

$$\sum_{s \geq 1} \frac{(-h^-)^{s-1}}{s!} D^s.$$

We rewrite this expression differently:

$$\frac{\delta}{\delta u} = \frac{\partial}{\partial u} - \frac{1}{h^+} \left(\frac{\partial}{\partial \underset{h}{u_x}} - \underset{-h}{S} \frac{\partial}{\partial \underset{h}{u_x}} \right)$$

$$= \frac{\partial}{\partial u} - \frac{h^-}{h^+ h^-} \left(\frac{\partial}{\partial \underset{h}{u_x}} - \underset{-h}{S} \frac{\partial}{\partial \underset{h}{u_x}} \right) = \frac{\partial}{\partial u} - \frac{h^-}{h^+} \underset{-h}{D} \left(\frac{\partial}{\partial \underset{h}{u_x}} \right).$$

Thus, the Euler equation on the nonuniform mesh can be written as

$$\frac{\partial \mathcal{L}}{\partial u} - \frac{h^-}{h^+} \underset{-h}{D} \left(\frac{\partial \mathcal{L}}{\partial \underset{h}{u_x}} \right) = 0. \tag{6.4}$$

The factor h^-/h^+ occurring in (6.4) characterizes the difference stencil proportions at a given point,

$$\frac{h^-}{h^+} = \varphi(x).$$

As was already shown, if the nonuniform mesh satisfies the invariance conditions, then this equation represents an invariant manifold in $\underset{h}{Z}$.

3. Consider the Euler operator in the *two-dimensional case*. We assume that the two-dimensional mesh is rectangular and uniform in each direction (with constant spacings h_1 and h_2, respectively). The variational derivative in $\tilde{Z}$ acquires the form

$$\frac{\delta}{\delta u} = \frac{\partial}{\partial u} + \sum_{s=1}^{\infty} (-1)^s D_{i_1} \cdots D_{i_s} \frac{\partial}{\partial u_{i_1 \ldots i_s}},$$

where $i_1 \dots i_s$ is the s-dimensional set of indices $(1, 2)$ and D_{i_s} is the complete ("continuous") differentiation in the respective direction,

$$D_1 = \frac{\partial}{\partial x^1} + u_1 \frac{\partial}{\partial u} + u_{11} \frac{\partial}{\partial u_1} + u_{21} \frac{\partial}{\partial u_2} + \cdots,$$

$$D_2 = \frac{\partial}{\partial x^2} + u_2 \frac{\partial}{\partial u} + u_{12} \frac{\partial}{\partial u_1} + u_{22} \frac{\partial}{\partial u_2} + \cdots.$$

We assume that the operator is applied in $\underset{h}{Z}$ to functions of the form

$$\mathcal{L} = \mathcal{L}(x^1, x^2, u, \underset{h}{u_1}, \underset{h}{u_2})$$

and obtain

$$\frac{\delta}{\delta u} = \frac{\partial}{\partial u} + \sum_{s \geq 1} (-1)^s \sum_{k+l=s} D_1^k D_2^l \frac{\partial}{\partial u_{1_1 \dots 1_k 2_1 \dots 2_l}}$$

$$= \frac{\partial}{\partial u} - \sum_{s \geq 1} (-1)^{s-1} \frac{h_1^{s-1}}{s!} D_1^s \frac{\partial}{\partial \underset{h}{u_1}} - \sum_{p \geq 1} (-1)^{p-1} \frac{h_2^{p-1}}{p!} D_2^p$$

$$= \frac{\partial}{\partial u} - \underset{-h}{D_1} \left(\frac{\partial}{\partial \underset{h}{u_1}} \right) - \underset{-h}{D_2} \left(\frac{\partial}{\partial \underset{h}{u_2}} \right).$$

Thus, the Euler equation on a two-dimensional uniform mesh acquires the form

$$\frac{\delta \mathcal{L}}{\delta u} = \frac{\partial \mathcal{L}}{\partial u} - \underset{-h}{D_1} \left(\frac{\partial \mathcal{L}}{\partial \underset{h}{u_1}} \right) - \underset{+h}{D_2} \left(\frac{\partial \mathcal{L}}{\partial \underset{h}{u_2}} \right) = 0, \qquad (6.5)$$

where $\underset{h}{u_1}$ and $\underset{h}{u_2}$ are the right difference derivatives in the directions (x^1, x^2), respectively.

4. In a similar way, we can obtain an expression for the Euler operator on a two-dimensional nonuniform rectangular mesh $\underset{h}{\omega}$ characterized by two local spacings, $h_1^{\pm}$ and $h_2^{\pm}$:

$$\frac{\delta \mathcal{L}}{\delta u} = \frac{\partial \mathcal{L}}{\partial u} - \frac{h_1^-}{h_1^+} \underset{-h}{D_1} \left(\frac{\partial \mathcal{L}}{\partial \underset{h}{u_1}} \right) - \frac{h_2^-}{h_2^+} \underset{-h}{D_2} \left(\frac{\partial \mathcal{L}}{\partial \underset{h}{u_2}} \right) = 0.$$

In what follows, considering difference functionals of specific forms, we also obtain various forms of the Euler operator.

EXAMPLE. On an orthogonal mesh $\underset{h}{\omega}$ uniform in each of the t- and x-directions with respective spacings τ and h, consider the Lagrangian

$$\mathcal{L} = e^{\frac{t}{\tau_0}} \left(\frac{k_0}{6} \underset{h}{u_x^3} - \frac{\tau_0}{4} u_t^2 - \frac{\tau_0}{4} \underset{\tau}{u_t^2} \right),$$

where k_0 and τ_0 are some positive constants. Then formula (6.5) gives the following difference Euler equation:

$$\tau_0 u_{t\bar{t}} + \frac{\tau_0}{2\tau}\{u_{\bar{t}}(1 - e^{-\tau/\tau_0}) + u_t(e^{\tau/\tau_0} - 1)\} = \frac{k_0}{2}u_{x\bar{x}}(u_x + u_{\bar{x}}). \qquad (6.6)$$

Equation (6.6) approximates the heat equation

$$\tau_0 u_{tt} + u_t = k_0 u_x u_{xx}$$

with heat flux relaxation taken into account up to $O(\tau^2 + h^2)$.

6.2. Criterion for the Invariance of Difference Functionals

1. Let the following finite-difference functional be given on a one-dimensional mesh $\underset{h}{\omega}$:

$$L = \sum_{\underset{h}{\Omega}} \mathcal{L}(x, u, \underset{h}{u_x})h^+, \qquad (6.7)$$

where the sum is taken over a finite or infinite domain $\underset{h}{\Omega} \subset \underset{h}{\omega}$. (In the latter case, we assume that $\mathcal{L}$ sufficiently rapidly decays at ∞.)

The functional (6.7) is defined on a difference mesh, uniform or nonuniform. If the mesh is uniform, then $h^+ = h^-$. A nonuniform mesh is introduced by a smooth function $\varphi(x)$, $h^+ = \varphi(x)$, so that $h^- = \varphi(x - h^-)$. It is also possible that φ depends on u, $h^+ = \varphi(x, u)$; i.e., the mesh can depend on the solution.

In the space $\underset{h}{Z} = (x, u, \underset{h}{u_x}, \underset{h}{u_{x\bar{x}}}, \dots, h^+)$ of difference variables, consider a one-parameter transformation group G_1 with operator

$$X = \xi\frac{\partial}{\partial x} + \eta\frac{\partial}{\partial u} + \zeta_1\frac{\partial}{\partial \underset{h}{u_x}} + \cdots + h^+ \underset{+h}{D}(\xi)\frac{\partial}{\partial h^+} + h^- \underset{-h}{D}(\xi)\frac{\partial}{\partial h^-}, \qquad (6.8)$$

where the functions $\xi, \eta, \underset{h}{\zeta_1}, \dots \in \underset{h}{A}$, $\underset{h}{\zeta_1}, \underset{h}{\zeta_2}, \dots$ are linear difference forms of (ξ, η) (see Chapter 1).

Under the transformations of the group G_1 generated by the operator (6.8), there are variations not only in the difference functional (6.7) but also in the difference mesh $\underset{h}{\omega}$ on which the operator is considered (and in the domain $\underset{h}{\Omega} \subset \underset{h}{\omega}$ along with the mesh). Therefore, in the definition of the functional transformation, it is necessary to introduce a transformation of the difference mesh $\underset{h}{\omega}$.

DEFINITION. An transformed value of the mesh functional (6.7) on a uniform mesh is defined to be the sum

$$L^* = \sum_{\underset{h}{\Omega^*}} \mathcal{L}^*(x^*, u^*, \underset{h}{u_x^*})h^{+*}, \qquad h^{+*} = \varphi(h^{-*}), \qquad (6.9)$$

where, in general, $\mathcal{L}^* \neq \mathcal{L}$ and $\varphi(h^{-*}) \neq h^{-*}$; the summation domain $\underset{h}{\Omega^*}$ is obtained from the domain $\underset{h}{\Omega}$ by the transformations of the group G_1.

Note that the transformed summation domain $\underset{h}{\Omega^*}$ may depend on the solution u if the transformed value x^* depends on u,

$$x^* = f(x, u, a) = x + a\xi(x, u) + \cdots .$$

DEFINITION. A difference functional L is said to be *invariant* under the group G_1 on a uniform mesh if the following relations hold for all transformations of the group G_1 and any summation domain $\underset{h}{\Omega}$:

$$\sum_{\underset{h}{\Omega}} \mathcal{L}(x, u, \underset{h}{u_x})h^+ = \sum_{\underset{h}{\Omega^*}} \mathcal{L}(x^*, u^*, \underset{h}{u_x^*})h^{+*}, \qquad h^{+*} = h^{-*}. \tag{6.10}$$

Let us find out under what conditions on the discrete Lagrangian $\mathcal{L}(x, u, \underset{h}{u_x})$ and for what classes of transformations conditions (6.10) are satisfied. We perform a change of variables in (6.10) so that the sum over the original domain $\underset{h}{\Omega}$ is on the right:

$$\sum_{\underset{h}{\Omega}} \mathcal{L}(x, u, \underset{h}{u_x})h^+ = \sum_{\underset{h}{\Omega}} \mathcal{L}(e^{aX}(x), e^{aX}(u), e^{aX}(\underset{h}{u_x}))e^{aX}(h^+), \qquad h^{+*} = h^{-*}.$$

Since the summation domain $\underset{h}{\Omega}$ is arbitrary, it follows that these relations are equivalent to the following relations for the elementary action:

$$\mathcal{L}(x, u, \underset{h}{u_x})h^+ = \mathcal{L}^*(x^*, u^*, \underset{h}{u_x^*})h^{-*}, \qquad h^{+*} = h^{-*}. \tag{6.11}$$

Relations (6.11) mean that the elementary action $\mathcal{L}(x, u, \underset{h}{u_1})h^+$ is an invariant of the transformation group G_1 in the space $\underset{h}{Z} = (x, u, \underset{h}{u_x}, \underset{h}{u_{x\bar{x}}}, \dots, h)$ on the invariant manifold $h^+ = h^-$.

We use the operator (6.8) to write out a necessary and sufficient invariance condition for the elementary action $\mathcal{L}(x, u, \underset{h}{u_x})h^+$ on the manifold $h^+ = h^-$. To this end, we apply the operation $\partial/\partial a\big|_{a=0}$ to relations (6.11):

$$\xi\frac{\partial\mathcal{L}}{\partial x} + \eta\frac{\partial\mathcal{L}}{\partial u} + [\underset{+h}{D}(\eta) - \underset{h}{u_x}\underset{+h}{D}(\xi)]\frac{\partial\mathcal{L}}{\partial \underset{h}{u_x}} + \mathcal{L}\underset{+h}{D}(\xi) = 0, \qquad \underset{-h+h}{D\,D}(\xi) = 0. \tag{6.12}$$

Thus, the following theorem holds.

THEOREM 6.1. *For the mesh functional (6.7) to be invariant on a uniform mesh under the one-parameter group G_1 with operator (6.8), it is necessary and sufficient that relations (6.12) be satisfied.*

The first relation in (6.12) is a difference analog of the first Noether theorem [104], and the second is of course absent in the continuum limit, because the equation $h^+ = h^-$ becomes the identity $0 = 0$.

2. Consider the case of a one-dimensional nonuniform mesh $\underset{h}{\omega}$ characterized in $\underset{h}{Z}$ by the relation

$$h^+ = \varphi(x, u), \tag{6.13}$$

where $\varphi \in A$ (i.e., the case in which the mesh depends on the solution).

The transformed value of the functional (6.7) on the mesh (6.13) is defined to be the expression

$$L^* = \sum_{\underset{h}{\Omega}} \mathcal{L}^*(x^*, u^*, \underset{h}{u_x^*})h^{+*}, \qquad h^{+*} = \varphi^*(x^*, u^*),$$

where, in general, $\varphi(z^*) \neq \varphi^*(z^*)$. Note that the invariant mesh is characterized by the same function $\varphi(z^*)$ for the spacing h^{+*} in the new variables.

DEFINITION. One says that the functional (6.7) is *invariant* on the nonuniform mesh (6.13) if the following conditions are satisfied:

$$\sum_{\underset{h}{\Omega}} \mathcal{L}(x, u, \underset{h}{u_x})h^+ = \sum_{\underset{h}{\Omega^*}} \mathcal{L}(x^*, u^*, \underset{h}{u_x^*})h^{+*}, \qquad h^{+*} = \varphi(x^*, u^*).$$

In this relation, we replace the sum over the points of the domain $\underset{h}{\Omega}$ by the equivalent sum over the powers of the shift operator $\underset{+h}{S}$:

$$\sum_{\alpha} \underset{+h}{S}{}^\alpha(\mathcal{L}(x, u, \underset{h}{u_x})h^+) = \sum_{\alpha} \mathcal{L}(\underset{+h}{S}{}^\alpha(x^*), \underset{+h}{S}{}^\alpha(u^*), \underset{+h}{S}{}^\alpha(\underset{h}{u_x^*}) \underset{+h}{S}{}^\alpha(h^{+*}),$$

$$h^{+*} = \varphi(x^*, u^*). \tag{6.14}$$

Since the arbitrariness of the domain $\underset{h}{\Omega}$ is equivalent to the arbitrariness of the domain of summation over the index α, it follows from (6.14) that

$$\mathcal{L}(x, u, \underset{h}{u_x})h^+ = \mathcal{L}(x^*, u^*, \underset{h}{u_x^*})h^{+*}, \qquad h^{+*} = \varphi(x^*, u^*). \tag{6.15}$$

By applying the operation $\frac{\partial}{\partial a}\big|_{a=0}$ to (6.15), we obtain an infinitesimal criterion for the functional (6.7) to be invariant on the nonuniform mesh (6.13):

$$\xi\frac{\partial \mathcal{L}}{\partial x} + \eta\frac{\partial \mathcal{L}}{\partial u} + \zeta_1 \frac{\partial \mathcal{L}}{\underset{h}{\partial u_x}} + \mathcal{L}\underset{+h}{D}(\xi) = 0, \qquad \underset{+h}{S}(\xi) - \xi(1 + \varphi_x) - \eta\varphi_u = 0, \tag{6.16}$$

where the operators $\underset{+h}{S}$ and $\underset{+h}{D}$ are taken on the mesh (6.13).

THEOREM 6.2. *For the mesh functional (6.9) to be invariant on a nonuniform mesh depending on the solution under the one-parameter group G_1 with operator (6.8), it is necessary and sufficient that relations (6.16) be satisfied.*

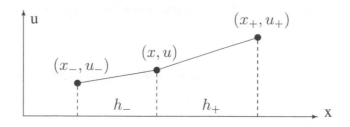

Figure 6.1

3. Let us present the above results for a finite-difference functional written in the different form

$$\mathbb{L}_h = \sum_{\Omega} \mathcal{L}(x, x_+, u, u_+)h_+, \tag{6.17}$$

where the mesh spacings $h_+ = x_+ - x$ can still depend on the solution,

$$h_+ = \varphi(x_+, x, u_+, u); \tag{6.18}$$

the notation is clear from Fig. 6.1.

We need the prolongation of the operator X by shifting the coefficients to the corresponding points (x_+, x, u_+, u)

$$X = \xi\frac{\partial}{\partial x} + \xi_-\frac{\partial}{\partial x_-} + \xi_+\frac{\partial}{\partial x_+} + \eta^i\frac{\partial}{\partial u^i} + \eta^i_-\frac{\partial}{\partial u^i_-}$$
$$+ \eta^i_+\frac{\partial}{\partial u^i_+} + (\xi_+ - \xi)\frac{\partial}{\partial h_+} + (\xi - \xi_-)\frac{\partial}{\partial h_-},$$

where the coefficients are given as follows:

$$\xi_- = \xi(x_-, u_-), \quad \eta^i_- = \eta^i(x_-, u_-), \quad \xi_+ = \xi(x_+, u_+), \quad \eta^i_+ = \eta^i(x_+, u_+).$$

The infinitesimal invariance condition for the functional (6.17) on the mesh (6.18) is given by the two equations

$$\left[\xi\frac{\partial \mathcal{L}}{\partial x} + \xi_-\frac{\partial \mathcal{L}}{\partial x_-} + \xi_+\frac{\partial \mathcal{L}}{\partial x_+} + \eta^i\frac{\partial \mathcal{L}}{\partial u^i} + \eta^i_-\frac{\partial \mathcal{L}}{\partial u^i_-} + \eta^i_+\frac{\partial \mathcal{L}}{\partial u^i_+} + \mathcal{L}\underset{+h}{D}(\xi)\right]\Bigg|_{h_+=\varphi} = 0,$$

$$\left[\underset{+h}{S}(\xi) - \xi - X(\varphi)\right]\Big|_{h_+=\varphi} = 0,$$

which hold on the lattice (6.18).

4. Consider the conditions for the mesh functional

$$L = \sum_{\substack{\Omega \\ h}} \mathcal{L}(x^1, x^2, u, \underset{h}{u_1}, \underset{h}{u_2}) h_1^+ h_2^+, \tag{6.19}$$

to be invariant on a two-dimensional uniform rectangular mesh $\underset{h_1}{\omega} \times \underset{h_2}{\omega}$, $h_i^+ = h_i^-$, characterized by two constant spacings h_1 and h_2. An argument similar to that in the one-dimensional case leads to the infinitesimal invariance criterion

$$X(\mathcal{L}) + \mathcal{L}(\underset{+h}{D_1}(\xi^1) + \underset{+h}{D_2}(\xi^2)) = 0,$$

$$\underset{+h}{D_1}\underset{-h}{D_1}(\xi^1) = 0, \qquad \underset{+h}{D_2}\underset{-h}{D_2}(\xi^2) = 0, \qquad \underset{\pm h}{D_1}(\xi^2) = -\underset{\pm h}{D_2}(\xi^1),$$

where $\underset{\pm h}{D_i}$ is the discrete differentiation in the ith direction, $\underset{h}{u_1}$ and $\underset{h}{u_2}$ are the right difference derivatives in the x^1- and x^2-directions, respectively,

$$X = \xi^1 \frac{\partial}{\partial x^1} + \xi^2 \frac{\partial}{\partial x^2} + \eta \frac{\partial}{\partial u} + \underset{h}{\zeta_1} \frac{\partial}{\partial u_1} + \underset{h}{\zeta_2} \frac{\partial}{\partial u_2}$$

$$+ \cdots + (\underset{+h}{S_1}(\xi^1) - \xi^1) \frac{\partial}{\partial h_1^+} + (\underset{+h}{S_2}(\xi^2) - \xi^2) \frac{\partial}{\partial h_2^+}$$

is the operator of the group G_1, and the $\underset{h}{\zeta_i}$ are linear difference forms of (ξ^1, ξ^2, η) obtained by the prolongation formulas derived in Chapter 1.

5. In the case of a nonuniform rectangular mesh depending on the solution,

$$h_1^+ = \varphi_1(x^1, x^2, u), \qquad h_2^+ = \varphi_2(x^1, x^2, u), \tag{6.20}$$

a necessary and sufficient condition for the functional (6.19) to be invariant on the mesh (6.20) is

$$X(\mathcal{L}) + \mathcal{L}(\underset{+h}{D_1}(\xi^1) + \underset{+h}{D_2}(\xi^2)) = 0, \qquad \underset{+h}{S_1}(\xi^1) - \xi^1 - X(\varphi_1) = 0,$$

$$\underset{+h}{S_2}(\xi^2) - \xi^2 - X(\varphi_2) = 0, \qquad \underset{\pm h}{D_1}(\xi^2) = -\underset{\pm h}{D_2}(\xi^1),$$

where the operators $\underset{+h}{S_i}$ and $\underset{+h}{D_i}$ are taken on the mesh (6.20).

6.3. Invariance of Difference Euler Equations

It is well known that the invariance of the Euler equations in the differential case is a consequence of the invariance of the corresponding variational functional [104]. Let us find out whether this situation is preserved in the finite-difference case.

For simplicity, consider the case of a one-dimensional uniform mesh. The finite-difference equation

$$\frac{\delta \mathcal{L}}{\delta u} \equiv \frac{\partial \mathcal{L}}{\partial u} - \underset{-h}{D}\left(\frac{\partial \mathcal{L}}{\partial \underset{h}{u_x}}\right) = 0 \tag{6.21}$$

for the mesh extremals in the case of a nondegenerate functional is a second-order difference equation written at three points $(x - h, x, x + h)$ of the mesh $\underset{h}{\omega}$. The invariance condition for the elementary Lagrangian action, which was obtained in the preceding section, contains only the first (right) derivative, i.e., is written at two points of the mesh $(x, x + h) \in \underset{h}{\omega}$. Therefore, to clarify the invariance conditions (6.21), we need to prolong the invariance condition for the elementary action to the left, i.e., to the point $x - h$. Now consider the condition that two terms in the sum of the difference functional are equal to each other:

$$\mathcal{L}(x^-, u^-, \underset{h}{u_{\bar{x}}})h^- + \mathcal{L}(x, u, \underset{h}{u_x})h^+ = \mathcal{L}(x^{-*}, u^{-*}, \underset{h}{u_{\bar{x}}^*})h_1^{-*} + \mathcal{L}(x^*, u^*, \underset{h}{u_x^*})h_1^{+*},$$

$$h^{-*} = h^{+*},$$

$$\tag{6.22}$$

where $x^{-*} = f(x^-, u^-, a)$, $u^{-*} = g(x^-, u^-, a)$, and

$$\underset{h}{u_x^*} = \underset{-h}{S}\left(\frac{\underset{+h}{D}(g(x, u, a))}{\underset{+h}{D}(f(x, u, a))}\right).$$

By applying the Euler operator to relations (6.22), we obtain

$$\frac{\partial \mathcal{L}}{\partial u} - \underset{-h}{D}\left(\frac{\partial \mathcal{L}}{\partial \underset{h}{u_x}}\right) = f_u\left(\frac{\partial \mathcal{L}(z^*)}{\partial x^*} + \underset{-h}{D^*}\left(\underset{h}{u_x^*}\frac{\partial \mathcal{L}(z^*)}{\partial \underset{h}{u_x^*}}\right)\right)$$

$$- \underset{-h}{D^*}(\mathcal{L}(z^*))) + g_u\left(\frac{\partial \mathcal{L}(z^*)}{\partial u^*} + \underset{-h}{D^*}\left(\frac{\partial \mathcal{L}(z^*)}{\partial \underset{h}{u_x^*}}\right)\right), \tag{6.23}$$

where $\underset{-h}{D^*}$ is the operator of left discrete differentiation in the new variables.

By applying the operation $\partial/\partial a\big|_{a=0}$ to relation (6.23), we obtain

$$\xi_u\left(\frac{\partial \mathcal{L}}{\partial x} + \underset{-h}{D}\left(\underset{h}{u_x}\frac{\partial \mathcal{L}}{\partial \underset{h}{u_x}}\right) - \underset{-h}{D}(\mathcal{L})\right) + \eta_u\left(\frac{\delta \mathcal{L}}{\delta u}\right)$$

$$+ \frac{\partial}{\partial a}\left(\frac{\partial \mathcal{L}(z^*)}{\partial u^*} - \underset{-h}{D^*}\left(\frac{\partial \mathcal{L}(z^*)}{\partial \underset{h}{u_x}}\right)\right)\Bigg|_{a=0} = 0,$$

where the last term is just the invariance condition for the Euler equation. Thus, for invariant functionals on an invariantly uniform mesh the action of the operator X

on the Euler equation gives

$$X\left(\frac{\delta\mathcal{L}}{\delta u}\right) = -\eta_u\left(\frac{\delta\mathcal{L}}{\delta u}\right) - \xi_u\left(\frac{\partial\mathcal{L}}{\partial x} + \underset{-h}{D}\left(u_x\frac{\partial\mathcal{L}}{\partial u_x} - \mathcal{L}\right)\right), \qquad \underset{+h-h}{D\,D}(\xi) = 0. \quad (6.24)$$

By substituting $\frac{\delta\mathcal{L}}{\delta u} = 0$ into (6.24), we obtain the following assertion.

THEOREM 6.3. *For the Euler equations* (6.21) *of the invariant functional* (6.11) *to be invariant on the uniform mesh* $\underset{h}{\omega}$, *it is necessary and sufficient that the following condition be satisfied on their solutions:*

$$\xi_u\left(\frac{\partial\mathcal{L}}{\partial x} + u_{\bar x}\frac{\partial\mathcal{L}}{\partial u} + u_{x\bar x}\frac{\partial\mathcal{L}}{\underset{h}{\partial u_x}} - \underset{-h}{D}(\mathcal{L})\right) = 0, \qquad \underset{-h+h}{D\,D}(\xi) = 0. \quad (6.25)$$

This condition is absent in the differential case, because the operator in parentheses becomes identically zero as $h \to 0$.

Conditions (6.25) are necessarily satisfied for degenerate functionals linearly depending on their variables and for x-autonomous transformation groups, for which

$$\xi_u = 0, \qquad \underset{-h+h}{D\,D}(\xi) = 0$$

on uniform meshes, which implies that $\xi(x) = Ax + B$, where A and B are constants.

Thus, it is only under conditions (6.25) that the situation in the difference case is similar to that in the differential case: the invariance of the functional implies the invariance of the corresponding Euler equation. But it is clear that rather wide classes of transformations and functionals do not satisfy conditions (6.25). This means that the group transformations can transform the Euler equation into some different equation without changing the difference functional.

The following question arises: For the solutions of what equation do the values of the functional remain constant under the group transformations? The answer to this question is given in the next section.

6.4. Variation of Difference Functional and Quasi-Extremal Equations

Consider the variation of the functional (6.7)

$$L = \sum_{\underset{h}{\Omega}} \mathcal{L}(x, u, \underset{h}{u_x})h_+$$

along a smooth curve

$$u = \Psi(x) \qquad (6.26)$$

passing through a given point (x, u). Let us calculate the increment of the functional (6.7) in terms of the variations δx and $\delta u = \Psi_x \delta x$. Since δx and δu are contained only in two neighboring terms of the sum (6.7), we omit the terms of smaller order than δx and obtain

$$\delta L = \left(\frac{\partial \mathcal{L}}{\partial x} + D_{-h} \left(u_{\underset{h}{x}} \frac{\partial \mathcal{L}}{\partial u_{\underset{h}{x}}} - \mathcal{L} \right) \right) \delta x + \left(\frac{\partial \mathcal{L}}{\partial u} - D_{-h} \left(\frac{\partial \mathcal{L}}{\partial u_{\underset{h}{x}}} \right) \right) \Psi_x \delta x. \qquad (6.27)$$

Thus, the values of the functional (6.7) remain constant in the case of variation along the curve (6.26) on the solutions of the equation

$$\frac{\partial \mathcal{L}}{\partial x} + D_{-h} \left(u_{\underset{h}{x}} \frac{\partial \mathcal{L}}{\partial u_{\underset{h}{1}}} - \mathcal{L} \right) + \Psi_x \left(\frac{\delta \mathcal{L}}{\delta u} \right) = 0.$$

The above-obtained expression

$$\frac{\delta}{\delta u} = \frac{\partial}{\partial u} - D_{-h} \left(\frac{\partial}{\partial u_{\underset{h}{x}}} \right)$$

for the Euler operator on the uniform mesh $\underset{h}{\omega}$ corresponds to the *vertical* variation in the functional (6.7). If $|\Psi_x| < C_0$ in a neighborhood of the point (x, u), then the variation is said to be *inclined*, and if $\Psi_x = 0$, then it is *horizontal*.

Now let the curve (6.26) be the orbit of the point (x, u) under transformations of the group G_1. In this case, the variations δx and δu are determined by the components of the operator of the group G_1,

$$\delta x = \xi(x, u)\delta a, \qquad \delta u = \eta(x, u)\delta a,$$

where δa is the variation of the group parameter (see Fig. 6.2). The corresponding extremal equation (6.27) is

$$\xi \left(\frac{\partial \mathcal{L}}{\partial x} + D_{-h} \left(u_{\underset{h}{x}} \frac{\partial \mathcal{L}}{\partial u_{\underset{h}{x}}} - \mathcal{L} \right) \right) + \eta \left(\frac{\partial \mathcal{L}}{\partial u} - D_{-h} \left(\frac{\partial \mathcal{L}}{\partial u_{\underset{h}{x}}} \right) \right) = 0. \qquad (6.28)$$

Equation (6.28) is said to be *quasi-extremal* (or called a *local extremal equation*), and any of its solutions are called *quasi-extremals*. Equation (6.28) can be obtained directly from transformations of the difference functional by applying the operation $\frac{\partial}{\partial a}\big|_{a=0}$ at the point (x, u).

Note that the slope characteristics $\xi(x, u)$ and $\eta(x, u)$ are contained in the quasi-extremal equation (6.28) on whose solutions the functional can be stationary in the general case. This means that the *equations for the quasi-extremals of one and the same invariant functional, in general, may have different form for different groups*.

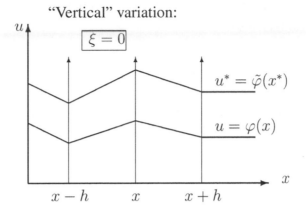

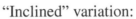

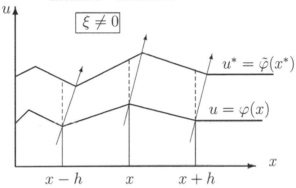

Figure 6.2

EXAMPLE 6.4. Consider the linear difference equation

$$u_{x\bar{x}} = 0 \qquad (6.29)$$

on the uniform mesh. In particular, this equation admits three translation operators

$$X_1 = \frac{\partial}{\partial x}, \qquad X_2 = \frac{\partial}{\partial u}, \qquad X_3 = \frac{\partial}{\partial x} + \frac{\partial}{\partial u}$$

preserving the mesh uniformity. The variational functional

$$L = \frac{1}{2} \sum_{\alpha} S^{\alpha}_{+h}(u_x^2 h) \qquad (6.30)$$

is invariant under X_1, X_2, and X_3, because so is the elementary action $\frac{1}{2}u_x^2 h$. Equation (6.29) is the Euler equation of the functional (6.30). All three operators X_1, X_2, and X_3 satisfy the invariance conditions for the Euler equation (6.29).

Now let us obtain the quasi-extremals of the functional (6.30) for the operators X_1, X_2, and X_3, respectively:

$$\underset{h}{u_{x\bar{x}}}\left(\frac{\underset{h}{u_x} + \underset{h}{u_{\bar{x}}}}{2}\right) = 0, \qquad \underset{h}{u_{x\bar{x}}} = 0, \qquad \underset{h}{u_{x\bar{x}}}\left(\frac{\underset{h}{u_x} + \underset{h}{u_{\bar{x}}}}{2} - 1\right) = 0.$$

Obviously, the solution of the extremal equation (6.29) is also a solution of each quasi-extremal equation, i.e., lies in the domain of intersection of quasi-extremals. But the converse is not true. Thus, on the set of solutions of Eq. (6.29), the stationary value of the functional (6.30) is attained in the case of "vertical" variation (X_2) and of "inclined" variation (X_3) and "horizontal" variation (X_1).

EXAMPLE 6.5. Consider the nonlinear difference equation

$$\underset{h}{u_{x\bar{x}}} = u^2 \tag{6.31}$$

on a uniform mesh, which obviously admits the translation operator

$$X = \frac{\partial}{\partial x}.$$

The variational functional with Lagrangian function

$$\mathcal{L} = \frac{1}{3}u^3 + \frac{1}{2}\underset{h}{u_x}^2 \tag{6.32}$$

has Eq. (6.31) as the Euler equation. But for (6.32) the horizontal variation along the orbit of the group with operator X gives the following equation for quasi-extremals:

$$\underset{h}{u_{x\bar{x}}} = \frac{2\underset{h}{u_{\bar{x}}}}{\underset{h}{u_x} + \underset{h}{u_{\bar{x}}}}\frac{u^2 + u_- u + u_-^2}{3}, \tag{6.33}$$

where $u_- = \underset{-h}{S}(u)$. Note that the nonlinear quasi-extremal equation (6.33) and the semi-linear Euler equation (6.31) are close to each other in the sense of the approximation order but still have different structures.

EXAMPLE 6.6. Consider the difference equation

$$\underset{h}{u_{x\bar{x}}} = e^u, \tag{6.34}$$

which obviously admits the translation operator $X = \partial/\partial x$ on the invariantly uniform mesh $\underset{h}{\omega}$. One can readily verify that Eq. (6.34) is the difference Euler equation of the difference functional with Lagrangian

$$\mathcal{L} = \frac{1}{2}\underset{h}{u_x}^2 + e^u. \tag{6.35}$$

The difference functional with the function $\mathcal{L}$ given by (6.35) also admits the operator $X = \frac{\partial}{\partial x}$. But the functional attains its stationary values not on the extremals (6.34) but on the quasi-extremals determined by the equation

$$u_{x\bar{x}} = e^u \left(\frac{2}{h} \frac{e^{\frac{hu_x}{h}} - 1}{u_x + u_{\bar{x}}} \right).$$

Thus, in difference variational problems, a variation of the functional leads to the Euler equation only in the case of vertical variations. The set of solutions on which the mesh functional attains stationary values depends on the variation direction, i.e., on the direction of the group orbit. If $\xi \neq 0$, then the group transformations change the independent variable and also the difference mesh. In this case, the action attains stationary values not on extremals but on quasi-extremals.

Each quasi-extremal depends on the group operator coordinates. Therefore, there are different equations of quasi-extremals for different symmetries. In the continuum limit, the difference between the quasi-extremals disappears, and all of them become the differential Euler equation.

Let us rewrite a finite-difference functional in the equivalent form

$$L = \sum_{\Omega} \mathcal{L}(x, x_+, u, u_+) h_+ \tag{6.36}$$

on some one dimensional lattice $\underset{h}{\omega}$

$$h_+ = \varphi(x_+, x, u_+, u)$$

with spacing $h_+ = x_+ - x$.

Let us take a variation of the difference functional (6.36) along some curve $u = \Psi(x)$ at some point (x, u). The variation will affect only two terms in the sum (6.36),

$$L = \cdots + \mathcal{L}(x_-, x, u_-, u) h_- + \mathcal{L}(x, x_+, u, u_+) h_+ + \cdots .$$

Thus, we obtain the following expression for the variation of the difference functional:

$$\delta L = \frac{\delta \mathcal{L}}{\delta x} \delta x + \frac{\delta \mathcal{L}}{\delta u} \delta u,$$

where $\delta u = \Psi' \delta x$ and

$$\frac{\delta \mathcal{L}}{\delta u} = h_+ \frac{\partial \mathcal{L}}{\partial u} + h_- \frac{\partial \mathcal{L}^-}{\partial u}, \qquad \frac{\delta \mathcal{L}}{\delta x} = h_+ \frac{\partial \mathcal{L}}{\partial x} + h_- \frac{\partial \mathcal{L}^-}{\partial x} + \mathcal{L}^- - \mathcal{L}$$

with $\mathcal{L} = \mathcal{L}(x, x_+, u, u_+)$ and $\mathcal{L}^- = \underset{-h}{S}(\mathcal{L}) = \mathcal{L}(x_-, x, u_-, u)$.

Now consider the variation of the functional (6.36) along the orbit of a group generated by the operator

$$X = \xi(x, u) \frac{\partial}{\partial x} + \eta(x, u) \frac{\partial}{\partial u} + \cdots . \tag{6.37}$$

Then, we have $\delta t = \xi \delta a$ and $\delta u = \eta \delta a$, where δa is the variation of the group parameter. A stationary value of the difference functional (6.36) along the flow generated by the vector field (6.37) is given by the equation

$$\xi \frac{\delta \mathcal{L}}{\delta t} + \eta \frac{\delta \mathcal{L}}{\delta u} = 0, \tag{6.38}$$

which is another representation of the quasi-extremal equation corresponding to the coefficients ξ, η of the operator (6.37).

If we have a Lie algebra of vector fields corresponding to two or more different coefficients ξ, η, then a stationary value of the difference functional (6.36) along the entire flow will be attained on the intersection of solutions of all quasi-extremal equations of type (6.38):

$$\frac{\delta \mathcal{L}}{\delta x} = 0, \qquad \frac{\delta \mathcal{L}}{\delta u} = 0. \tag{6.39}$$

This intersection will be referred as *global extremal equations*. Note that the variational equations (6.39) can be obtained by the action of the discrete variational operators

$$\frac{\delta}{\delta x} = \frac{\partial}{\partial x} + S_{-h} \frac{\partial}{\partial x_+}, \qquad \frac{\delta}{\delta u} = \frac{\partial}{\partial u} + S_{-h} \frac{\partial}{\partial u_+}$$

on the discrete elementary action $\mathcal{L}(x, x_+, u, u_+)h_+$.

Remark 6.7. We shall also use the global extremal equations (6.39) in the modified form

$$\frac{\partial \mathcal{L}}{\partial x} + \frac{h_-}{h_+} \frac{\partial \mathcal{L}^-}{\partial x} - D_{+h}(\mathcal{L}^-) = 0, \qquad \frac{\partial \mathcal{L}}{\partial u} + \frac{h_-}{h_+} \frac{\partial \mathcal{L}^-}{\partial u} = 0, \tag{6.40}$$

which is obtained by division by h_+.

Thus, for an arbitrary curve, a stationary value of a difference functional is given by a solution of Eq. (6.40).

EXAMPLE 6.8. Consider the difference model of the ordinary differential equation

$$u'' = u^{-3} \tag{6.41}$$

from the standpoint of Lagrangian formalism. For the difference analog of the Lagrangian function $u^{-2} - u_x^2$ we take the expression

$$\mathcal{L} = \frac{1}{uu_+} - \left(\frac{u_+ - u}{h_+} \right)^2, \tag{6.42}$$

which is defined at two points of a mesh $\underset{h}{\omega}$. First, let us verify the variational invariance of the Lagrangian (6.42) under the original group admitted by Eq. (6.41):

$$X_1 = \frac{\partial}{\partial x}, \qquad X_2 = 2x \frac{\partial}{\partial x} + u \frac{\partial}{\partial u}, \qquad X_3 = x^2 \frac{\partial}{\partial x} + xu \frac{\partial}{\partial u}. \tag{6.43}$$

Obviously, the difference functional

$$L = \sum_{\Omega} \left(\frac{1}{uu_+} - \left(\frac{u_+ - u}{h_+} \right)^2 \right) h_+ = \sum_{\Omega} \left(\frac{h_+}{uu_+} - \frac{(u_+ - u)^2}{h_+} \right), \tag{6.44}$$

where $h_+ = x_+ - x$, admits the translation X_1. The invariance of the elementary action $\mathcal{L}h_+$,

$$X\mathcal{L} + \mathcal{L}\underset{+h}{D}(\xi) = 0,$$

can also readily be verified for the dilation X_2. In the case of the operator X_3, we obtain the "divergence invariance"; i.e., the action of the operator X_3 gives the finite difference of some non-zero expression,

$$X_3\mathcal{L} + \mathcal{L}\underset{+h}{D}(x^2) = \frac{u_+{}^2 - u^2}{h_+} \equiv \underset{+h}{D}(u^2).$$

Let us write out the extremal difference equation

$$2\underset{h}{u_x} - 2\underset{h}{u_{\bar{x}}} - \frac{1}{u^2} \left(\frac{h_+}{u_+} + \frac{h_-}{u_-} \right) = 0 \tag{6.45}$$

corresponding to the functional (6.44), where, as usual, $\underset{h}{u_x} = (u_+ - u)/h_+$, $\underset{h}{u_{\bar{x}}} = (u - u_-)/h_-$. Note that to write out the extremal equation, we need the two terms of the sum (6.44):

$$\mathcal{L}(u, u_+)h_+ + \mathcal{L}(u_-, u)h_-.$$

Note also that the mesh $\underset{h}{\omega}$ remains undefined,

$$h_+ = \varphi(z). \tag{6.46}$$

Since the group G_3 is x-autonomous, it follows that the Euler equation (6.45) admits the operators (6.43) *under the condition that the mesh (6.46) is invariant as well*.

Now consider the quasi-extremals of the functional (6.44). In the case of an "inclined" variation (6.44) along the orbits of subgroups corresponding to the operators (6.43), we also need to take into account two terms of the sum (6.44). The operator X_1 gives us the following quasi-extremal:

$$\frac{1}{u} \left(\frac{1}{u_-} - \frac{1}{u_+} \right) + \underset{h}{u_x}(u_+ - u) - \underset{h}{u_{\bar{x}}}(u - u_-) = 0. \tag{6.47}$$

The quasi-extremal corresponding to the dilation operator X_2 is determined by the equation

$$\frac{x}{u} \left(\frac{1}{u_-} - \frac{1}{u_+} \right) + \frac{1}{u} \left(\frac{x - h_-}{u_-} - \frac{x + h_+}{u_+} \right) + 2x(\underset{h}{u_{\bar{x}}}{}^2 - \underset{h}{u_x}{}^2) + 2u(\underset{h}{u_x} - \underset{h}{u_{\bar{x}}}) = 0. \tag{6.48}$$

The third quasi-extremal equation on whose solutions the functional (6.44) takes stationary values under the transformations X_3 acquires the form

$$x^2\left(\frac{1}{uu_-} - \frac{1}{uu_+} - u_x{}^2 + u_{\bar{x}}{}^2\right) + xu\left(2u_x - 2u_{\bar{x}} - \frac{h_+}{h} - \frac{h_-}{u^2u_+} - \frac{h_-}{u^2u_-}\right) = 0.$$
(6.49)

Note that all quasi-extremal equations can be written in the unified form

$$\frac{\delta\mathcal{L}}{\delta x} = 0, \qquad 2x\left(\frac{\delta\mathcal{L}}{\delta x}\right) + \left(\frac{\delta\mathcal{L}}{\delta u}\right) = 0, \qquad x^2\left(\frac{\delta\mathcal{L}}{\delta x}\right) + xu\left(\frac{\delta\mathcal{L}}{\delta u}\right) = 0, \quad (6.50)$$

where

$$\frac{\delta\mathcal{L}}{\delta x} = \frac{\partial\mathcal{L}}{\partial x}h_+ + \frac{\partial\mathcal{L}^-}{\partial x}h_- + \mathcal{L}^- - \mathcal{L}, \qquad \frac{\delta\mathcal{L}}{\delta u} = h_+\frac{\partial\mathcal{L}}{\partial u} + h_-\frac{\partial\mathcal{L}^-}{\partial u}, \qquad \mathcal{L}^- = \underset{-h}{S}(\mathcal{L}).$$

One can readily verify that the representation of quasi-extremals in the form (6.50) completely corresponds to the already introduced form based on the use of differentiation with respect to the difference derivatives. Obviously, the quasi-extremal equations have the common domain of intersection, i.e., the global extremal

$$\frac{\delta\mathcal{L}}{\delta x} = 0, \qquad \frac{\delta\mathcal{L}}{\delta u} = 0.$$
(6.51)

Each solution of system (6.51) is a solution of system (6.50). It is remarkable that the set of solutions of system (6.51) admits the same three-parameter group as the original functional.

6.5. Invariance of Global Extremal Equations and Properties of Quasi-Extremal Equations

Now consider the invariant properties of quasi-extremal equations and Eqs. (6.51) determining the global extremal. The global extremal equations satisfy the following theorem.

THEOREM 6.9. *Let a difference functional*

$$L = \sum \mathcal{L}(x, x_+, u, u_+)h_+$$
(6.52)

invariant under the group G_1 with operator

$$X = \xi\frac{\partial}{\partial x} + \eta\frac{\partial}{\partial u} + \xi_+\frac{\partial}{\partial x_+} + \eta_+\frac{\partial}{\partial u_+}$$
(6.53)

be given. Then system (6.51) admits the same group G_1.

Proof. We express the invariance of the functional (6.52) as the equality of the two corresponding series, in which we preserve only two terms of the series:

$$\mathcal{L}(x, x_+, u, u_+)h_+ + \mathcal{L}^-(x_-, x, u_-, u)h_-$$
$$= \mathcal{L}(x^*, x^*_+, u^*, u^*_+)h^*_+ + \mathcal{L}^-(x^*_-, x^*, u^*_-, u^*)h^*_-.$$

After differentiating the last relation with respect to x and u, we apply the operation $\partial/\partial a\big|_{a=0}$ to the result and obtain

$$\frac{\partial}{\partial a}\left(\frac{\partial \mathcal{L}}{\partial x^*}h^*_+ + \frac{\partial \mathcal{L}^-}{\partial x^*}h^*_- + \mathcal{L}^- - \mathcal{L}\right)\bigg|_{a=0} + \xi_x \frac{\delta \mathcal{L}}{\delta x} + \eta_x \frac{\delta \mathcal{L}}{\delta u} = 0,$$

$$\frac{\partial}{\partial a}\left(\frac{\partial \mathcal{L}}{\partial u^*}h^*_+ + \frac{\partial \mathcal{L}^-}{\partial u^*}h^*_-\right)\bigg|_{a=0} + \xi_u \frac{\delta \mathcal{L}}{\delta x} + \eta_u \frac{\delta \mathcal{L}}{\delta u} = 0.$$

The substitution of Eqs. (6.51) into the above relations completes the proof of the theorem,

$$X\left(\frac{\delta \mathcal{L}}{\delta x}\right)\bigg|_{(6.51)} = 0, \quad X\left(\frac{\delta \mathcal{L}}{\delta u}\right)\bigg|_{(6.51)} = 0. \qquad \square$$

COROLLARY. *If the intersection (6.51) of quasi-extremals corresponds to a functional (6.52) invariant under an r-parameter group G_r, then system (6.51) is also invariant under G_r.*

Note that system (6.51), as well as any invariant manifold, can be written in terms of difference invariants of the group G_r. We also note that we are not yet interested in the set of solutions of system (6.51), which may be empty (i.e., the quasi-extremal equations may be inconsistent).

Another property of quasi-extremals of an invariant functional is given by the following theorem.

THEOREM 6.10. *Let the quasi-extremal equation*

$$\xi \frac{\delta \mathcal{L}}{\delta x} + \eta \frac{\delta \mathcal{L}}{\delta x} = 0 \qquad (6.54)$$

correspond to the stationary values of the functional (6.52) under variations along the orbit of the subgroup G_1 with operator (6.53). Then Eq. (6.54) admits the same subgroup G_1.

In other words, each quasi-extremal equation is invariant under "its own" subgroup. This fact is almost obvious. Indeed, assume that Eq. (6.54) does not admit the operator X (with the same coordinates (ξ, η)). Then this equation under the action of transformations of the corresponding subgroup G_1 becomes a certain different equation (6.54)*, and the functional remains unchanged in this case.

But Eq. (6.54) was obtained as the set of *all* solutions on which (6.52) is stationary. Therefore, Eq. (6.54)* cannot produce an additional manifold on which the functional (6.52) is stationary. Just this contradiction proves the invariance of the quasi-extremal equation (6.54).

Thus, the quasi-extremal equations are invariant under their "own" subgroups, and their intersection admits the entire set of symmetries of the invariant functional.

The following natural question arises: How does a "foreign" subgroup associated with another extremal act on the given quasi-extremal equation?

Rather often, this question can be answered by using the following relation, which holds for the invariant Lagrangian and permits writing out the quasi-extremal equation in divergence form:

$$\xi\left(\frac{\partial\mathcal{L}}{\partial x} + \frac{h_-}{h_+}\frac{\partial\mathcal{L}^-}{\partial x} - \underset{+h}{D}(\mathcal{L}^-)\right) + \eta\left(\frac{\partial\mathcal{L}}{\partial u} + \frac{h_-}{h_+}\frac{\partial\mathcal{L}^-}{\partial u}\right)$$
$$+ \underset{+h}{D}\left(h^-\eta\frac{\partial\mathcal{L}^-}{\partial u} + h^-\xi\frac{\partial\mathcal{L}^-}{\partial x} + \xi\mathcal{L}^-\right) = 0. \quad (6.55)$$

One can readily verify that relation (6.55) is equivalent to the invariance condition for the functional (6.52):

$$\xi\frac{\partial\mathcal{L}}{\partial x} + \xi_+\frac{\partial\mathcal{L}}{\partial x_+} + \eta\frac{\partial\mathcal{L}}{\partial u} + \eta_+\frac{\partial\mathcal{L}}{\partial u_+} + \mathcal{L}\underset{+h}{D}(\xi) = 0.$$

Relation (6.55) allows us to prove the following assertion.

THEOREM 6.11. *Suppose that the quasi-extremal equation (6.54) of the invariant functional (6.52) corresponds to a group G_1 with operator X. Let there exist an operator $\bar{X}$ commuting on the solutions of (6.54) with the discrete differentiation operator,*

$$[\bar{X}, \underset{+h}{D}] = 0. \quad (6.56)$$

Then the action of the operator $\bar{X}$ transforms the quasi-extremal (6.54) into another quasi-extremal of the same functional.

COROLLARY. *The operators of the adjoint Lie algebra admitted by (6.52) have the property (6.56). Thus, a new additional operation, namely, the action of the adjoint algebra, is introduced on the set of all quasi-extremals of the invariant functional. In this connection, one can to introduce the following new notion: a basis of quasi-extremal equations, which is a minimal set of quasi-extremals from which all the other quasi-extremal equations can be obtained by the action of the operators $\bar{X}$.*

Note that all the above-listed properties of quasi-extremal equations are of difference character; i.e., in the continuum limit all the quasi-extremal equations become a single Euler equation, which is invariant under the entire set of symmetries of the corresponding Lagrangian.

EXAMPLE. Now let us illustrate all these properties by the example of the equation

$$u'' = u^{-3},$$

which was considered in the preceding section. Under the transformations (6.43), the invariant Lagrangian (6.42) preserves its constant values on the solutions of the corresponding quasi-extremal equations (6.47), (6.48), and (6.49). For these equations, system (6.51) acquires the form

$$\frac{1}{uu_-} - \frac{1}{uu_+} + u_x{}^2 - u_{\bar{x}}^2 = 0, \qquad 2u_x - 2u_{\bar{x}} - \frac{h_+}{u^2 u_+} - \frac{h_-}{u^2 u_-} = 0. \qquad (6.57)$$

One can readily verify that system (6.57) admits the entire symmetry of (6.43).

The quasi-extremal equations (6.50) were obtained in the preceding section. They permit verifying the second property of the quasi-extremals. One can see that the first equation in system (6.50) admits the operator X_1, the second equation admits the operator X_2, and the third equation admits the operator X_3; i.e., each quasi-extremal is invariant under its respective subgroup.

We also easily see the third property of the quasi-extremals: the operator $\partial/\partial x$ takes the third equation to the second, and the action of the operator $\frac{1}{2}\partial/\partial x$ transforms the second equation into the first equation. Thus, the third equation in system (6.50) forms a basis of quasi-extremals. As we shall see later, the same relation holds for the conservation laws.

An invariant system determines both an invariant mesh and an invariant difference equation for u. In particular, this system contains equations obtained earlier by the method of difference invariants. Indeed, by substituting

$$h_+ = \varepsilon uu_+, \qquad h_- = \varepsilon uu_- \qquad (6.58)$$

into system (6.57), we obtain

$$(u_+ - u)u_- - (u - u_-)u_+ = \varepsilon^2 u_+ u_-, \qquad (u_+ - u)^2 u_- - (u - u_-)^2 u_+ = \varepsilon(u_- - u_+).$$

Both of the last equations are equivalent to the already known mapping

$$u_+ u_- (2 - \varepsilon^2) = u(u_+ + u_-),$$

which approximates Eq. (6.41) up to second-order terms on the mesh (6.58).

6.6. Conservation Laws for Difference Equations

Let finite-difference equations

$$F_\alpha(x, u, \underset{h}{u_x}, \dots, h) = 0, \qquad \alpha = 1, 2, \dots, m, \qquad (6.59)$$

be given on a difference mesh

$$\Omega_\beta(x, u, \dots, h) = 0, \quad \beta = 1, \dots, n, \qquad F_\alpha, \Omega_\beta \in \underset{h}{A}. \qquad (6.60)$$

DEFINITION. We say that a system of difference equations has a *conservation law* if there exists a vector $\mathbf{A}$ with components $A^i = A^i(x, u, \underset{h}{u_x}, \ldots, h)$, $i = 1, \ldots, n$, $A^i \in \underset{h}{A}$, such that, on any solution $u = \varphi(x)$ of system (6.59)–(6.60), it satisfies the condition

$$\mathrm{DIV}\,\mathbf{A} \equiv \sum_{i=1}^{n} \underset{+h}{D_i}(A^i) = 0, \qquad (6.61)$$

where $\underset{+h}{D_i}$ denotes the discrete differentiation in the ith direction.

If there exist r vectors $\mathbf{A}$, satisfying condition (6.61) and *linearly independent with constant coefficients*, then one says that system (6.59)–(6.60) has r independent conservation laws.

Note that in the one-dimensional case, where $\mathbf{A} \equiv A^1 = A(x, u, \underset{h}{u_x}, \ldots, h)$, the condition

$$\underset{+h}{D}(A(z))\big|_{(6.59),(6.60)} = 0 \qquad (6.62)$$

for system (6.59)–(6.60) to be conservative is equivalent to the condition saying that $A(x, u, \underset{h}{u_1})$ is a *mesh invariant* on the solutions of (6.59)–(6.60),

$$A(z)\big|_{F=0} = \underset{\pm h}{S^{\alpha}}(A(z))\big|_{(6.59),(6.60)}, \qquad \alpha = 1, 2, \ldots, \qquad (6.63)$$

where the $\underset{\pm h}{S}$ are the discrete shift operators. The conservation law taken in the form (6.63) is a *first integral* of the one-dimensional version of system (6.59)–(6.60) and is a finite algebraic expression on the mesh $\underset{h}{\omega}$.

In the case of arbitrary nonuniform meshes $\underset{h}{\omega}$, it should be remembered that the operators $\underset{\pm h}{S_i}$ and $\underset{\pm h}{D_i}$ in (6.61) and (6.63) are of "local" character; i.e., they depend on the spacings h_i^+ and h_i^- at a given point of the mesh.

Note also that condition (6.62) can be rewritten in infinitesimal form using the fact that the shift operator $\underset{\pm h}{S}$ was obtained with the use of the tangent field of the Taylor group:

$$\underset{+h}{D^{\pm}}(A(z))\big|_{(6.59),(6.60)} = 0, \qquad \underset{+h}{D^{\pm}} = \frac{\partial}{\partial x} + \underset{\pm h}{\tilde{D}}(u)\frac{\partial}{\partial u} + \underset{\pm h}{\tilde{D}}(\underset{h}{u_1})\frac{\partial}{\partial u_x} + \cdots, \qquad (6.64)$$

where $\underset{\pm h}{\tilde{D}}$ is the Lagrangian representation of the differentiation operator on the difference mesh,

$$\underset{\pm h}{\tilde{D}} = \sum_{n=1}^{\infty} \frac{(\mp h)^{n-1}}{n}\underset{\pm h}{D^n}.$$

Note that the use of the conservation law in the form (6.64) requires the use of all points of the difference mesh.

Let us show that the difference analog of the differential form of the conservation law (6.61) implies a difference analog of the integral form of the conservation law. Let the last coordinate of the vector $(x^1, x^2, \ldots, x^n)$ be the time $x^n = t$. Take the cylindrical domain

$$\underset{h}{\Omega} = \left\{ x \underset{h}{\in} \omega : \sum_{i=1}^{n-1} (x^i)^2 = r_0^2, \quad t_1 \leq t \leq t_2 \right\},$$

where $x = (x^1, x^2, \ldots, x^{n-1}, t)$ and r_0, t_1, and t_2 are constants.

Then it follows from (6.61) that

$$\sum_s A l_n h_1 h_2 \ldots h_{n-1} \tau = \sum_{\underset{h}{\Omega}} \mathrm{DIV}\ \mathbf{A} = 0, \tag{6.65}$$

where S is the surface bounding the domain $\underset{h}{\Omega}$ and l_n is the unit exterior normal vector on S.

If A^i decays sufficiently rapidly at spatial infinity for the solutions of system (6.59) (6.60), then, by setting $r_0 \to \infty$, we can omit the summation over the cylindrical surface. (If $A^i = 0$ on it, then the situation is similar.)

We have

$$A l_n = -A^n \big|_{t=t_1}$$

on the lower base of the cylinder $\underset{h}{\Omega}$ and

$$A l_n = A^n \big|_{t=t_2}$$

on the upper base. It follows from (6.65) that for any solution $u = u(x)$ of system (6.59)–(6.60) the function $A^n(x, u(x)\underset{h}{u}(x)_1, \ldots)$ satisfies the equation

$$\sum_{R^{n-1}} A^n h_1 h_2 \cdots h_{n-1} \bigg|_{t=t_1} = \sum_{R^{n-1}} A^n h_1 h_2 \cdots h_{n-1} \bigg|_{t=t_2},$$

which implies that the variable

$$E = \sum_{R^{n-1}} A^n(x, u(x), \underset{h}{u_x}(x), \ldots) h_1 h_2 \cdots h_{n-1}$$

is independent of time on the solution of system (6.59)–(6.60),

$$\underset{+\tau}{D} E \big|_{(6.59)-(6.60)} = 0, \tag{6.66}$$

where

$$\underset{+\tau}{D} = \frac{1}{\tau} (\underset{+\tau}{S} - 1)$$

is the discrete differentiation with respect to time on the mesh $\underset{h}{\omega}$.

Note that for $n = 1$ conditions (6.62) and (6.66) coincide and are equivalent to the existence of a first integral.

The difference schemes with conservation laws of the form (6.61) are said to be *conservative*. Such schemes ensure that there are no fictitious sources in the computational domain, which is of high importance in numerical implementations (see [122]).

6.7. Noether-Type Identities and Difference Analog of Noether's Theorem

In this section, we develop difference analogs of the Noether identity (see the Introduction). This identity plays a crucial role for formulation of a difference analog of Noether's theorem for difference equations. The approach based on an operator identity provides a simple, clear way to construct first integrals (conservation laws) for quasi-extremal and global extremal equations just by means of algebraic manipulations.

LEMMA 6.12. *The operator identity* ([30, 36, 39])

$$
X(\mathcal{L}) + \mathcal{L}\underset{+h}{D}(\xi) \equiv \xi\left(\frac{\partial\mathcal{L}}{\partial x} + \frac{h_-}{h_+}\frac{\partial\mathcal{L}^-}{\partial x} - \underset{+h}{D}(\mathcal{L}^-)\right)
$$
$$
+ \eta\left(\frac{\partial\mathcal{L}}{\partial u} + \frac{h_-}{h_+}\frac{\partial\mathcal{L}^-}{\partial u}\right) + \underset{+h}{D}\left(h_-\eta\frac{\partial\mathcal{L}^-}{\partial u} + h_-\xi\frac{\partial\mathcal{L}^-}{\partial x} + \xi\mathcal{L}^-\right) \quad (6.67)
$$

holds for any function $\mathcal{L} = \mathcal{L}(x, x_+, u, u_+)$ *and any vector field* X.

The left hand-side of this identity can be written as follows:

$$
X(\mathcal{L}) + \mathcal{L}\underset{+h}{D}(\xi) = \xi\frac{\partial\mathcal{L}}{\partial x} + \xi + \frac{\partial\mathcal{L}}{\partial x} + \eta\frac{\partial\mathcal{L}}{\partial u} + \eta\frac{\partial\mathcal{L}}{\partial u} + \mathcal{L}\underset{+h}{D}(\xi).
$$

This identity is a discrete analog of the Noether identity and will be called the *discrete* (or *difference*) *Noether identity*.

We rewrite the discrete Noether identity for the Lagrangian in the equivalent form $\mathcal{L} = \underset{h}{\mathcal{L}}(x, u, u_x)$:

a. For a one-dimensional regular mesh, one has

$$\xi\frac{\partial\mathcal{L}}{\partial x} + \eta\frac{\partial\mathcal{L}}{\partial u} + [\underset{+h}{D}(\eta) - u_x\underset{h}{D}(\xi)]\frac{\partial\mathcal{L}}{\partial\underset{h}{u_x}} + \mathcal{L}\underset{+h}{D}(\xi)$$

$$\equiv \xi\left[\frac{\partial\mathcal{L}}{\partial x} + \underset{h}{u_{\bar{x}}}\frac{\partial\mathcal{L}}{\partial u} + \underset{h}{u_{x\bar{x}}}\frac{\partial\mathcal{L}}{\partial\underset{h}{u_x}} - \underset{-h}{D}(\mathcal{L})\right]$$

$$+ (\eta - \xi\underset{h}{u_{\bar{x}}})\left[\frac{\partial\mathcal{L}}{\partial u} - \underset{-h}{D}\left(\frac{\partial\mathcal{L}}{\partial\underset{h}{u_x}}\right)\right] + \underset{+h}{D}\left[\xi\mathcal{L}^- + (\eta - \xi\underset{h}{u_{\bar{x}}})\left(\frac{\partial\mathcal{L}^-}{\partial\underset{h}{u_{\bar{x}}}}\right)\right], \quad (6.68)$$

where $\mathcal{L}^- = \mathcal{L}(x_-, u_-, \underset{h}{u_{\bar{x}}})$.

b. For a one-dimensional irregular mesh, one has

$$\xi\frac{\partial\mathcal{L}}{\partial x} + \eta\frac{\partial\mathcal{L}}{\partial u} + [\underset{+h}{D}(\eta) - u_x\underset{h}{D}(\xi)]\frac{\partial\mathcal{L}}{\partial\underset{h}{u_x}} + \mathcal{L}\underset{+h}{D}(\xi)$$

$$\equiv \xi\left[\frac{\partial\mathcal{L}}{\partial x} + \underset{h}{u_{\bar{x}}}\frac{\partial\mathcal{L}}{\partial u} + \frac{h^-}{h^+}\underset{-h}{D}(u_x)\frac{\partial\mathcal{L}}{\partial\underset{h}{u_{\bar{x}}}} - \frac{h^-}{h^+}\underset{-h}{D}(\mathcal{L})\right]$$

$$+ (\eta - \xi\underset{h}{u_{\bar{x}}})\left[\frac{\partial\mathcal{L}}{\partial u} - \frac{h^-}{h^+}\underset{-h}{D}\left(\frac{\partial\mathcal{L}}{\partial\underset{h}{u_x}}\right)\right] + \underset{+h}{D}\left[\xi\mathcal{L} + (\eta - \xi\underset{h}{u_{\bar{x}}})\left(\frac{\partial\mathcal{L}^-}{\partial\underset{h}{u_{\bar{x}}}}\right)\right]. \quad (6.69)$$

The following operator identity holds for the two-dimensional case if the Lagrangian has the form $\mathcal{L} = \mathcal{L}(x^1, x^2, u, \underset{h}{u_1}, \underset{h}{u_2})$ on a two-dimensional orthogonal regular mesh:

$$\xi^1\frac{\partial\mathcal{L}}{\partial x^1} + \xi^2\frac{\partial\mathcal{L}}{\partial x^2} + \eta\frac{\partial\mathcal{L}}{\partial u} + \mathcal{L}\left(\underset{+h}{D_1}(\xi^1) + \underset{+h}{D_2}(\xi^2)\right)$$

$$+ [\underset{+h}{D_1}(\eta) - u_1\underset{h}{D_1}(\xi^1) - \underset{+h}{S_1}(u_2)\underset{+h}{D_1}(\xi^2)]\frac{\partial\mathcal{L}}{\partial\underset{h}{u_1}}$$

$$+ [\underset{+h}{D_2}(\eta) - u_2\underset{h}{D_2}(\xi^2) - \underset{+h}{S_2}(u_1)\underset{+h}{D_2}(\xi^1)]\frac{\partial\mathcal{L}}{\partial\underset{h}{u_2}}$$

$$\equiv \xi^1\left[\frac{\partial\mathcal{L}}{\partial x^1} + \underset{-h}{D_1}\left(u_1\frac{\partial\mathcal{L}}{\partial\underset{h}{u_1}}\right) - \underset{-h}{D_1}(\mathcal{L}) + \underset{-h}{D_2}\left(\underset{+h}{S_2}(u_1)\frac{\partial\mathcal{L}}{\partial\underset{h}{u_2}}\right)\right]$$

$$+ \xi^2\left[\frac{\partial\mathcal{L}}{\partial x^2} + \underset{-h}{D_2}\left(u_2\frac{\partial\mathcal{L}}{\partial\underset{h}{u_2}}\right) - \underset{-h}{D_2}(\mathcal{L}) + \underset{-h}{D_1}\left(\underset{+h}{S_1}(u_2)\frac{\partial\mathcal{L}}{\partial\underset{h}{u_1}}\right)\right]$$

$$+ \eta\left[\frac{\partial\mathcal{L}}{\partial u} - \underset{-h}{D_1}\left(\frac{\partial\mathcal{L}}{\partial\underset{h}{u_1}}\right) - \underset{-h}{D_2}\left(\frac{\partial\mathcal{L}}{\partial\underset{h}{u_2}}\right)\right]$$

$$+ \underset{+h}{D_1}\left[\xi^1\underset{-h}{S_1}(\mathcal{L}) + (\eta - \xi^1\underset{h}{u_{\bar{1}}} - \xi^2 u_2)\underset{-h}{S_1}\left(\frac{\partial\mathcal{L}}{\partial\underset{h}{u_1}}\right)\right]$$

$$+ \underset{+h}{D_2}\left[\xi^2 \underset{-h}{S}_2(\mathcal{L}) + (\eta - \xi^1 \underset{h}{u_{\bar{1}}} - \xi^2 u_2) \underset{-h}{S}_2\left(\frac{\partial \mathcal{L}}{\partial \underset{h}{u_2}}\right)\right], \quad (6.70)$$

where $\underset{h}{u_1}$ and $\underset{h}{u_2}$ are the right difference derivatives in the x^1- and x^2-directions, respectively, and u_1 and u_2 are the continuous partial derivatives in discrete representation,

$$u_i = \sum_{n=1}^{\infty} \frac{(-h_i)^{n-1}}{n} \underset{+h}{D}_i^n(u), \qquad i = 1, 2.$$

Proof. Identities (6.67)–(6.70) can be proved by a straightforward verification. $\quad\square$

Remark. In the continuous limit $h \to 0$, identities (6.67)–(6.70) tend to Noether's differential identities of appropriate dimensions.

From this relations, we obtain the following theorem, which we state just for the case of identity (6.67). (For all other cases, one can state the theorem in a similar way.)

THEOREM 6.13. *Let an element X of the Lie algebra of the group G give the quasi-extremal equation*

$$\xi\left(\frac{\partial \mathcal{L}}{\partial x} + \frac{h_-}{h_+}\frac{\partial \mathcal{L}^-}{\partial x} - \underset{+h}{D}(\mathcal{L}^-)\right) + \eta\left(\frac{\partial \mathcal{L}}{\partial u} + \frac{h_-}{h_+}\frac{\partial \mathcal{L}^-}{\partial u}\right) = 0.$$

This quasi-extremal equation possesses a first integral of the form

$$\mathcal{I} = \left(h_-\eta\frac{\partial \mathcal{L}^-}{\partial u} + h_-\xi\frac{\partial \mathcal{L}^-}{\partial x} + \xi\mathcal{L}^-\right)$$

if and only if the Lagrangian function $\mathcal{L}$ is invariant under the Lie group G of local point transformations generated by vector fields X of the form

$$X = \xi\frac{\partial}{\partial x} + \xi_+\frac{\partial}{\partial x_+} + \eta\frac{\partial}{\partial u} + \eta\frac{\partial}{\partial u_+}.$$

Proof. The assertion follows from identity (6.67). $\quad\square$

Remark 6.14. If the Lagrangian density $\mathcal{L}$ is divergence invariant under a Lie group of local point transformations, i.e., if

$$X(\mathcal{L}) + \mathcal{L}\underset{+h}{D}(\xi) = \underset{+h}{D}(V)$$

for some function $V(x, u)$, then each element X of the Lie algebra of G provides the first integral

$$\mathcal{I} = h_-\eta\frac{\partial \mathcal{L}^-}{\partial u} + h_-\xi\frac{\partial \mathcal{L}^-}{\partial x} + \xi\mathcal{L}^- - V$$

of the quasi-extremal equations.

Remark 6.15. As we know, the invariance of a difference functional leads to the invariance of the corresponding quasi-extremal equation. In the case of a multi-parameter group G_r the invariance of a difference functional gives r distinct quasi-extremal equations. Each of these quasi-extremal equations is invariant under its own subgroup and, in accordance with the above theorem, possesses its own first integral.

EXAMPLE 6.16. Consider the functional with $\mathcal{L} = 3u_{\underset{h}{x}}{}^2 + 2u^3$ on a uniform mesh and the translation operator $X = \partial/\partial x$. The function $\mathcal{L}$ satisfies the invariance condition, just as does the Euler equation

$$u_{\underset{h}{x\bar{x}}} = u^2.$$

The quasi-extremal equation

$$u_{\underset{h}{x\bar{x}}} = \frac{2u_{\underset{h}{\bar{x}}}}{u_{\underset{h}{x}} + u_{\underset{h}{\bar{x}}}} \frac{u^2 + uu_- + u_-{}^2}{3}$$

is invariant under translations in x and has the first integral

$$I = 2u^3 - 3u_{\underset{h}{x}}{}^2 = C_0, \qquad C_0 = \text{const},$$

which implies the exact solution of the nonlinear difference equation in the recursive form

$$u_+ = u \pm h\sqrt{\frac{2}{3}u^3 + C_0}.$$

EXAMPLE 6.17. With the same translation operator on the uniform mesh $\underset{h}{\omega}$, consider the functional with $\mathcal{L} = 0.5u_{\underset{h}{x}}^2 + e^u$, which, as well as the Euler equation, has the following difference equation:

$$u_{\underset{h}{x\bar{x}}} = e^u.$$

The quasi-extremal equation

$$u_{\underset{h}{x\bar{x}}} = \frac{2}{h}e^u\left(\frac{1 - e^{-\frac{hu_{\bar{x}}}{h}}}{u_{\underset{h}{x}} + u_{\underset{h}{\bar{x}}}}\right)$$

of this functional corresponding to the operator $X = \partial/\partial x$ has the conservation law

$$\underset{+h}{D}\left(0.5u_{\underset{h}{x\bar{x}}}^2 - e^{u_-}\right) = 0,$$

which is equivalent to the existence of the first integral for the quasi-extremal

$$u_x^2 - 2e^u = C_0, \qquad C_0 = \text{const},$$

and which implies the solution in the recursive form

$$u_+ = u \pm h\sqrt{e^u + C_0}.$$

Theorem 6.13 has an important practical consequence, which can be viewed as a *difference analog of Noether's theorem.*

THEOREM 6.18 ([37]). *Let a nondegenerate functional with Lagrangian function* $\mathcal{L} = \mathcal{L}(x, x_+, u, u_+)$ *admit an r-parameter group* G_r. *(That is, assume that the functional has a variational symmetry or a divergence invariance.) Assume that there are r quasi-extremal equations*

$$\xi^\alpha \frac{\delta \mathcal{L}}{\delta x} + \eta^\alpha \frac{\delta \mathcal{L}}{\delta u} = 0, \qquad \alpha = 1, 2, \ldots, r. \tag{6.71}$$

Then the system of global extremals

$$\frac{\delta \mathcal{L}}{\delta x} = 0, \qquad \frac{\delta \mathcal{L}}{\delta u} = 0$$

lying in the intersection of the quasi-extremals (6.71) *admits the group* G_r *and has r conservation laws of the form*

$$\mathcal{I} = \left(h_- \eta^\alpha \frac{\partial \mathcal{L}^-}{\partial u} + h_- \xi^\alpha \frac{\partial \mathcal{L}^-}{\partial x} + \xi \mathcal{L}^- \right) \qquad \alpha = 1, 2, \ldots, r.$$

Theorems for Lagrangians of different form and for meshes of different dimension can be stated in a similar way. Note that the invariant mesh is either determined in the process of finding the first integrals or can be added independently to the system of global extremals.

EXAMPLE 6.19 (see [48, 49]). Let us consider how the above-obtained theorems can be used to compose a conservative difference model of the ordinary differential equation

$$u'' = u^{-3}.$$

For the difference Lagrangian function we take (Sec. 6.4) the function

$$\mathcal{L} = \frac{1}{uu_+} - \left(\frac{u_+ - u}{h_+} \right)^2,$$

which is defined at two mesh points. The variational invariance

$$X_1 \mathcal{L} + \mathcal{L} \underset{+h}{D}(\xi_1) = 0, \qquad X_2 \mathcal{L} + \mathcal{L} \underset{+h}{D}(\xi_2) = 0, \qquad X_3 \mathcal{L} + \mathcal{L} \underset{+h}{D}(\xi_3) = \underset{+h}{D}(y^2)$$

of the Lagrangian implies the three quasi-extremal equations

$$\frac{\delta \mathcal{L}}{\delta x} = 0, \qquad 2x\frac{\delta \mathcal{L}}{\delta x} + \frac{\delta \mathcal{L}}{\delta u} = 0, \qquad x^2\frac{\delta \mathcal{L}}{\delta x} + xu\frac{\delta \mathcal{L}}{\delta u} = 0$$

and the corresponding system of global extremals

$$
\begin{aligned}
\frac{\delta \mathcal{L}}{\delta u} &: \quad 2(\underset{h}{u_x} - \underset{h}{u_{\bar{x}}}) = \frac{h_+}{u^2 u_+} + \frac{h_-}{u^2 u_-}, \\
\frac{\delta \mathcal{L}}{\delta x} &: \quad (\underset{h}{u_x})^2 + \frac{1}{u u_+} - (\underset{h}{u_{\bar{x}}})^2 - \frac{1}{u u_-} = 0,
\end{aligned}
\tag{6.72}
$$

where, as usual, $\underset{h}{u_x} = (u_+ - u)/h_+$ and $\underset{h}{u_{\bar{x}}} = (u - u_-)/h_-$.

Application of Theorem 6.18 gives the following three functionally independent first integrals:

$$I_1 = \underset{h}{u_x^2} + \frac{1}{u u_+} = A, \quad I_2 = \frac{2x + h^+}{2}\underset{h}{u_x^2} + \frac{2x + h^+}{2 u u^+} - \frac{u + u^+}{2}\underset{h}{u_x} = 2B,$$

$$I_3 = \frac{x(x + h^+)}{u u^+} + \left(\frac{u + u^+}{2} - \frac{2x + h^+}{2}\underset{h}{u_x} \right)^2 = C.$$

$$\tag{6.73}$$

By analogy with the continuous case, the discrete first integrals obey the relations

$$X_1(I_3) = I_2, \qquad X_1(I_2) = 2I_1, \qquad X_3(I_1) = -I_2, \qquad X_3(I_2) = -2I_3,$$

and consequently, each first integral can be taken as a basic integral.

It can be verified the three first integrals satisfy the relation

$$\frac{1}{4}\left(\frac{h_+}{u u_+} \right)^2 = 1 + \frac{I_2^2}{4} - I_1 I_3,$$

which means that $h_+ (yy_+)^{-1}$ is a constant on the solutions of the global extremal equations. This allows us to introduce the special case

$$\frac{h_-}{y_- y} = \frac{h_+}{y y_+} = \varepsilon, \qquad \varepsilon = \text{const}, \quad 0 < \varepsilon \ll 1, \tag{6.74}$$

of the invariant mesh.

Then the general solution of the discrete model (6.72) can be found with the help of the first integrals $I_1 = A$ and $I_2 = 2B$ as

$$Au^2 = (Ax + B)^2 + 1 - \frac{\varepsilon^2}{4} \tag{6.75}$$

just by algebraic manipulations.

Note that the solution (6.75) contains three parameters and a third parameter ε arises from the mesh where (6.75) is defined. This solution differs from the solution of the underlying continuous equation by $\varepsilon^2/4$, and the estimate is uniform. The solution for the mesh points x^n and u^n, $n = 0, 1, 2, \ldots$, was obtained in [48, 49].

The difference integrals (6.73) are partly inherited in an appropriate mapping. The substitution of the mesh (6.74) into system (6.72) of global extremal equations reduces the latter to the equation

$$u_+ u_- (2 - \varepsilon^2) = u(u_+ + u_-),$$

which is a one-dimensional mapping. This mapping has the only first integral

$$\tilde{I}_1 = \left(\frac{u_+ - u}{\varepsilon u u_+} \right)^2 + \frac{1}{u u_+},$$

which is the first integral I_1 inherited from the system of global extremals.

Thus, we have developed the invariant difference model

$$\frac{\frac{u_x}{h} - \frac{u_{\bar{x}}}{h}}{h_-} = \frac{1}{u^2 u_-}, \qquad \frac{h_+}{u u_+} = \frac{h_-}{u u_-} = \varepsilon, \qquad \varepsilon = \text{const}, \quad 0 < \varepsilon \ll 1,$$

of the original ODE, which possesses the same Lagrangian structure. The complete set of first integrals allows us to write out an invariant lattice and completely integrate the discrete model.

It should be noted that not all quasi-extremal equations are consistent, and one cannot find invariant lattices from the first integrals in all cases. Several other cases will be considered in Sec. 6.9.

6.8. Necessary and Sufficient Conditions for Global Extremal Equations to Be Invariant

It has been shown in Sec. 6.5 that if the functional $\mathcal{L}$ is invariant under some group G, then the global extremal equations

$$\frac{\delta \mathcal{L}}{\delta x} = 0, \qquad \frac{\delta \mathcal{L}}{\delta u} = 0 \tag{6.76}$$

are invariant with respect to G as well. If the Lagrangian $\mathcal{L}$ is divergence invariant, then so are the global extremal equations (6.76). This follows from the fact that the total finite differences belong to the kernel of the discrete variational operators.

As in the continuous case, the global extremal equations can be invariant with respect to a larger group than the corresponding Lagrangian.

Now we are in a position to establish a *necessary and sufficient* condition for the invariance of global extremal equations. We present new identities and a new theorem [45].

LEMMA 6.20. *The following identities hold for any smooth function $\mathcal{L}(t, t_+, u, u_+)$:*

$$\frac{\delta}{\delta u}((X(\mathcal{L}) + \mathcal{L}\underset{+h}{D}(\xi))h_+) \equiv X\left(\frac{\delta\mathcal{L}}{\delta u}\right) + \frac{\partial\eta}{\partial u}\frac{\delta\mathcal{L}}{\delta u} + \frac{\partial\xi}{\partial u}\frac{\delta\mathcal{L}}{\delta x},$$

$$\frac{\delta}{\delta x}((X(\mathcal{L}) + \mathcal{L}\underset{+h}{D}(\xi))h_+) \equiv X\left(\frac{\delta\mathcal{L}}{\delta x}\right) + \frac{\partial\eta}{\partial x}\frac{\delta\mathcal{L}}{\delta u} + \frac{\partial\xi}{\partial x}\frac{\delta\mathcal{L}}{\delta x}.$$

Proof. The identities can be proved by a straightforward verification. □

Lemma 6.20 allows one to obtain not only a sufficient but also a necessary and sufficient condition for the invariance of the global extremal equations.

THEOREM 6.21. *The global extremal equations (6.76) are invariant with respect to a symmetry generator X if and only if the following conditions are true on the solutions of Eqs. (6.76):*

$$\frac{\delta}{\delta u}((X(\mathcal{L}) + \mathcal{L}\underset{+h}{D}(\xi))h_+)\Big|_{(6.76)} = 0,$$

$$\frac{\delta}{\delta x}((X(\mathcal{L}) + \mathcal{L}\underset{+h}{D}(\xi))h_+)\Big|_{(6.76)} = 0.$$

Proof. The assertion follows from the identities in Lemma 6.20. □

EXAMPLE. Consider the difference functional with Lagrangian

$$\mathcal{L} = \frac{u_{\overset{x}{h}}^2}{2} + \frac{u^3}{3} = \frac{(u_+ - u)^2}{2(x_+ - x)^2} + \frac{u^3}{3}$$

on a regular lattice, which is invariant with respect to the operators

$$X_1 = \frac{\partial}{\partial x}, \qquad X_2 = x\frac{\partial}{\partial x} - 2u\frac{\partial}{\partial u}.$$

Then the global extremal system is

$$\frac{\delta\mathcal{L}}{\delta u} = u_{\overset{x\bar{x}}{h}} - u^2 = 0,$$

$$\frac{\delta\mathcal{L}}{\delta x} = u_{\overset{x\bar{x}}{h}} - \frac{2u_{\overset{\bar{x}}{h}}}{u_{\overset{x}{h}} + u_{\overset{\bar{x}}{h}}}\frac{u^2 + uu_- + u_-^2}{3} = 0. \tag{6.77}$$

The invariance of the difference Lagrangian is satisfied for the operator X_1,

$$X_1(\mathcal{L}) + \mathcal{L}\underset{+h}{D}(\xi^1) = 0,$$

while the application of the operator X_2 to the Lagrangian action gives

$$X_2(\mathcal{L}) + \mathcal{L}\underset{+h}{D}(\xi^2) = -5\mathcal{L} \neq 0.$$

Meanwhile, an application of Theorem 6.21 indicates the invariance of both Eqs. (6.77) with respect to the operators X_1 and X_2:

$$\frac{\delta}{\delta u}((X_1(\mathcal{L}) + \mathcal{L}\underset{+h}{D}(\xi^1))h_+)\Big|_{(6.77)} = 0,$$

$$\frac{\delta}{\delta x}((X_2(\mathcal{L}) + \mathcal{L}\underset{+h}{D}(\xi^2))h_+)\Big|_{(6.77)} = 0.$$

An alternative approach to conservation laws for difference equations is developed in [68, 70].

6.9. Applications of Lagrangian Formalism to Second-Order Difference Equations

In this section, we give examples of construction of conservative second-order difference equations approximating second-order ODE and admitting the same transformation group [48, 49]. The example of an invariant conservative scheme for second-order ODE considered in the preceding section is, in a sense, the simplest example; in this example, we can find a difference analog of the invariant Lagrangian, which gives a first integral for all symmetries. Moreover, the use of first integrals permits construction of an invariant mesh. In this section, we consider more complicated situations. We use a modification of the above-obtained constructions of the difference Lagrangian formalism. It turns out that in the difference case not only different Lagrangian functions can be used for variations along the orbits of different subgroups (just as in the continuous case) but also *different approximations* to the same invariant Lagrangian for different subgroups (which, of course, do not exist in the continuous case) can be involved.

In the case where second-order ODEs have two and more symmetries, they can be integrated completely. The methods for finding the general solution can be different, but most of them employ the mathematical apparatus of integration. This integration technique is in fact absent in the difference case. (The technique of exact integration of ordinary difference equations, including nonlinear equations, is required.) But the use of first integrals permits obtaining the general solution with the use of only algebraic operations. Therefore, to obtain analytic solutions of difference schemes, we use a technique based on the difference analog of the Noether theorem.

6.9.1. Equations corresponding to Lagrangians invariant under one-dimensional groups

First, consider the simplest case of a one-dimensional group. By an appropriate change of variables it can be transformed into

$$X_1 = \frac{\partial}{\partial y}.$$

The most general second-order ODE invariant under X_1 is

$$y'' = F(x, y'), \tag{6.78}$$

where F is an arbitrary given function.

An invariant Lagrangian function should have the form $L = L(x, y')$, and the Euler equation is

$$\frac{\partial^2 L}{\partial x \partial y'} + y'' \frac{\partial^2 L}{\partial y'^2} = 0.$$

By substituting y'' from Eq. (6.78), we obtain a linear partial differential equation for $L(x, y')$. This, of course, has infinitely many solutions. Let us assume that we know a solution $L(x, y)$ explicitly. Then the Noether theorem provides the first integral

$$\frac{\partial L}{\partial y'}(x, y') = A. \tag{6.79}$$

If we can solve Eq. (6.79) for y' as a function of x (and A), then the general solution can be obtained by a quadrature,

$$y' = \phi(x, A), \qquad y(x) = y_0 + \int_0^x \phi(x, A)\, dt.$$

In the discrete case, the situation is similar. Assume that we know a Lagrangian $\mathcal{L}(x, x_+, u, u_+)$ invariant under the group of transformations of u generated by X_1. (We replace y by u to distinguish the discrete case from the continuous one.) It has the form

$$\mathcal{L} = \mathcal{L}(x, x_+, u_{\underset{h}{x}}), \qquad u_{\underset{h}{x}} = \frac{u_+ - u}{x_+ - x}.$$

The corresponding global extremal equations are

$$\frac{\delta \mathcal{L}}{\delta u} = -\frac{\partial \mathcal{L}}{\partial u_{\underset{h}{x}}}(x, x_+, u_{\underset{h}{x}}) + \frac{\partial \mathcal{L}}{\partial u_{\underset{h}{\bar{x}}}}(x_-, x, u_{\underset{h}{\bar{x}}}) = 0, \tag{6.80}$$

$$\frac{\delta \mathcal{L}}{\delta x} = h_+ \frac{\partial \mathcal{L}}{\partial x}(x, x_+, u_{\underset{h}{x}}) + u_{\underset{h}{x}} \frac{\partial \mathcal{L}}{\partial u_{\underset{h}{x}}}(x, x_+, u_{\underset{h}{x}}) - \mathcal{L}(x, x_+, u_{\underset{h}{x}})$$

$$+ h_- \frac{\partial \mathcal{L}}{\partial x}(x_-, x, u_{\underset{h}{\bar{x}}}) - u_{\underset{h}{\bar{x}}} \frac{\partial \mathcal{L}}{\partial u_{\underset{h}{\bar{x}}}}(x_-, x, u_{\underset{h}{\bar{x}}}) + \mathcal{L}(x_-, x, u_{\underset{h}{\bar{x}}}) = 0. \tag{6.81}$$

The first integral can be obtained from Eq. (6.80); it is given by

$$\frac{\partial \mathcal{L}}{\partial u_x}(x, x_+, \underset{h}{u_x}) = A. \tag{6.82}$$

We can solve Eq. (6.82) for $\underset{h}{u_x}$ and by shifting down obtain $\underset{h}{u_{\bar{x}}}$,

$$\underset{h}{u_x} = \phi(x, x_+, A), \qquad \underset{h}{u_{\bar{x}}} = \phi(x_-, x, A).$$

By substituting this into the global extremal equation (6.81), we obtain a relation between x_+, x_-, and x, i.e., a single three-point relation for the variable x. For u we then obtain the two-point equation

$$u_+ - u = (x_+ - x)\phi(x, x_+, A). \tag{6.83}$$

Equation (6.83) is a discrete analog of quadrature; it is a first-order inhomogeneous linear equation for u. Note that any lattice $h_+ = f(h_-, x)$ will be invariant.

6.9.2. Equations corresponding to Lagrangians invariant under two-dimensional groups

A. The Abelian Lie algebra with unconnected basis elements

$$X_1 = \frac{\partial}{\partial x}, \qquad X_2 = \frac{\partial}{\partial y} \tag{6.84}$$

corresponds to the invariant ODE

$$y'' = F(y'), \tag{6.85}$$

where F is an arbitrary function.

The equation can be obtained from the Lagrangian

$$L = y + G(y'), \qquad F(y') = \frac{1}{G''(y')}. \tag{6.86}$$

The Lagrangian admits the symmetries X_1 and X_2,

$$X_1 L + L D(\xi_1) = 0, \qquad X_2 L + L D(\xi_2) = 1 = D(x). \tag{6.87}$$

With the help of Noether's theorem, we obtain the following first integrals:

$$J_1 = y + G(y') - y'G'(y'), \qquad J_2 = G'(y') - x. \tag{6.88}$$

It suffices to have two first integrals to write out the general solution of a second-order ODE without quadratures. We can solve the second equation (6.88) for y' in terms of x and obtain

$$y' = H(J_2 + x), \qquad H(J_2 + x) = [G']^{-1}(J_2 + x).$$

By substituting this into the first equation, we obtain the general solution in the form

$$y(x) = J_1 - G[H(J_2 + x)] + (J_2 + x)H(J_2 + x). \qquad (6.89)$$

Now we show how one can find a discrete model and its first integrals by means of Lagrange-type technique. Let us choose a difference Lagrangian in the form

$$\mathcal{L} = \frac{u + u_+}{2} + G(\underset{h}{u_x}); \qquad (6.90)$$

then we verify the invariance conditions

$$X_1\mathcal{L} + \mathcal{L}\underset{+h}{D}(\xi_1) = 0, \qquad X_2\mathcal{L} + \mathcal{L}\underset{+h}{D}(\xi_2) = 1 = \underset{+h}{D}(x).$$

The variations of $\mathcal{L}$ yield the following global extremal equations:

$$\frac{\delta\mathcal{L}}{\delta u}: \qquad G'(\underset{h}{u_x}) - G'(\underset{h}{u_{\bar{x}}}) = \frac{h_+ + h_-}{2}, \qquad (6.91)$$

$$\frac{\delta\mathcal{L}}{\delta x}: \qquad -\frac{u + u_+}{2} - G(\underset{h}{u_x}) + u_x G'(\underset{h}{u_x}) + \frac{u + u_-}{2}$$
$$+ G(\underset{h}{u_{\bar{x}}}) - u_{\bar{x}} G'(\underset{h}{u_{\bar{x}}}) = 0. \qquad (6.92)$$

The difference analog of Noether's theorem yields two first integrals

$$I_1 = u + G(\underset{h}{u_x}) - u_x G'(\underset{h}{u_x}) + \frac{x_+ - x}{2}u_x, \qquad (6.93)$$

$$I_2 = G'(\underset{h}{u_x}) - \frac{x + x_+}{2}. \qquad (6.94)$$

We can solve Eq. (6.94) for u_x to obtain

$$\underset{h}{u_x} = \Phi_1(I_2, x + x_+). \qquad (6.95)$$

By substituting this into the equation for I_1, we obtain

$$u = \Phi_2(I_1, I_2, x, x_+). \qquad (6.96)$$

Calculating $\underset{h}{u_x}$ from Eq. (6.96) and setting it equal to (6.95), we obtain a three-point recursion relation for x. Solving it, we turn Eq. (6.96) into an explicit general solution of the difference scheme (6.91), (6.92).

EXAMPLE. Consider the case in which

$$\mathcal{L} = \frac{u + u_+}{2} + \exp(\underset{h}{u_x}).$$

The two first integrals (6.93) and (6.94) in this case are the following:

$$I_1 = u + \exp(\underset{h}{u_x}) - u_x \exp(\underset{h}{u_x}) + \frac{x_{n+1} - x_n}{2} \underset{h}{u_x}, \qquad I_2 = \exp(\underset{h}{u_x}) - \frac{x_{n+1} + x_n}{2}.$$

Equations (6.95) and (6.96) are reduced to

$$\underset{h}{u_x} = \ln\left(I_2 + \frac{x_{n+1} + x_n}{2}\right),$$

$$u = I_1 - I_2 - \frac{x_{n+1} + x_n}{2} + (I_2 + x_n)\ln\left(I_2 + \frac{x_{n+1} + x_n}{2}\right).$$

The recursion relation for x reads

$$\frac{-x_{n+1} + x_{n-1}}{2} + (I_2 + x_n)\big[\ln(2I_2 + x_{n+1} + x_n) - \ln(2I_2 + x_n + x_{n-1})\big].$$

The last equation provides the lattice, but it is difficult to solve. We have however reduced a system of two three-point equations to a single three-point equation. We shall return to this case later, applying an alternative method.

B. The non-Abelian Lie algebra with unconnected elements

$$X_1 = \frac{\partial}{\partial y}, \qquad X_2 = x\frac{\partial}{\partial x} + y\frac{\partial}{\partial y} \qquad (6.97)$$

gives the invariant ODE

$$y'' = \frac{1}{x}F(y'). \qquad (6.98)$$

We define a function $G(y')$ by the equation

$$F(y') = \frac{G'(y')}{G''(y')}.$$

Then the ODE (6.98) is the Euler equation for the Lagrangian

$$L = \frac{1}{x}G(y'),$$

which admits X_1 and X_2 as variational symmetries,

$$X_1 L + LD(\xi_1) = 0, \qquad X_2 L + LD(\xi_2) = 0.$$

Noether's theorem provides two first integrals

$$J_1 = \frac{1}{x}G'(y'), \qquad J_2 = G(y') + \left(\frac{y}{x} - y'\right)G'(y'),$$

which are sufficient to integrate the ODE.

Let us take the difference Lagrangian

$$\mathcal{L} = \frac{2}{x + x_+} \underset{h}{G}(u_x),$$

which satisfies the invariance conditions

$$X_1 \mathcal{L} + \mathcal{L} \underset{+h}{D}(\xi_1) = 0, \qquad X_2 \mathcal{L} + \mathcal{L} \underset{+h}{D}(\xi_2) = 0.$$

Then the variations of $\mathcal{L}$ give the following global extremal equations:

$$\frac{\delta \mathcal{L}}{\delta u} : \quad \frac{2}{x + x_+} \underset{h}{G'}(u_x) - \frac{2}{x_- + x} \underset{h}{G'}(u_{\bar{x}}) = 0,$$

$$\frac{\delta \mathcal{L}}{\delta x} : \quad -\frac{2h_+}{(x + x_+)^2} \underset{h}{G}(u_x) + \frac{2}{(x + x_+)} \underset{h}{G'}(u_x) \underset{h}{u_x} - \frac{2}{(x + x_+)} \underset{h}{G}(u_x) \qquad (6.99)$$

$$- \frac{2h_-}{(x_- + x)^2} \underset{h}{G}(u_{\bar{x}}) - \frac{2}{(x_- + x)} \underset{h}{G'}(u_{\bar{x}}) \underset{h}{u_{\bar{x}}} + \frac{2}{(x_- + x)} \underset{h}{G}(u_{\bar{x}}) = 0.$$

Since the Lagrangian is invariant with respect to the operators X_1 and X_2, we find the first integrals

$$I_1 = \frac{2 \underset{h}{G'}(u_x)}{x + x_+}, \qquad I_2 = \frac{4xx_+}{(x + x_+)^2} \underset{h}{G}(u_x) + \frac{2 \underset{h}{G'}(u_x)}{x + x_+} \left(u - x \underset{h}{u_x} \right)$$

for the solutions of (6.99).

As in the previous case, we can express $\underset{h}{u_x}$ from the integral I_1 and obtain

$$u_x = \Phi_1(I_1, x + x_+).$$

The second integral allows us to express u as a function of x and x_+,

$$u = x\Phi_1 + \frac{I_2}{I_1} - \frac{4xx_+}{I_1(x + x_+)^2} G(\Phi_1).$$

Remark. The Lie algebra of operators

$$X_1 = \frac{\partial}{\partial x}, \qquad X_2 = (n - 1)x\frac{\partial}{\partial x} - 2y\frac{\partial}{\partial y}, \qquad n \neq -1, 1,$$

which is isomorphic to the algebra (6.97), was considered in [40], where an invariant difference equation and an invariant lattice were constructed for the approximation of the invariant ODE

$$y'' = y^n.$$

6.9.3. Equations corresponding to Lagrangians invariant under three-dimensional groups

We have already considered in Sec. 6.7 one invariant difference model for ODE that has three-dimensional symmetry groups. In this section, we consider another two cases. Both of them come from Lagrangians that have three-dimensional symmetry groups as well.

A. First, consider the family of solvable Lie algebras

$$X_1 = \frac{\partial}{\partial x}, \qquad X_2 = \frac{\partial}{\partial y}, \qquad X_3 = x\frac{\partial}{\partial x} + ky\frac{\partial}{\partial y}, \qquad k \neq 0, \frac{1}{2}, 1, 2, \quad (6.100)$$

depending on one constant k. The invariant equation has the form

$$y'' = y'^{\frac{k-2}{k-1}}. \tag{6.101}$$

This equation can be obtained by the usual variational procedure from the Lagrangian

$$L = \frac{(k-1)^2}{k}(y')^{\frac{k}{k-1}} + y,$$

which admits the operators X_1 and X_2 for any parameter k,

$$X_1 L + LD(\xi_1) = 0, \qquad X_2 L + LD(\xi_2) = 1 = D(x),$$

and the operator X_3 for $k = -1$,

$$X_3 L + LD(\xi_3) = (k+1)L.$$

One can show that there is no Lagrangian function $L(x, y, y')$ that produces Eq. (6.101) with $k \neq -1$ as its Euler equation and is divergence invariant for all three symmetries (6.100).

For arbitrary k there are two first integrals

$$J_1 = \frac{(1-k)}{k}(y')^{\frac{k}{k-1}} + y = A^0, \qquad J_2 = (k-1)(y')^{\frac{1}{k-1}} - x = B^0.$$

By eliminating y', we find the general solution

$$y = \frac{1}{k}\left(\frac{1}{k-1}\right)^{k-1}(x + B^0)^k + A^0. \tag{6.102}$$

For $k = -1$, we have yet another first integral corresponding to the symmetry X_3,

$$J_3 = \frac{2}{\sqrt{y'}}(y - xy') + xy = C^0.$$

It is functionally dependent on J_1 and J_2, since a second-order ODE can possess only two functionally independent first integrals. Note that the integral J_3 is the basic integral,

$$J_1 = X_1(J_3), \qquad J_2 = -X_2(J_3),$$

since

$$[X_1, X_3] = X_1, \qquad [X_2, X_3] = kX_2.$$

In this case, we have the following relation:

$$4 - J_1 J_2 - J_3 = 0. \tag{6.103}$$

Thus, the integral J_3 is not independent and is of no use in the present context.

B. Now let us proceed to the discrete case and consider only $k = -1$. (The other values of k will be considered later on under a different approach.) Take

$$\mathcal{L} = -4\sqrt{\underset{h}{u_x}} + \frac{u + u_+}{2}$$

as the discrete Lagrangian, which is invariant with respect to X_1 and X_3 and divergence invariant with respect to X_2,

$$X_1\mathcal{L} + \mathcal{L}\underset{+h}{D}(\xi_1) = 0, \qquad X_2\mathcal{L} + \mathcal{L}\underset{+h}{D}(\xi_2) = 1 = \underset{+h}{D}(x), \qquad X_3\mathcal{L} + \mathcal{L}\underset{+h}{D}(\xi_3) = 0.$$

From the Lagrangian, we obtain the global extremal equations

$$\frac{\delta\mathcal{L}}{\delta u} : \quad -\frac{4}{h_- + h_+}\left(\frac{1}{\sqrt{\underset{h}{u_x}}} - \frac{1}{\sqrt{\underset{h}{u_{\bar{x}}}}}\right) = 1,$$

$$\frac{\delta\mathcal{L}}{\delta x} : \quad 4\left(\sqrt{\underset{h}{u_x}} - \sqrt{\underset{h}{u_{\bar{x}}}}\right) - \frac{u + u_+}{2} + \frac{u_- + u}{2} = 0. \tag{6.104}$$

This system of equations is invariant with respect to all three operators (6.100). Application of the difference analog of the Noether theorem gives us three first integrals

$$I_1 = -2\sqrt{\underset{h}{u_x}} + \frac{u + u_+}{2} = A, \qquad I_2 = -\frac{2}{\sqrt{\underset{h}{u_x}}} - \frac{x + x_+}{2} = B,$$

$$I_3 = \frac{2(x_+ u - u_+ x)}{h_+ \sqrt{\underset{h}{u_x}}} + \frac{x_+ u + u_+ x}{2} = C.$$

In contrast to the continuous case, these three first integrals I_1, I_2, and I_3 are functionally independent, and instead of Eq. (6.103) we have the relation

$$4 - I_1 I_2 - I_3 = \frac{1}{4}h_+^2 \underset{h}{u_x} = \frac{4\varepsilon^2}{(\varepsilon + 2)^2}.$$

This coincides with Eq. (6.103) in the continuous limit $\varepsilon \to 0$. We see that the expression $h^2_{+}u_{\underset{h}{x}}$ is a first integral of (6.104) as well. This allows us to introduce a convenient lattice, namely,

$$\frac{1}{4}h^2_{-}u_{\underset{h}{\bar{x}}} = \frac{1}{4}h^2_{+}u_{\underset{h}{x}} = \frac{4\varepsilon^2}{(\varepsilon + 2)^2}, \qquad \varepsilon = \text{const}, \quad 0 < \varepsilon \ll 1. \tag{6.105}$$

By substituting $u_{\underset{h}{x}}$ from Eq. (6.105) into I_2, we obtain a two-term recursion relation for x,

$$x_{n+1} - (1 + \varepsilon)x_n - \varepsilon B = 0 \tag{6.106}$$

or

$$-(1 + \varepsilon)x_{n+1} + x_n - \varepsilon B = 0,$$

depending on the sign choice for $\sqrt{u_{\underset{h}{x}}}$. These equations can be solved, and we obtain a lattice satisfying

$$x_n = (x_0 + B)(1 + \varepsilon)^n - B, \qquad x_0 > -B, \tag{6.107}$$

for the first equation and a lattice satisfying

$$x_n = (x_0 + B)(1 + \varepsilon)^{-n} - B, \qquad x_0 < -B, \tag{6.108}$$

for the second equation. Using the expressions for I_1, we obtain the general solution for u (the same for both lattices (6.107) and (6.108)) in the form

$$u_n = A - \frac{4}{x_n + B}\frac{1 + \varepsilon}{(1 + \varepsilon/2)^2}. \tag{6.109}$$

We have used the three integrals I_1, I_2, and I_3 to obtain the general solution of the difference scheme (6.104). Indeed, the solution (6.107), (6.109) for x_n and u_n depends on the four constants (A, B, x_0, ε), as it should.

The difference scheme is not consistent with a regular lattice but requires an exponential one, as in Eq. (6.107). Note that the only nonalgebraic step in the integration was the solution of a linear two-point equation with constant coefficients (6.106), which is known from the theory of difference schemes (e.g., see [122]).

6.9.4. Integration of difference equation with two variational symmetries: The method of perturbed Lagrangians

It was mentioned that a two-dimensional group of Lagrangian symmetries is always sufficient to reduce the original system of two three-point equations to a single three-point equation for the independent variable alone. Using a different approach, we shall actually obtain a complete solution of a difference scheme approximating a differential equation with a Lagrangian invariant under a two-dimensional symmetry group.

The case we shall consider is Eq. (6.85) and hence the two-dimensional Abelian group corresponding to the algebra (6.84). We shall make use of the fact that the Lagrangian is not unique. In fact, we consider three different Lagrangians, all having the same continuous limit (6.86). Instead of writing out the Lagrangian (6.90) in the discrete case, we shall use a family of Lagrangians parametrized by two constants α and β,

$$\mathcal{L} = \alpha G(\underset{h}{u_x}) + \beta u + (1 - \beta)u_+, \qquad \alpha \approx 1, \quad 0 \le \beta \le 1.$$

Each Lagrangian provides its own global extremal system

$$\alpha\left[-G'(\underset{h}{u_x}) + G'(\underset{h}{u_{\bar{x}}})\right] + \beta h_+ + (1 - \beta)h_- = 0, \tag{6.110}$$

$$\alpha\left[\underset{h}{u_x}G'(\underset{h}{u_x}) - \underset{h}{u_{\bar{x}}}G'(\underset{h}{u_{\bar{x}}}) - G(\underset{h}{u_x}) + G(\underset{h}{u_{\bar{x}}})\right]$$
$$- \beta(u - u_-) - (1 - \beta)(u_+ - u) = 0. \tag{6.111}$$

We shall view one Lagrangian, with $\alpha_3 = 1$ and $\beta_3 = 0.5$, as the basic one, and the other two as its perturbations.

Each Lagrangian in the family is divergence invariant under $X_1 = \frac{\partial}{\partial x}$ and $X_2 = \frac{\partial}{\partial u}$ and hence provides two first integrals of the corresponding global extremal equations (6.110) and (6.111),

$$\alpha\left[\underset{h}{u_x}G'(\underset{h}{u_x}) - G(\underset{h}{u_x})\right] + u + (1 - \beta)h_+\underset{h}{u_x} = A, \tag{6.112}$$

$$\alpha G'(\underset{h}{u_x}) - x - \beta h_+ = B. \tag{6.113}$$

Let us now choose three different pairs (α_i, β_i). They provide six integrals (and six global extremal equations). We shall show that, by appropriately fine tuning the constants α_i and β_i and by choosing some of the constants A_i and B_i, we can manufacture a consistent difference system representing both the equation and the lattice. Moreover, we can explicitly integrate the equations in a manner that approximates the exact solution obtained in the continuous limit.

Let us take one equation of the form (6.112) and two of the form (6.113). In these three equations, we choose $\alpha_3 = 1$, $\beta_3 = 0.5$, and $B_2 = B_3 = B$. We then take the difference between the two equations involving B and finally obtain the following system of three two-point equations:

$$\alpha_1\left[-\underset{h}{u_x}G'(\underset{h}{u_x}) + G(\underset{h}{u_x})\right] + u + (1 - \beta_1)h_+\underset{h}{u_x} = A, \tag{6.114}$$

$$G'(\underset{h}{u_x}) - x - \frac{1}{2}h_+ = B, \tag{6.115}$$

$$(1 - \alpha_2)G'(\underset{h}{u_x}) - \left(\frac{1}{2} - \beta_2\right)h_+ = 0. \tag{6.116}$$

From Eq. (6.115) and (6.116), we have

$$G'(\underset{h}{u_x}) = \frac{x_+ + x + 2B}{2},\tag{6.117}$$

$$x_+ - (1+\varepsilon)x - \varepsilon B = 0,\tag{6.118}$$

where we have set

$$\varepsilon = \frac{2(1-\alpha_2)}{\alpha_2 - 2\beta_2}.\tag{6.119}$$

The continuous limit will correspond to $\varepsilon \to 0$.

Equation (6.118) coincides with Eq. (6.106) obtained by using three Lagrangian symmetries in a special case. Here it appears in a much more general setting. The general solution

$$x_n = (x_0 + B)(1+\varepsilon)^n - B\tag{6.120}$$

of Eq. (6.118) depends on one integration constant x_0. This solution gives a lattice satisfying $h_- > 0$ and $h_+ > 0$ for $x_0 > -B$ if $\varepsilon > 0$ and for $x_0 < -B$ if $\varepsilon < 0$. For the other cases, namely for $x_0 < -B$ if $\varepsilon > 0$ and for $x_0 > -B$ if $\varepsilon < 0$, formula (6.120) gives a lattice with reverse order of points, $h_- < 0$ and $h_+ < 0$.

Using (6.120) and (6.117), we can express $\underset{h}{u_x}$ via x. We have

$$G'(\underset{h}{u_x}) = \left(1 + \frac{\varepsilon}{2}\right)(B + x).$$

Denoting the inverse function of $G'(u_x)$ by H, we have

$$\underset{h}{u_x} = H\left[\left(1 + \frac{\varepsilon}{2}\right)(B + x)\right].\tag{6.121}$$

Using (6.114) and (6.121), we can now write out the general solution of system (6.114)–(6.116) as

$$u(x) = A - \alpha_1 G(H) + (x + B)H,\tag{6.122}$$

where we have set

$$\alpha_1\left(1 + \frac{\varepsilon}{2}\right) - (1 - \beta_1)\varepsilon = 1.$$

The value of α_1, still occurring in the solution (6.122), must be chosen so as to obtain a consistent scheme. Indeed, x_n and u_n given in Eq. (6.120) and (6.122) will satisfy system (6.114)–(6.116). We should however ensure that $\underset{h}{u_x}$ in Eq. (6.121) and $\underset{h}{u_x} = (u_{n+1} - u_n)/(x_{n+1} - x_n)$ coincide. A simple computation shows that this equation requires that α_1 should satisfy

$$\alpha_1 = (1 + \varepsilon)^{n+1}(x_0 + B)\frac{H_{n+1} - H_n}{G(H_{n+1}) - G(H_n)}.\tag{6.123}$$

This equation is consistent only if the right-hand side is a constant (independent on n). The constants α_i and β_i can depend on the constant ε, and for $\varepsilon \to 0$ we should have $\alpha_1, \alpha_2 \to 1$ and $\beta_1, \beta_2 \to 0.5$.

From Eq. (6.116), we have

$$\frac{h_+}{\underset{h}{G'(u_x)}} = \frac{2(1 - \alpha_2)}{1 - 2\beta_2}.$$

This expression must vanish as $\varepsilon \to 0$. To achieve this while respecting Eq. (6.119), we set

$$\alpha_2 = 1 + \varepsilon^2, \qquad \beta_2 = \frac{1}{2} + \varepsilon + \frac{\varepsilon^2}{2}.$$

Equation (6.119) is satisfied exactly, and we have

$$\frac{h_+}{\underset{h}{G'(u_x)}} = \frac{2\varepsilon}{\varepsilon + 2}. \tag{6.124}$$

We can view Eqs. (6.120) and (6.122) as the general solution of the three-point difference scheme

$$G'(u_x) - G'(u_{\bar{x}}) - \frac{x_+ - x_-}{2} = 0, \qquad \frac{h_+}{\underset{h}{G'(u_x)}} = \frac{h_-}{\underset{h}{G'(u_{\bar{x}})}}. \tag{6.125}$$

System (6.125) is invariant under the group corresponding to (6.84). Strictly speaking, this is not a global extremal system, since it cannot be derived from any single Lagrangian. The arbitrary constants A, B, and ε come from the three first integrals (6.114), (6.115), and (6.124), which are associated with three different Lagrangians.

Thus, the ODE (6.85) obtained from the Lagrangian (6.86) can be approximated by the difference system (6.125). If α_1 in Eq. (6.123) is constant, then the general solution of this system is given by

$$x_n = (x_0 + B)(1 + \varepsilon)^n - B, \qquad u(x_n) = A - \alpha_1 G(H_n) + (x_n + B)H_n, \tag{6.126}$$

where A, B, ε, and x_0 are arbitrary constants. As $\varepsilon \to 0$, $u(x_n)$ agrees with the solution (6.89) of the ODE (6.85).

We have not proved that Eq. (6.123) is consistent for arbitrary functions $G(u_x)$. We shall however show that the above integration scheme is consistent in at least two interesting special cases. In both cases, the Lagrangian is only divergence invariant under a two-dimensional subgroup.

EXAMPLE 6.22 (A polynomial nonlinearity).

$$X_1 = \frac{\partial}{\partial x}, \qquad X_2 = \frac{\partial}{\partial u}, \qquad X_3 = x\frac{\partial}{\partial x} + ku\frac{\partial}{\partial u}, \qquad k \neq 0, \frac{1}{2}, \pm 1, 2.$$

This algebra for $k = -1$ was treated earlier, and now we consider the general case. We take

$$G(\underset{h}{u_x}) = \frac{(k-1)^2}{k} \underset{h}{u_x^{\frac{k}{k-1}}}$$

and hence

$$G'(\underset{h}{u_x}) = (k-1)\underset{h}{u_x^{\frac{1}{k-1}}} = \left(1 + \frac{\varepsilon}{2}\right)(x + B).$$

Equation (6.121) is reduced to

$$\underset{h}{u_x} = H_n(x) = \left(\frac{x+B}{k-1}\right)^{k-1}\left(1 + \frac{\varepsilon}{2}\right)^{k-1},$$

and we have

$$G(H_n) = \frac{(k-1)^2}{k}\left(\frac{x+B}{k-1}\right)^k\left(1 + \frac{\varepsilon}{2}\right)^k.$$

Substituting this into (6.123), we find

$$\alpha_1 = \frac{k(1+\varepsilon)((1+\varepsilon)^{k-1} - 1)}{(k-1)(1+\frac{\varepsilon}{2})((1+\varepsilon)^k - 1)},$$

so that $\alpha_1 = 1 + O(\varepsilon^2)$.

Thus, α_1 is a constant close to $\alpha_1 = 1$ for $\varepsilon \ll 1$. The solution u_n of (6.126) specializes to

$$u_n = A + \frac{(x+B)^k}{(k-1)^{k-1}}\frac{\varepsilon(1+\varepsilon/2)^{k-1}}{(1+\varepsilon)^k - 1}. \tag{6.127}$$

This agrees with the solution (6.102) of the ODE (6.101) up to $O(\varepsilon^2)$.

It is of interest to note that α_1 becomes independent on ε for $k = -1$ and we obtain $\alpha_1 = 1$ and $\beta_1 = 0.5$. The solution (6.127) provides us with the solution (6.109), which was obtained in Sec. 6.4 with the help of a different method.

EXAMPLE 6.23 (An exponential nonlinearity). Consider another three-dimensional group and its Lie algebra

$$X_1 = \frac{\partial}{\partial x}, \qquad X_2 = \frac{\partial}{\partial y}, \qquad X_3 = x\frac{\partial}{\partial x} + (x+y)\frac{\partial}{\partial y}.$$

The corresponding invariant ODE is

$$y'' = \exp(-y') \tag{6.128}$$

and can be obtained from the Lagrangian

$$L = \exp(y') + y.$$

We have

$$X_1 L + LD(\xi_1) = 0, \qquad X_2 L + LD(\xi_2) = 1 = D(x).$$

The corresponding first integrals of Eq. (6.128) are

$$\exp(y')(1 - y') + y = A, \qquad \exp(y') - x = B.$$

Finally, the general solution of Eq. (6.128) is

$$y = (x + B)(\ln(x + B) - 1) + A. \tag{6.129}$$

Now consider the discrete case, following the method of Sec. 6.9.4. We have

$$G(\underset{h}{u_x}) = \exp(\underset{h}{u_x})$$

and hence

$$G'(\underset{h}{u_x}) = \exp(\underset{h}{u_x}) = (x_n + B)\left(1 + \frac{\varepsilon}{2}\right), \qquad H_n = \underset{h}{u_x} = \ln(x_n + B) + \ln\left(1 + \frac{\varepsilon}{2}\right).$$

Substituting this into Eq. (6.123), we find

$$\alpha_1 = \frac{(1 + \varepsilon)\ln(1 + \varepsilon)}{\varepsilon\,(1 + \varepsilon/2)},$$

so that α_1 is indeed a constant, and moreover, $\alpha_1 = 1 + O(\varepsilon^2)$.

The solution $u(x)$ on the lattice given in Eq. (6.126) is

$$u_n = A + (x_n + B)\ln(x_n + B) + (x_n + B)\left[\ln\left(1 + \frac{\varepsilon}{2}\right) - \frac{(1 + \varepsilon)\ln(1 + \varepsilon)}{\varepsilon}\right].$$

This agrees with the solution (6.129) of the ODE (6.128) up to $O(\varepsilon^2)$.

We see that variational symmetries and the first integrals they provide play a crucial role in the study of exact solutions of invariant difference schemes, much more so than in the theory of ordinary differential equations.

The procedure followed in this section can be reformulated as follows. We start from the continuous case, where we know a Lagrangian density $L(x, y, y')$ invariant under a local point transformation group G_0. We hence also know the corresponding Euler–Lagrange equation invariant under the same group or a larger group containing G_0 as a subgroup.

Then we approximate this Lagrangian by a discrete Lagrangian $L(x, u, x_+, u_+)$ invariant under the same group G_0. Even in the absence of any symmetry group, the Lagrangian will provide the global extremal equations (or the discrete Euler–Lagrange system).

If the Lagrangian is invariant under a one-dimensional symmetry group, then we can reduce the global extremal system to a three-point relation for x alone plus a "discrete quadrature" for u. If the symmetry group of the Lagrangian is two dimensional, then we can always reduce the global extremal system to one three-point equation for x alone and write out the solution $u_n(x)$ directly.

If the invariance group of the Lagrangian is three dimensional, then we can integrate the system explicitly.

It was shown that if the symmetry group of the Lagrangian is two dimensional but the global extremal system has a third (non-Lagrangian) symmetry, then we can also integrate the difference scheme explicitly.

6.10. Moving Mesh Schemes for the Nonlinear Schrödinger Equation

In this section, we apply the Lagrangian formalism with the conservation of Lie point symmetries to partial differential equations and construct several conservative difference schemes. The object of our study is the nonlinear Schrödinger equation (NLS) [19, 22, 38]. In one dimension, this equation is integrable and there exist many numerical schemes (e.g., see [1, 4, 100, 137]) which are based on either preserving this integrability (often through a direct discretization of the underlying Lagrangian) and/or preserving the mass or energy (Hamiltonian) of the solution. When studying blow-up phenomena in higher-dimensional NLS, it is much less clear whether this is a good idea. In particular, it can lead to meshes that are relatively sparse in the blow-up region [20]. An alternative approach [20,23], which has proved efficient for a number of blow-up problems, is to construct moving mesh numerical methods which preserve the scaling symmetries close to the blow-up point. While these may not strictly conserve mass or energy, they have proved efficient in resolving the blow-up structure. A key test of this is whether they can accurately reproduce the self-similar evolution behavior which is known [131] to describe the asymptotic behavior of the blow-up. The purpose of this section is to determine how feasible it is to combine these three approaches, namely, a discretization of the Lagrangian, preserving the symmetry, and a moving mesh numerical method.

6.10.1. The cubic nonlinear Schrödinger equation: Symmetry, Lagrangian structure, and conservation laws

Consider the radially symmetric cubic nonlinear Schrödinger equation

$$i\frac{\partial u}{\partial t} + \frac{\partial^2 u}{\partial r^2} + \frac{n-1}{r}\frac{\partial u}{\partial r} + u|u|^2 = 0, \tag{6.130}$$

where n is the number of space dimensions.

This equation describes many physical situations, including some phenomena in plasma physics and nonlinear optics (see [131]). For the case of $n = 1$, the equation is integrable and the solution exists globally. We shall consider the non-integrable case $n \geq 2$, in which singularities are observed to develop for suitable initial data. A motivation for considering the radially symmetric form of the non-linear Schrödinger equation is that it has been observed in numerical experiments reported in [131] that singularities in the NLS when posed as a problem in three dimensions are highly symmetric close to the singular point.

Let us substitute the polar representation

$$u(r, t) = A e^{i\Phi},$$

where $A = A(r, t)$ and $\Phi = \Phi(r, t)$ are real functions, into Eq. (6.130); we then obtain the following two equations:

$$A_t + A\Phi_{rr} + 2A_r\Phi_r + \frac{n-1}{r}A\Phi_r = 0, \qquad (6.131)$$

$$A\Phi_t + A\Phi_r^2 - A_{rr} - \frac{n-1}{r}A_r - A^3 = 0. \qquad (6.132)$$

Lie group analysis yields the symmetries of system (6.131)–(6.132), and for $n \geq 2$ the admitted Lie algebra of operators is the following:

$$X_1 = \frac{\partial}{\partial t}, \qquad X_2 = \frac{\partial}{\partial \Phi}, \qquad X_3 = 2t\frac{\partial}{\partial t} + r\frac{\partial}{\partial r} - A\frac{\partial}{\partial A}, \qquad (6.133)$$

which describe translations in time and phase and scaling.

To apply Noether's theorem to (6.131)–(6.132), consider the functional

$$L = \int\limits_{\Omega} \mathcal{L}(t, r, u, u_t, u_r) r^{n-1} \, dr \, dt, \qquad (6.134)$$

where $\mathcal{L}$ is some Lagrangian function.

The invariance of $\mathcal{L}$ under the action of a symmetry group is connected via Noether's theorem with the existence of conservation laws for the Euler equations, which give the stationary value of the functional (6.134).

We have generalized the Noether-type identity to the case of radially symmetric solutions in dimension n:

$$\xi^t \frac{\partial \mathcal{L}}{\partial t} + \xi^r \frac{\partial \mathcal{L}}{\partial r} + \eta \frac{\partial \mathcal{L}}{\partial u} + [D_t(\eta) - u_t D_t(\xi^t) - u_r D_t(\xi^r)] \frac{\partial \mathcal{L}}{\partial u_t}$$

$$+ [D_r(\eta) - u_t D_r(\xi^t) - u_r D_r(\xi^r)] \frac{\partial \mathcal{L}}{\partial u_r}$$

$$+ \mathcal{L}[D_t(\xi^t) + D_r(\xi^r)] + \frac{n-1}{r} \xi^r \mathcal{L}$$

$$\equiv (\eta - \xi^t u_t - \xi^r u_r) \left[\frac{\partial \mathcal{L}}{\partial u} - D_t \left(\frac{\partial \mathcal{L}}{\partial u_t} \right) - \frac{1}{r^{n-1}} D_r \left(r^{n-1} \frac{\partial \mathcal{L}}{\partial u_r} \right) \right]$$

$$+ D_t \left[\xi^t \mathcal{L} + (\eta - \xi^t u_t - \xi^r u_r) \frac{\partial \mathcal{L}}{\partial u_t} \right]$$

$$+ \frac{1}{r^{(n-1)}} D_r \left[r^{n-1} \left(\xi^r \mathcal{L} + (\eta - \xi^t u_t - \xi^r u_r) \frac{\partial \mathcal{L}}{\partial u_r} \right) \right], \quad (6.135)$$

where

$$u = (A, \Phi), \qquad D_t = \frac{\partial}{\partial t} + u_t \frac{\partial}{\partial u} + \cdots, \qquad D_r = \frac{\partial}{\partial r} + u_r \frac{\partial}{\partial u} + \cdots.$$

The operator identity (6.135) makes it obvious that there is a connection between the invariance of the functional (6.134) and the conservation law

$$D_t \left[\xi^t \mathcal{L} + (\eta - \xi^t u_t - \xi^r u_r) \frac{\partial \mathcal{L}}{\partial u_t} \right]$$

$$+ \frac{1}{r^{(n-1)}} D_r \left[r^{n-1} \left(\xi^r \mathcal{L} + (\eta - \xi^t u_t - \xi^r u_r) \frac{\partial \mathcal{L}}{\partial u_r} \right) \right] = 0$$

for any solution of the Euler equations

$$\frac{\partial \mathcal{L}}{\partial u} - D_t \left(\frac{\partial \mathcal{L}}{\partial u_t} \right) - \frac{1}{r^{n-1}} D_r \left(r^{n-1} \frac{\partial \mathcal{L}}{\partial u_r} \right) = 0.$$

For the NLS problem, one can readily verify that the Lagrangian

$$\mathcal{L} = A_r{}^2 + A^2 \Phi_r{}^2 + A^2 \Phi_t - \frac{1}{2} A^4 \qquad (6.136)$$

has system (6.131)–(6.132) as the Euler equations. The Lagrangian (6.136) is invariant with respect to X_1 and X_2, and according to the Noether theorem (see identity (6.135) with $u = (A, \Phi)$) equips system (6.131)–(6.132) with the following conservation laws:

$$D_t\{A^2\} + \frac{1}{r^{n-1}} D_r\{r^{n-1} 2A^2 \Phi_r\} = 0,$$

$$D_t\{0.5A^4 - A_r^2 - A^2 \Phi_r^2\} + \frac{1}{r^{n-1}} D_r\{r^{n-1}(2A_t A_r + 2A^2 \Phi_t \Phi_r)\} = 0,$$

which are the well-known laws of conservation of mass and Hamiltonian for the NLS system.

6.10.2. Intermediate Lagrange coordinate system

Now we shall change the coordinate system to allow for the possible motion of points in the mesh, which we shall analyze later. Let us prolong the symmetry operators (6.133) to the subspace

$$\{t, r, A, \Phi; dt; dr; dA; d\Phi\},$$

which contains the differentials $dt, dr, dA, d\Phi$, so that

$$X_1 = \frac{\partial}{\partial t}, \qquad X_2 = \frac{\partial}{\partial \Phi},$$

$$X_3 = 2t\frac{\partial}{\partial t} + r\frac{\partial}{\partial r} - A\frac{\partial}{\partial A} + 2dt\frac{\partial}{\partial(dt)} + dr\frac{\partial}{\partial(dr)} - dA\frac{\partial}{\partial(dA)}.$$

(6.137)

By solving the system of linear partial differential equations

$$X_i(J_k) = 0, \qquad i = 1, 2, 3, \quad k = 1, 2, 3, 4, 5,$$

we obtain the following complete set of differential invariants:

$$J_1 = rA, \qquad J_2 = dt(dA)^2, \qquad J_3 = \frac{(dr)^2}{dt}, \qquad J_4 = d\Phi, \qquad J_5 = A\,dr.$$

This set gives us the possibility of finding the most general form for the evolution of r which preserves the symmetry (6.137):

$$\frac{dr}{dt} = \frac{1}{dr}F(A\,dr; dt(dA)^2; d\Phi; rA).$$

This result provides means for the evolution of a computational mesh: if $F = 0$, then we obtain an orthogonal coordinate system (on a fixed mesh); if $F \neq 0$, then we have a moving coordinate system with an invariant evolution of r.

For simplicity, we choose the following invariant evolution of r:

$$F = k(d\Phi), \qquad \frac{dr}{dt} = k\Phi_r,$$

where $k > 0$ is a control parameter (depending on n), which can be chosen to control the form of the mesh obtained in the numerical calculations. For example, it can prevent the mesh from becoming too sparse in certain regions.

Since r varies, we should prolong the time derivative to involve the following Lagrangian derivative:

$$\frac{d}{dt} = D_t + k\Phi_r D_r.$$

Significantly, this operation does not commute with D_r:

$$\left[\frac{d}{dt}, D_r\right] \neq 0.$$

Rewriting system (6.131)–(6.132) in the Lagrangian coordinates then gives

$$
\frac{dr}{dt} = k\Phi_r,
$$
$$
\frac{dA}{dt} = -A\Phi_{rr} + (k-2)\Phi_r A_r - \frac{n-1}{r}A\Phi_r, \tag{6.138}
$$
$$
A\frac{d\Phi}{dt} = A_{rr} + \frac{n-1}{r}A_r + (k-1)A\Phi_r{}^2 + A^3.
$$

One can readily show that the prolonged system (6.138) admits the symmetry operators (6.137) prolonged for the partial derivatives $\Phi_r, \Phi_{rr}, A_r, A_{rr}, \frac{dr}{dt}, \frac{dA}{dt}, \frac{d\Phi}{dt}$:

$$
X_1 = \frac{\partial}{\partial t}, \qquad X_2 = \frac{\partial}{\partial \Phi},
$$
$$
X_3 = 2t\frac{\partial}{\partial t} + r\frac{\partial}{\partial r} - A\frac{\partial}{\partial A} - 2A_r\frac{\partial}{\partial A_r} - 3A_{rr}\frac{\partial}{\partial A_{rr}} - \Phi_r\frac{\partial}{\partial \Phi_r}
$$
$$
- 2\Phi_{rr}\frac{\partial}{\partial \Phi_{rr}} - \frac{dr}{dt}\frac{\partial}{\partial(dr/dt)} - 3\frac{dA}{dt}\frac{\partial}{\partial(dA/dt)} - 2\frac{d\Phi}{dt}\frac{\partial}{\partial(d\Phi/dt)}.
$$

6.10.3. The "substantive" Lagrange coordinate system

Now we will rewrite system (6.138) in an orthogonal coordinate system by changing independent variables (t, r) to (t, s) and involving the new dependent variable ρ as follows:

$$
D_s = \frac{1}{\rho r^{n-1}}D_r. \tag{6.139}
$$

The purpose of this procedure is to recover the orthogonal differentiation property satisfied by a fixed coordinate mesh, so that in the revised coordinate system one has

$$
\left[\frac{d}{dt}, D_s\right] = 0, \tag{6.140}
$$

where

$$
\frac{d}{dt} = D_t + k\Phi_r D_r.
$$

From (6.140), we have the following equation for ρ:

$$
\rho_t + (k\rho\Phi_r)_r + \frac{n-1}{r}k\rho\Phi_r = 0. \tag{6.141}
$$

From (6.141) and the relation

$$
\frac{d\rho}{dt} = \rho_t + k\Phi_r\rho_r,
$$

we obtain one more equation, which gives the evolution for ρ in the form

$$\frac{d\rho}{dt} = -\frac{k\rho}{r^{n-1}}(r^{n-1}\Phi_r)_r. \tag{6.142}$$

Let us find the connection between s and t, r. From (6.139), we have

$$s_r = \rho r^{n-1}.$$

From the orthogonality of the coordinates (t, s),

$$\frac{ds}{dt} = 0,$$

we obtain

$$s_t = -k\rho\Phi_r r^{n-1}.$$

Thus, we have a contact transformation of the independent variables from (t, r) to (t, s),

$$\bar{t} = t, \qquad ds = \rho r^{n-1}\, dr - k\rho r^{n-1}\Phi_r\, dt. \tag{6.143}$$

We also should add $\rho > 0$ to (6.143), which implies the absence of a "vacuum gap" in the Lagrange coordinate system.

Remark. It is worth drawing a link at this stage between this approach, based on Lagrangian coordinate system, and the method of equidistributed meshes (see [23]). In this procedure a time-independent *computational variable* s is used for all calculations, and r is expressed in terms of s. To determine r, a *monitoring* function M is used, so that $\partial s/\partial r = M$. It is straightforward that this approach is equivalent to the orthogonal Lagrangian approach we consider provided that we set $M = \rho r^{n-1}$.

Note that the differential form (6.143) is total (complete); moreover, it is possible to start from the differential form

$$ds = \rho r^{n-1}\, dr - k\rho r^{n-1}\Phi_r\, dt$$

and then require the completeness of it; i.e.,

$$D_t(\rho r^{n-1}) = -D_r(k\rho r^{n-1}\Phi_r),$$

which is equivalent to (6.141).

Now, rewriting system (6.138)–(6.142) in terms of the orthogonal Lagrange coordinates (t, s), we obtain the system

$$\frac{dr}{dt} = k\rho r^{n-1}\Phi_s,$$

$$\frac{dA}{dt} = r^{n-1}\left(-A\rho(\rho r^{n-1}\Phi_s)_s + (k-2)\rho^2 r^{n-1}\Phi_s A_s - (n-1)A\rho r^{-1}\Phi_s\right),$$

$$A\frac{d\Phi}{dt} = \rho r^{n-1}(\rho r^{n-1}A_s)_s + (n-1)r^{n-2}\rho A_s + (k-1)A\rho^2\Phi_s^2 r^{2(n-1)} + A^3,$$

$$\frac{d\rho}{dt} = -k\rho^2(\rho r^{2(n-1)}\Phi_s)_s.$$

$$\tag{6.144}$$

6.10.4. Conservation laws in the Lagrangian coordinate system

We can now derive conservation laws for system (6.144) by using the conservation of differential forms. We denote the conservation laws for system (6.131)–(6.132) by

$$D_t\{A_0\} + D_r\{B_0\} = 0, \tag{6.145}$$

where A_0 (the density of conservation law), for instance, for the first (mass) conservation law is

$$A_0 = r^{n-1}A^2.$$

Equation (6.145) is equivalent to the existence of the total differential form

$$d\Omega = A_0\,dr - B_0\,dt. \tag{6.146}$$

If we now transform the differential form (6.146) to the new set of independent variables (6.143), then we have

$$d\Omega = A_1\,ds - B_1\,dt = A_1(\rho r^{n-1}\,dr - k\rho r^{n-1}\Phi_r\,dt) - B_1\,dt.$$

It follows that

$$A_1 = \frac{A_0}{\rho r^{n-1}}, \qquad B_1 = B_0 - kA_0\Phi_r. \tag{6.147}$$

We can rewrite this as a conservation law in the new coordinate system to obtain

$$\frac{d}{dt}\{A_1\} + D_s\{B_1\} = 0.$$

In accordance with (6.147), system (6.144) has the following conservation laws:

$$\frac{d}{dt}\{\rho^{-1}A^2\} + D_s\{r^{2(n-1)}A^2\rho\Phi_s(2-k)\} = 0,$$

$$\frac{d}{dt}\{\rho^{-1}(0.5A^4 - \rho^2 A_s^2 r^{2(n-1)} - \rho^2 A^2 r^{2(n-1)}\Phi_s^2)\}$$
$$+ D_s\{\rho r^{2(n-1)}\big[2A_s(\dot{A} - k\rho^2 r^{2(n-1)}A_s\Phi_s) + 2A^2\Phi_s(\dot{\Phi} - k\rho^2 r^{2(n-1)}\Phi_s^2)$$
$$- k\Phi_s(0.5A^4 - \rho^2 r^{2(n-1)}A_s^2 - A^2\rho^2 r^{2(n-1)}\Phi_s^2)\big]\} = 0,$$

where

$$\dot{A} = \frac{dA}{dt}, \qquad \dot{\Phi} = \frac{d\Phi}{dt}.$$

Interestingly, system (6.144) has the additional conservation law

$$\frac{d}{dt}\{\rho^{-1}\} + D_s\{-k\rho r^{2(n-1)}\Phi_s\} = 0,$$

which does not stem from the invariant Lagrange structure and is a continuity equation for the mesh density. Being transformed into the space (s, t, A, Φ, ρ, r), the symmetry operators (6.133) become

$$X_1 = \frac{\partial}{\partial t}, \qquad X_2 = \frac{\partial}{\partial \Phi},$$

$$X_3 = 2t\frac{\partial}{\partial t} + r\frac{\partial}{\partial r} - A\frac{\partial}{\partial A} + s\frac{\partial}{\partial s} + (1 - n)\rho\frac{\partial}{\partial \rho}. \tag{6.148}$$

System (6.144) admits an infinite-dimensional symmetry algebra. Indeed, in addition to the algebra (6.148), it admits the symmetries generated by

$$X_4 = f(s)\frac{\partial}{\partial s} + \rho f_s\frac{\partial}{\partial \rho},$$

where $f = f(s)$ is an arbitrary function.

6.10.5. The discretization procedure

Having considered various coordinate transformations of the NLS, we now turn our attention to discretization of this system in terms of these coordinates. We start our study of such discretization by considering NLS in the original variables (6.131)–(6.132) and then proceed to construct an invariant difference scheme with all appropriate conservation laws. The first question we need to address is finding the mesh geometry appropriate for the discretization procedure (see [19]).

One can readily verify that all operators (6.133) preserve the mesh orthogonality and regularity in both directions. Thus, we shall initially use the simplest invariant mesh that is orthogonal and regular in both directions (see Fig. 6.3) with constant steps h and τ. We note at this stage that while this mesh has good symmetry properties, it is not necessarily ideal for problems with associated small time and length scales—this leads to discretization on regular meshes in the Lagrangian variables, which we consider in subsequent sections.

On this mesh, we can consider the discrete version of the Lagrangian functional given by

$$L = \sum_i \mathcal{L}_i(A, A_r, \Phi_r, \Phi_t)h\tau,$$

derived over an appropriate domain Ω. For the discrete Lagrangian we take

$$\mathcal{L} = A_r{}^2 + A^2\Phi_r{}^2 + A^2\Phi_t - 0.5A^4, \tag{6.149}$$

where

$$A_r = \frac{A^+ - A}{h} = \underset{+h}{D}(A), \quad \Phi_r = \frac{\Phi^+ - \Phi}{h} = \underset{+h}{D}(\Phi), \quad \Phi_t = \frac{\hat{\Phi} - \Phi}{\tau} = \underset{+\tau}{D}(\Phi)$$

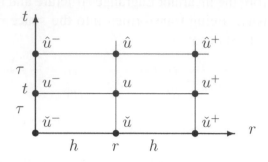

Figure 6.3: The orthogonal regular mesh in the original variables

are the corresponding right difference derivatives. (We omit the subscripts h and τ under difference derivatives so as not to overload formulas.) Our motivation is now to use this to derive discretization schemes with correct conservation laws. To apply the difference Noether theorem to (6.149), we generalize the discrete Noether-type operator identity [30] to give

$$
\begin{aligned}
&\xi^t \frac{\partial \mathcal{L}}{\partial t} + \xi^r \frac{\partial \mathcal{L}}{\partial r} + \eta \frac{\partial \mathcal{L}}{\partial u} + \left[\underset{+\tau}{D}(\eta) - u_t \underset{+\tau}{D}(\xi^t) - u_r \underset{+\tau}{D}(\xi^r) \right] \frac{\partial \mathcal{L}}{\partial u_t} \\
&\quad + \left[\underset{+h}{D}(\eta) - u_t \underset{+h}{D}(\xi^t) - u_r \underset{+h}{D}(\xi^r) \right] \frac{\partial \mathcal{L}}{\partial u_r} + \mathcal{L}\left[\underset{+\tau}{D}(\xi^t) + \underset{+h}{D}(\xi^r) \right] + \frac{n-1}{r} \xi^r \mathcal{L} \\
&\equiv \xi^t \left\{ \frac{\partial \mathcal{L}}{\partial t} + \underset{-\tau}{D}\left(u_t \frac{\partial \mathcal{L}}{\partial u_t} - \mathcal{L} \right) + \frac{1}{r^{n-1}} \underset{-h}{D}\left(r^{n-1} u_t \frac{\partial \mathcal{L}}{\partial u_r} \right) \right\} \\
&+ \xi^r \left\{ \frac{\partial \mathcal{L}}{\partial r} + \underset{+\tau}{D}\left(\check{u}_r \left(\frac{\partial \mathcal{L}}{\partial u_t} \right) \right) + \frac{1}{r^{(n-1)}} \underset{+h}{D}\left\{ (r^-)^{n-1} \left[u_r^- \left(\frac{\partial \mathcal{L}}{\partial u_r} \right)^- - \mathcal{L}^- \right] + \frac{n-1}{r} \mathcal{L} \right\} \\
&+ \eta \left\{ \frac{\partial \mathcal{L}}{\partial u} - \underset{-\tau}{D}\left(\frac{\partial \mathcal{L}}{\partial u_t} \right) - \frac{1}{r^{n-1}} \underset{-h}{D}\left(r^{n-1} \frac{\partial \mathcal{L}}{\partial u_r} \right) \right\} + \underset{+\tau}{D}\left\{ \xi^t \check{\mathcal{L}} + (\eta - \xi^t \check{u}_t - \xi^r \check{u}_r)\left(\frac{\partial \mathcal{L}}{\partial u_t} \right) \right\} \\
&\quad + \frac{1}{r^{(n-1)}} \underset{+h}{D}\left\{ (r^-)^{n-1} \left[\xi^r \mathcal{L}^- + (\eta - \xi^t u_t^- - \xi^r u_r^-)\left(\frac{\partial \mathcal{L}}{\partial u_r} \right)^- \right] \right\}, \quad (6.150)
\end{aligned}
$$

where $u = (A, \Phi)$ and all derivatives are difference derivatives. Identity (6.150) can be proved by straightforward computations.

From identity (6.150), we have the difference Euler equation

$$
\frac{\partial \mathcal{L}}{\partial u} - \underset{-\tau}{D}\left(\frac{\partial \mathcal{L}}{\partial u_t} \right) - \frac{1}{r^{n-1}} \underset{-h}{D}\left(r^{n-1} \frac{\partial \mathcal{L}}{\partial u_r} \right) = 0,
$$

but this equation possesses conservation laws only if $\xi^t = \xi^r = 0$.

If the left hand-side of the Eq. (6.150) is zero, then the quasi-extremal equation

$$\xi^t\left\{\frac{\partial\mathcal{L}}{\partial t} + \underset{-\tau}{D}\left(u_t\frac{\partial\mathcal{L}}{\partial u_t} - \mathcal{L}\right) + \frac{1}{r^{n-1}}\underset{-h}{D}\left(r^{n-1}u_t\frac{\partial\mathcal{L}}{\partial u_r}\right)\right\}$$

$$+\xi^r\left\{\frac{\partial\mathcal{L}}{\partial r} + \underset{+\tau}{D}\left(\breve{u}_r\left(\frac{\partial\mathcal{L}}{\partial u_t}\right)\right) + \frac{1}{r^{(n-1)}}\underset{+h}{D}\left\{(r^-)^{n-1}\left[u_r^-\left(\frac{\partial\mathcal{L}}{\partial u_r}\right)^- - \mathcal{L}^-\right] + \frac{n-1}{r}\mathcal{L}\right\}$$

$$+\eta\left\{\frac{\partial\mathcal{L}}{\partial u} - \underset{-\tau}{D}\left(\frac{\partial\mathcal{L}}{\partial u_t}\right) - \frac{1}{r^{n-1}}\underset{-h}{D}(r^{n-1}\frac{\partial\mathcal{L}}{\partial u_r})\right\} = 0$$

possesses the conservation law

$$\underset{+\tau}{D}\left\{\xi^t\check{\mathcal{L}} + (\eta - \xi^t\breve{u}_t - \xi^r\breve{u}_r)\left(\frac{\breve{\partial\mathcal{L}}}{\partial u_t}\right)\right\}$$

$$+ \frac{1}{r^{(n-1)}}\underset{+h}{D}\left\{(r^-)^{n-1}\left[\xi^r\mathcal{L}^- + (\eta - \xi^tu_t^- - \xi^ru_r^-)\left(\frac{\partial\mathcal{L}}{\partial u_r}\right)^-\right]\right\} = 0.$$

Now let us apply the difference Noether theorem to the Lagrangian (6.149). The Lagrangian (6.149) is invariant under the actions of X_1 and X_2. This leads to the conservation laws

$$\underset{-\tau}{D}\{A^2\} + \frac{1}{r^{n-1}}\underset{-h}{D}\{2r^{n-1}A^2\Phi_r\} - 0, \quad (6.151)$$

$$\underset{-\tau}{D}\{0.5A^4 - A^2\Phi_r^2 - A_r^2\} + \frac{1}{r^{n-1}}\underset{-h}{D}\{2r^{n-1}[A_rA_t + A^2\Phi_t\Phi_r]\} = 0 \quad (6.152)$$

for the global extremal difference equations. Note that in the underlying case we already have the invariant mesh and only need two difference equations for the solution (A, Φ). For such equations we can take two equations (6.151) and (6.152). Thus, the difference equations (6.151)–(6.152) form an invariant scheme on an orthogonal regular mesh and thus coincide with the difference conservation laws.

Note that this model is not unique, because some other equations can be obtained by the same procedure starting from some other invariant Lagrangian.

6.10.6. The total difference form

Now consider the difference analog of a total differential form on the orthogonal difference mesh in the computational variable s in accordance with Fig. 6.4.

By doing this, we can derive discretizations of the NLS more appropriate for problems with increasingly small length scales. On the right upper box, we consider the difference form

$$\Delta s = s_rh + s_t^+\tau = \hat{s}_rh + s_t\tau, \quad (6.153)$$

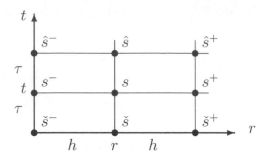

Figure 6.4: The orthogonal mesh in the computational variables

where the difference operators on s are as follows:

$$s_r = \frac{s^+ - s}{h}, \qquad \hat{s}_r = \frac{\hat{s}^+ - \hat{s}}{h}, \qquad s_t = \frac{\hat{s} - s}{\tau}, \qquad s_t^+ = \frac{\hat{s}^+ - s^+}{\tau}.$$

It follows from (6.153) that

$$\underset{+h}{D}(s_t) = \underset{+\tau}{D}(s_r), \tag{6.154}$$

leading to the completeness of the difference form (6.153).

Let us restate the following difference derivatives of the computational variable s:

$$s_r = \rho r^{n-1}, \qquad s_t = -k\rho r^{n-1}\Phi_r.$$

The completeness condition (6.154) then gives

$$\underset{+\tau}{D}(\rho r^{n-1}) = -k\underset{+h}{D}(\rho r^{n-1}\Phi_r). \tag{6.155}$$

Relation (6.155) can readily be shifted to any desired mesh point.

Now we introduce the new discrete differentiation operators of Lagrange type:

$$\frac{d}{dt}_+ = \underset{+\tau}{D} + k\Phi_r\underset{+h}{D}, \qquad \frac{d}{dt}_- = \underset{-\tau}{D} + k\check{\Phi}_r\underset{+h}{\check{D}},$$

where

$$\underset{+h}{\check{D}} = \underset{-\tau+h}{S}\underset{}{D} = \underset{+h-\tau}{D}\underset{}{S}, \qquad \check{\Phi}_r = \underset{-\tau}{S}\Phi_r.$$

We also invoke a couple of difference operators corresponding to right and left differentiation in the s-direction as follows:

$$\rho r^{n-1}\underset{+h}{D}_s = \underset{+h}{D}, \qquad \rho^-(r^-)^{n-1}\underset{-h}{D}_s = \underset{-h}{D}.$$

One can readily verify that the above-stated definitions give the orthogonality of the new mesh in the computational (t, s)-coordinate system,

$$\frac{ds}{dt}_+ = 0, \qquad \frac{ds}{dt}_- = 0.$$

6.10.7. Difference conservation laws in the Lagrange coordinate system

To study discretization in the new coordinate system, we transform the difference conservation laws (6.151)–(6.152) for the Lagrange coordinate system in the same manner as was done in the continuous case by using the conservation of differential forms.

We represent the conservation law (6.151)–(6.152) as

$$\underset{+\tau}{D}\{A_0\} + \underset{+h}{D}\{B_0\} = 0,$$

which is equivalent to the existence of the difference form

$$\Delta_0 = A_0 h - B_0^+ \tau = \hat{A}_0 h - B_0 \tau. \tag{6.156}$$

Now we transform the difference form (6.156) by the change of independent variables:

$$\bar{t} = t, \qquad \Delta s = \rho r^{n-1} h - k \rho^+ (r^+)^{n-1} \Phi_r^+ \tau = \hat{\rho}\hat{r}^{n-1} h - k \rho r^{n-1} \Phi_r \tau.$$

In this derivation, we have the "new" spatial step $h_s = \Delta s$ in the computational variable.

The difference form (6.156) can be represented in the (t, s)-coordinate system by

$$\Delta_0 = A_1 h_s - B_1^+ \tau = \hat{A}_1 h_s - B_1 \tau = A_1(\rho r^{(n-1)} h - k\rho^+(r^+)^{n-1}\Phi_r^+ \tau) - B_1^+ \tau$$
$$= \hat{A}_1(\hat{\rho}\hat{r}^{n-1} h - k\rho r^{n-1}\Phi_r \tau) - B_1 \tau.$$

Then we have

$$A_1 = \frac{A_0}{\rho r^{n-1}}, \qquad B_1 = B_0 - kA_0 \Phi_r \frac{\rho r^{n-1}}{\hat{\rho}\hat{r}^{n-1}}, \tag{6.157}$$

which we can rewrite as the difference conservation law

$$\underset{+\tau}{D}\{A_1\} + \underset{+h}{D_s}\{B_1\} = 0 \tag{6.158}$$

in the new coordinate system.

In accordance with (6.157) and (6.158), we can finally rewrite the conservation laws in the following form:

$$\underset{+}{\frac{d}{dt}}\left(\check{A}^2 \frac{\check{r}^{n-1}}{\rho r^{n-1}}\right) + \underset{+h}{D_s}\left(2(r^-)^{2(n-1)}(A^-)^2\rho^- \Phi_s^- - kA^2 \frac{\rho^2 r^{3(n-1)}}{\hat{\rho}\hat{r}^{n-1}}\Phi_s\right) = 0, \tag{6.159}$$

$$\frac{d}{dt_+}\left\{\frac{\check{r}^{n-1}}{\rho r^{n-1}}\left[0.5\check{A}^4 - \check{\rho}^2\check{A}^2\check{\Phi}_s^2\check{r}^{2(n-1)} - \check{\rho}^2\check{r}^{2(n-1)}\check{A}_s^2\right]\right\}$$

$$+ \underset{+h}{D_s}\left\{2(r^-)^{2(n-1)}\rho^-\left[A_s^-\left(\dot{A}^- - k(\rho^-)^2(r^-)^{2(n-1)}A_s^-\Phi_s^-\right)\right.\right.$$

$$+ (A^-)^2\Phi_s^-\left(\dot{\Phi}^- - k(\rho^-)^2(r^-)^{2(n-1)}(\Phi_s^-)^2\right)\right]$$

$$- k\Phi_s\frac{\rho^2 r^{3(n-1)}}{\hat{\rho}\hat{r}^{n-1}}\left[0.5A^4 - \rho^2 A^2\Phi_s^2 r^{2(n-1)} - \rho^2 r^{2(n-1)}A_s^2\right]\right\} = 0. \quad (6.160)$$

This system allows us to evolve the discrete solution. We should also allow for the evolution of the mesh points given by the following two equations for the evolution of r and ρ:

$$\frac{dr}{dt_+} = k\rho r^{n-1}\Phi_s, \quad (6.161)$$

$$\frac{d}{dt_+}\left(\rho r^{n-1}\right) = -k\rho^+(r^+)^{n-1}\rho r^{n-1}(\rho r^{n-1}\Phi_s)_s. \quad (6.162)$$

Thus, Eqs. (6.159)–(6.162) form an invariant difference scheme on the orthogonal mesh in the (t, s)-plane, which can be implemented to calculate solutions of the NLS as it evolves toward a singularity.

6.10.8. The blow-up invariant solution

Finally, consider the application of the discretization (6.159) in the context of solutions with developing singularities. First, let us transform the symmetry operators (6.133) into the space (s, t, A, Φ, ρ, r):

$$X_1 = \frac{\partial}{\partial t}, \qquad X_2 = \frac{\partial}{\partial\Phi},$$

$$X_3 = 2t\frac{\partial}{\partial t} + r\frac{\partial}{\partial r} - A\frac{\partial}{\partial A} + s\frac{\partial}{\partial s} + (1-n)\rho\frac{\partial}{\partial\rho}. \quad (6.163)$$

We have shown that system (6.144), together with (6.163), has one more additional symmetry

$$X^* = f(s)\frac{\partial}{\partial s} + \rho f_s\frac{\partial}{\partial\rho},$$

where $f = f(s)$ is an arbitrary function.

System (6.159)–(6.162) possesses the same symmetries (6.163) and has the additional symmetry

$$X^* = f(s)\frac{\partial}{\partial s} + \rho \underset{+h}{D_s}(f)\frac{\partial}{\partial\rho}. \quad (6.164)$$

Now consider the symmetry subalgebra

$$X = -2T_0 X_1 + X_3 = 2(T_0 - t)\frac{\partial}{\partial(T_0 - t)} + r\frac{\partial}{\partial r} - A\frac{\partial}{\partial A} + s\frac{\partial}{\partial s} + (1-n)\rho\frac{\partial}{\partial\rho},$$

$$(6.165)$$

where T_0 is some positive constant. Then to (6.165) we add the special case of the operator (6.164),

$$X^{**} = \gamma \left(s \frac{\partial}{\partial s} + \rho \frac{\partial}{\partial \rho} \right),$$

which gives the subalgebra

$$\hat{X} = 2(T_0 - t) \frac{\partial}{\partial(T_0 - t)} + r \frac{\partial}{\partial r} - A \frac{\partial}{\partial A} + s(1+\gamma) \frac{\partial}{\partial s} + (1-n+\gamma)\rho \frac{\partial}{\partial \rho}, \quad (6.166)$$

where γ is some "monitoring" parameter. One can readily see that $\gamma = -1$ corresponds to the situation in which s is an invariant of the subalgebra (6.166). The corresponding symmetry operator is the following:

$$\hat{X}^* = 2(T_0 - t) \frac{\partial}{\partial(T_0 - t)} + r \frac{\partial}{\partial r} - A \frac{\partial}{\partial A} - n\rho \frac{\partial}{\partial \rho}.$$

Let us write out the invariant representation of the solution in this case:

$$A = \bar{A}(\lambda)(T_0 - t)^{-1/2}, \qquad \Phi = \bar{\Phi}(\lambda), \qquad \rho = \bar{\rho}(\lambda)(T_0 - t)^{-n/2},$$
$$s = \bar{s}(\lambda), \qquad \lambda = r(T_0 - t)^{-1/2}. \qquad (6.167)$$

The ordinary differential system and the corresponding ordinary difference system can readily be obtained by substituting the invariant representation (6.167) into system (6.144) and (6.159)–(6.162).

This solution has the desired property of having a self-similar form and of becoming singular in finite time T_0 with amplitude proportional to $(T_0 - t)^{-1/2}$ while evolving on a length scale proportional to $(T_0 - t)^{1/2}$. Thus, if such a solution exists for the underlying problem, it is admitted by the discretization. As was noted earlier, this is a significant feature of such a method, as it is known [131] that if $n > 2$, then the stable form of singularity evolution is that of a monotone decreasing self-similar solution.

Since s is an invariant of this solution, there is no movement of waves in the s-direction for the invariant solution of the form (6.167), and any distinctive point of the solution (6.167) in λ (gradient maximum or zero point for example) does not move in the s-direction. Thus, s is a true computational variable, in the sense that a computationally "difficult" problem when expressed in terms of r has been transformed into a more "regular" problem in s allowing for a more straightforward discretization.

Chapter 7

Hamiltonian Formalism for Difference Equations: Symmetries and First Integrals

In this chapter, the relationship between symmetries and first integrals for difference Hamiltonian equations is considered. These results are built upon those for the continuous canonical Hamiltonian equations

$$\dot{q}^i = \frac{\partial H}{\partial p_i}, \quad \dot{p}_i = -\frac{\partial H}{\partial q^i}, \qquad i = 1, \ldots, n, \tag{7.1}$$

considered in the Introduction. It was shown there that the continuous Hamiltonian equations can be obtained by the variational principle from action functionals. On the basis of Noether-type identities, there was developed a Noether-type theorem for the canonical Hamiltonian equations.

Now we shall develop a similar mathematical formalism for their discrete counterparts, i.e., for difference Hamiltonian equations [43–45]. The approach based on symmetries of discrete functionals provides a simple, clear way to construct first integrals of difference Hamiltonian equations by means of purely algebraic manipulations. It can be used to preserve the structure properties of the underlying differential equations under the discretization procedure; this is useful for numerical implementation.

7.1. Discrete Legendre Transform

Consider difference Hamiltonian equations at some lattice point $(t, \mathbf{q}, \mathbf{p})$. The notation is given in Fig. 7.1. Generally, the lattice in not regular. Using the analogy with the continuous case, we can construct discrete Hamiltonian equations on the basis of the discrete equations in the Lagrangian framework. We use the slightly modified version [83] of the *Legendre transform*. The discrete Legendre transform of $\mathcal{L}(t, t_+, \mathbf{q}, \mathbf{q}_+)$ with respect to $\mathbf{q}_+$ is the function

$$\mathcal{H}(t, t_+, \mathbf{q}, \mathbf{p}^+) = p_i^+ \underset{+h}{D}(q^i) - \mathcal{L}(t, t_+, \mathbf{q}, \mathbf{q}_+), \tag{7.2}$$

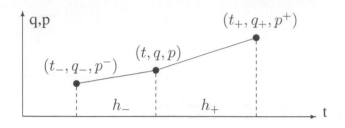

Figure 7.1

where $D_{+h}(q^i) = (q^i_+ - q^i)/h_+$ and $\mathbf{q}_+$ is defined implicitly by

$$\mathbf{p}^+ = h_+ \frac{\partial \mathcal{L}}{\partial \mathbf{q}_+}. \tag{7.3}$$

Remark. Alternatively, one can consider the discrete Legendre transform with respect to $\mathbf{q}$. Then

$$\mathcal{H}(t, t_+, \mathbf{q}_+, \mathbf{p}) = p_i \underset{+h}{D}(q^i) - \mathcal{L}(t, t_+, \mathbf{q}, \mathbf{q}_+), \tag{7.4}$$

where $\mathbf{q}$ is found from

$$\mathbf{p} = -h_+ \frac{\partial \mathcal{L}}{\partial \mathbf{q}}. \tag{7.5}$$

For the discrete Legendre transform (7.2), (7.3) we obtain the following relations for the derivatives of the Hamiltonian function:

$$\frac{\partial \mathcal{H}}{\partial \mathbf{p}^+} = \underset{+h}{D}(\mathbf{q}), \qquad\qquad \frac{\partial \mathcal{H}}{\partial \mathbf{q}} = -\frac{\mathbf{p}^+}{h_+} - \frac{\partial \mathcal{L}}{\partial \mathbf{q}},$$

$$\frac{\partial \mathcal{H}}{\partial t} = \frac{p_i^+}{h_+} \underset{+h}{D}(q^i) - \frac{\partial \mathcal{L}}{\partial t}, \qquad \frac{\partial \mathcal{H}}{\partial t_+} = -\frac{p_i^+}{h_+} \underset{+h}{D}(q^i) - \frac{\partial \mathcal{L}}{\partial t_+}.$$

By using these relations as well as relations (7.2) and (7.3), we can transform the $n + 1$ global extremal equations for the Lagrangian $\mathcal{L}(t, t_+, \mathbf{q}, \mathbf{q}_+)$ into the Hamiltonian framework. We arrive at the system of $2n + 1$ equations

$$\underset{+h}{D}(q^i) = \frac{\partial \mathcal{H}}{\partial p_i^+}, \quad \underset{+h}{D}(p_i) = -\frac{\partial \mathcal{H}}{\partial q^i}, \qquad i = 1, \dots, n,$$

$$\frac{\partial \mathcal{H}}{\partial t} + \frac{h_-}{h_+} \frac{\partial \mathcal{H}^-}{\partial t} - \underset{+h}{D}(\mathcal{H}^-) = 0, \tag{7.6}$$

where $\mathcal{H} = \mathcal{H}(t, t_+, \mathbf{q}, \mathbf{p}_+)$ and $\mathcal{H}^- = \underset{-h}{S}(\mathcal{H}) = \mathcal{H}(t_-, t, \mathbf{q}_-, \mathbf{p})$. We refer to these equations as the *difference* (or *discrete*) *Hamiltonian* equations. Although these equations have been introduced in terms of the discrete Legendre transform, they can be considered independently of the Lagrangian framework (see the next section).

Note that the first $2n$ equations in (7.6) are first-order discrete equations, which correspond to the canonical Hamiltonian equations (7.1) in the continuous limit. These (and equivalent) equations were considered in a number of papers [3, 52, 53, 129]. The last equation is of second order. It defines the lattice on which the canonical Hamiltonian equations are discretized. In the continuous limit, the lattice equation itself disappears. Being a second-order difference equation, it needs one more initial value (the first spacing of the lattice) to state the initial-value problem.

Remark. The second version (7.4), (7.5) of the discrete Legendre transform yields the discrete Hamiltonian equations

$$\underset{+h}{D}(q^i) = \frac{\partial \mathcal{H}}{\partial p_i}, \quad \underset{+h}{D}(p_i) = -\frac{\partial \mathcal{H}}{\partial q_+^i}, \qquad i = 1, \dots, n,$$

$$\frac{\partial \mathcal{H}}{\partial t} + \frac{h_-}{h_+}\frac{\partial \mathcal{H}^-}{\partial t} - \underset{+h}{D}(\mathcal{H}^-) = 0. \tag{7.7}$$

where $\mathcal{H} = \mathcal{H}(t, t_+, \mathbf{q}^+, \mathbf{p})$ and $\mathcal{H}^- = \underset{-h}{S}(\mathcal{H}) = \mathcal{H}(t_-, t, \mathbf{q}, \mathbf{p})$.

7.2. Variational Statement of the Difference Hamiltonian Equations

The discrete Hamiltonian equations (7.6), which were obtained by an application of the discrete Legendre transform to the discrete Euler–Lagrange equations, can be obtained from a variational principle. Indeed, consider the finite-difference functional

$$\mathbb{H}_h = \sum_{\Omega} (p_i^+ \underset{+h}{D}(q^i) - \mathcal{H}(t, t_+, \mathbf{q}, \mathbf{p}_+))h_+$$

$$= \sum_{\Omega} (p_i^+(q_+^i - q^i) - \mathcal{H}(t, t_+, \mathbf{q}, \mathbf{p}_+)h_+). \tag{7.8}$$

The variation of this functional along a curve $q^i = \phi_i(t)$, $p_i = \psi_i(t)$, $i = 1, \dots, n$, at some point $(t, \mathbf{q}, \mathbf{p})$ will affect only two terms in the sum (7.8),

$$\mathbb{H}_h = \cdots + p_i(q^i - q_-^i) - \mathcal{H}(t_-, t, \mathbf{q}_-, \mathbf{p})h_-$$
$$+ p_i^+(q_+^i - q^i) - \mathcal{H}(t, t_+, \mathbf{q}, \mathbf{p}_+)h_+ + \cdots .$$

Therefore, we obtain the expression

$$\delta \mathbb{H}_h = \frac{\delta \mathcal{H}}{\delta p_i}\delta p_i + \frac{\delta \mathcal{H}}{\delta q^i}\delta q^i + \frac{\delta \mathcal{H}}{\delta t}\delta t$$

for the variation of the functional, where $\delta q^i = \phi_i' \delta t$, $\delta p_i = \psi_i' \delta t$, $i = 1, \ldots, n$, and

$$\frac{\delta \mathcal{H}}{\delta p_i} = q^i - q_-^i - h_- \frac{\partial \mathcal{H}^-}{\partial p_i}, \qquad i = 1, \ldots, n,$$

$$\frac{\delta \mathcal{H}}{\delta q^i} = -\left(p_i^+ - p_i + h_+ \frac{\partial \mathcal{H}}{\partial q^i} \right), \qquad i = 1, \ldots, n, \tag{7.9}$$

$$\frac{\delta \mathcal{H}}{\delta t} = -\left(h_+ \frac{\partial \mathcal{H}}{\partial t} - \mathcal{H} + h_- \frac{\partial \mathcal{H}^-}{\partial t} + \mathcal{H}^- \right).$$

For the stationary value of the finite-difference functional (7.8), we obtain the system of $2n + 1$ equations

$$\frac{\delta \mathcal{H}}{\delta p_i} = 0, \quad \frac{\delta \mathcal{H}}{\delta q^i} = 0, \quad i = 1, \ldots, n, \qquad \frac{\delta \mathcal{H}}{\delta t} = 0, \tag{7.10}$$

which are equivalent to the discrete Hamiltonian equations (7.6).

Note that the variational equations (7.9) can be derived by applying the variational operators

$$\frac{\delta}{\delta p_i} = \frac{\partial}{\partial p_i} + S_{-h} \frac{\partial}{\partial p_i^+}, \qquad i = 1, \ldots, n, \tag{7.11}$$

$$\frac{\delta}{\delta q^i} = \frac{\partial}{\partial q^i} + S_{-h} \frac{\partial}{\partial q_+^i}, \qquad i = 1, \ldots, n, \tag{7.12}$$

$$\frac{\delta}{\delta t} = \frac{\partial}{\partial t} + S_{-h} \frac{\partial}{\partial t_+} \tag{7.13}$$

to the discrete Hamiltonian elementary action $p_i^+(q_+^i - q^i) - \mathcal{H}(t, t_+, \mathbf{q}, \mathbf{p}^+)h_+$.

For the variation of the functional (7.8) along an orbit of the group generated by the operator

$$X = \xi(t, \mathbf{q}, \mathbf{p}) \frac{\partial}{\partial t} + \eta^i(t, \mathbf{q}, \mathbf{p}) \frac{\partial}{\partial q^i} + \zeta_i(t, \mathbf{q}, \mathbf{p}) \frac{\partial}{\partial p_i}, \tag{7.14}$$

we have $\delta t = \xi \delta a$, $\delta q^i = \eta^i \delta a$, and $\delta p_i = \zeta_i \delta a$, $i = 1, \ldots, n$, where δa is the variation of the group parameter. The stationary value of the finite-difference functional (7.8) along the flow generated by this vector field is given by the equation

$$\zeta_i \frac{\delta \mathcal{H}}{\delta p_i} + \eta^i \frac{\delta \mathcal{H}}{\delta q^i} + \xi \frac{\delta \mathcal{H}}{\delta t} = 0,$$

which explicitly depends on the coefficients of the generator and is the Hamiltonian counterpart of the quasi-extremal equation in the Lagrangian framework.

If we have a Lie algebra of vector fields of dimension $\geq 2n+1$, then the stationary value of the functional (7.8) along the entire flow is attained on the solutions of the system (7.10).

Remark. In a similar way, one can show that the discrete Hamiltonian equations (7.7) can be obtained from variations of the finite-difference functional

$$\mathbb{H}_h = \sum_{\Omega} (p_i \underset{+h}{D}(q^i) - \mathcal{H}(t, t_+, \mathbf{q}_+, \mathbf{p}))h_+$$

$$= \sum_{\Omega} (p_i(q^i_+ - q^i) - \mathcal{H}(t, t_+, \mathbf{q}_+, \mathbf{p})h_+).$$

In forthcoming sections, we consider the invariance and conservation properties of discrete Hamiltonian equations. For simplicity, we restrict ourselves to the version (7.6) of such equations. All results can be equivalently stated for the other version, i.e., for the discrete Hamiltonian equations (7.7).

7.3. Symplecticity of Difference Hamiltonian Equations

The canonical Hamiltonian equations generate symplectic transformations in the phase space $(\mathbf{q}, \mathbf{p})$. For the solution $(\mathbf{q}(t), \mathbf{p}(t))$ of system (7.1) with initial data $\mathbf{q}(t_0) = \mathbf{q}_0$, $\mathbf{p}(t_0) = \mathbf{p}^0$, this property can be expressed as the conservation of the 2-form

$$dp_i \wedge dq^i = dp_i^0 \wedge dq_0^i.$$

This property is used to select symplectic numerical integrators [84, 128] as numerical schemes with the property

$$dp_i^{n+1} \wedge dq_{n+1}^i = dp_i^n \wedge dq_n^i, \qquad n = 0, 1, \dots . \tag{7.15}$$

Definition (7.15) of conservation of symplecticity cannot be used for discretization on solution-dependent meshes such as the discrete Euler–Lagrange equations and the discrete Hamiltonian equations (7.6). Generally, the variations of the dependent variables involve the variations of the lattice points. This is clearly seen from the variational equations

$$dq_+^i - dq^i = \frac{\partial^2(\mathcal{H}h_+)}{\partial p_i^+ \partial t} dt + \frac{\partial^2(\mathcal{H}h_+)}{\partial p_i^+ \partial t_+} dt_+ + \frac{\partial^2(\mathcal{H}h_+)}{\partial p_i^+ \partial q^j} dq^j + \frac{\partial^2(\mathcal{H}h_+)}{\partial p_i^+ \partial p_j^+} dp_j^+,$$

$$i = 1, \dots, n,$$

$$dp_i^+ - dp_i = -\frac{\partial^2(\mathcal{H}h_+)}{\partial q^i \partial t} dt - \frac{\partial^2(\mathcal{H}h_+)}{\partial q^i \partial t_+} dt_+ - \frac{\partial^2(\mathcal{H}h_+)}{\partial q^i \partial q^j} dq^j - \frac{\partial^2(\mathcal{H}h_+)}{\partial q^i \partial p_j^+} dp_j^+,$$

$$i = 1, \dots, n,$$

$$\frac{\partial^2(\mathcal{H}h_+)}{\partial t^2} dt + \frac{\partial^2(\mathcal{H}h_+)}{\partial t \partial t_+} dt_+ + \frac{\partial^2(\mathcal{H}h_+)}{\partial t \partial q^j} dq^j + \frac{\partial^2(\mathcal{H}h_+)}{\partial t \partial p_j^+} dp_j^+$$

$$+ \frac{\partial^2(\mathcal{H}^-h_-)}{\partial t \partial t_-} dt_- + \frac{\partial^2(\mathcal{H}^-h_-)}{\partial t^2} dt + \frac{\partial^2(\mathcal{H}^-h_-)}{\partial t \partial q_-^j} dq_-^j + \frac{\partial^2(\mathcal{H}^-h_-)}{\partial t \partial p_j} dp_j = 0$$

for system (7.6). These equations are a system of $2n + 1$ linear algebraic equations for the variations dt_+, $d\mathbf{q}_+$, and $d\mathbf{p}^+$ at the next lattice point. Thus, the variational equations considered in the phase space (without the variations of the independent variable) form an overdetermined system of $2n + 1$ equations for $2n$ variables, which has only the trivial solution in the general case.

Therefore, we are forced to look for symplecticity in the extended phase space $(t, \mathbf{q}, \mathbf{p})$. (See also the general considerations for the continuous case in [25].)

THEOREM. *The difference Hamiltonian equations* (7.6) *preserve symplecticity*:

$$dp_i^+ \wedge dq_+^i - d\mathcal{E}_+ \wedge dt_+ = dp_i \wedge dq^i - d\mathcal{E} \wedge dt,$$

where

$$\mathcal{E}_+ = \mathcal{H} + h_+ \frac{\partial \mathcal{H}}{\partial t_+}, \qquad \mathcal{E} = \mathcal{H}^- + h_- \frac{\partial \mathcal{H}^-}{\partial t}$$

are the discrete energies for the lattice points t_+ and t.

Proof. From the first $2n$ variational equations, we obtain

$$dp_i^+ \wedge dq_+^i - dp_i \wedge dq^i = \frac{\partial^2 (\mathcal{H} h_+)}{\partial p_i^+ \partial t} dp_i^+ \wedge dt + \frac{\partial^2 (\mathcal{H} h_+)}{\partial p_i^+ \partial t_+} dp_i^+ \wedge dt_+$$
$$+ \frac{\partial^2 (\mathcal{H} h_+)}{\partial q^i \partial t} dq^i \wedge dt + \frac{\partial^2 (\mathcal{H} h_+)}{\partial q^i \partial t_+} dq^i \wedge dt_+. \quad (7.16)$$

With the help of the relations

$$d\mathcal{E}_+ = \frac{\partial^2 (\mathcal{H} h_+)}{\partial t \partial t_+} dt + \frac{\partial^2 (\mathcal{H} h_+)}{\partial t_+^2} dt_+ + \frac{\partial^2 (\mathcal{H} h_+)}{\partial q^j \partial t_+} dq^j + \frac{\partial^2 (\mathcal{H} h_+)}{\partial p_j^+ \partial t_+} dp_j^+$$

$$d\mathcal{E} = \frac{\partial^2 (\mathcal{H}^- h_-)}{\partial t_- \partial t} dt_- + \frac{\partial^2 (\mathcal{H}^- h_-)}{\partial t^2} dt + \frac{\partial^2 (\mathcal{H}^- h_-)}{\partial q_-^j \partial t} dq_-^j + \frac{\partial^2 (\mathcal{H}^- h_-)}{\partial p_j \partial t} dp_j$$

$$= -\frac{\partial^2 (\mathcal{H} h_+)}{\partial t^2} dt - \frac{\partial^2 (\mathcal{H} h_+)}{\partial t \partial t_+} dt_+ - \frac{\partial^2 (\mathcal{H} h_+)}{\partial t \partial q^j} dq^j - \frac{\partial^2 (\mathcal{H} h_+)}{\partial t \partial p_j^+} dp_j^+$$

for the variations (where the last variational equation was used), we obtain

$$d\mathcal{E}_+ \wedge dt_+ - d\mathcal{E} \wedge dt = \frac{\partial^2 (\mathcal{H} h_+)}{\partial p_i^+ \partial t} dp_i^+ \wedge dt + \frac{\partial^2 (\mathcal{H} h_+)}{\partial p_i^+ \partial t_+} dp_i^+ \wedge dt_+$$
$$+ \frac{\partial^2 (\mathcal{H} h_+)}{\partial q^i \partial t} dq^i \wedge dt + \frac{\partial^2 (\mathcal{H} h_+)}{\partial q^i \partial t_+} dq^i \wedge dt_+. \quad (7.17)$$

By comparing the right-hand sides of (7.16) and (7.17), we arrive at the statement of the theorem. □

7.4. Invariance of the Hamiltonian Action

To consider discrete Hamiltonian equations, we need three lattice points. The prolongation of the Lie group operator (7.14) to the neighboring points $(t_-, \mathbf{q}_-, \mathbf{p}^-)$ and $(t_+, \mathbf{q}_+, \mathbf{p}^+)$ is as follows:

$$X = \xi \frac{\partial}{\partial t} + \xi_- \frac{\partial}{\partial t_-} + \xi_+ \frac{\partial}{\partial t_+} + \eta^i \frac{\partial}{\partial q^i} + \eta^i_- \frac{\partial}{\partial q^i_-} + \eta^i_+ \frac{\partial}{\partial q^i_+}$$

$$+ \zeta_i \frac{\partial}{\partial p_i} + \zeta^-_i \frac{\partial}{\partial p^-_i} + \zeta^+_i \frac{\partial}{\partial p^+_i} + (\xi_+ - \xi) \frac{\partial}{\partial h_+} + (\xi - \xi_-) \frac{\partial}{\partial h_-}, \quad (7.18)$$

where

$$\xi_- = \xi(t_-, \mathbf{q}_-, \mathbf{p}^-), \qquad \eta^i_- = \eta^i(t_-, \mathbf{q}_-, \mathbf{p}^-), \qquad \zeta^-_i = \zeta_i(t_-, \mathbf{q}_-, \mathbf{p}^-),$$
$$\xi_+ = \xi(t_+, \mathbf{q}_+, \mathbf{p}^+), \qquad \eta^i_+ = \eta^i(t_+, \mathbf{q}_+, \mathbf{p}^+), \qquad \zeta^+_i = \zeta_i(t_+, \mathbf{q}_+, \mathbf{p}^+).$$

Consider the functional (7.8) on some lattice given by equation

$$\Omega(t, t_-, t_+, \mathbf{q}, \mathbf{p}, \mathbf{q}_-, \mathbf{p}_-, \mathbf{q}_+, \mathbf{p}_+) = 0. \tag{7.19}$$

DEFINITION 7.1. We say that the discrete Hamiltonian function $\mathcal{H}$ considered on the mesh (7.19) is *invariant* with respect to the group generated by the operator (7.14) if the action functional (7.8) considered on the mesh (7.19) is an invariant of the group.

THEOREM 7.2. *The Hamiltonian function considered together with the mesh (7.19) is invariant with respect to the group generated by the operator (7.14) if and only if*

$$\left[\zeta^+_i \underset{+h}{D}(q^i) + p^+_i \underset{+h}{D}(\eta^i) - X(\mathcal{H}) - \mathcal{H} \underset{+h}{D}(\xi) \right]\Big|_{\Omega=0} = 0, \qquad X\Omega\big|_{\Omega=0} = 0. \tag{7.20}$$

Proof. The invariance condition readily follows from the action of X on the functional:

$$X(\mathbb{H}_h) = X\left(\sum_\Omega (p^+_i(q^i_+ - q^i) - \mathcal{H}h_+) \right)$$

$$= \sum_\Omega (\zeta^+_i(q^i_+ - q^i) + p^+_i(\eta^i_+ - \eta^i) - X(\mathcal{H})h_+ - \mathcal{H}(\xi_+ - \xi))$$

$$= \sum_\Omega (\zeta^+_i \underset{+h}{D}(q^i) + p^+_i \underset{+h}{D}(\eta^i) - X(\mathcal{H}) - \mathcal{H} \underset{+h}{D}(\xi))h_+ = 0.$$

It should be supplemented with the invariance of the mesh, which is obtained by the action of the symmetry operator on the mesh equation (7.19). $\quad\square$

In the general case, the lattice is provided by the discrete Hamiltonian equations (7.6). Therefore, we need to require their invariance to consider the invariance of the Hamiltonian function.

7.5. Difference Hamiltonian Identity and Noether-Type Theorem for Difference Hamiltonian Equations

Just as in the continuous case, the invariance of a difference Hamiltonian on a specified mesh yields first integrals of the discrete Hamiltonian equations.

LEMMA 7.3. *For any smooth function* $\mathcal{H} = \mathcal{H}(t, t_+, \mathbf{q}, \mathbf{p}_+)$, *one has the identity*

$$
\zeta_i^+ \underset{+h}{D}(q^i) + p_i^+ \underset{+h}{D}(\eta^i) - X(\mathcal{H}) - \mathcal{H} \underset{+h}{D}(\xi)
$$

$$
\equiv -\xi \left(\frac{\partial \mathcal{H}}{\partial t} + \frac{h_-}{h_+} \frac{\partial \mathcal{H}^-}{\partial t} - \underset{+h}{D}(\mathcal{H}^-) \right) - \eta^i \left(\underset{+h}{D}(p_i) + \frac{\partial \mathcal{H}}{\partial q^i} \right)
$$

$$
+ \zeta_i^+ \left(\underset{+h}{D}(q^i) - \frac{\partial \mathcal{H}}{\partial p_i^+} \right) + \underset{+h}{D} \left[\eta^i p_i - \xi \left(\mathcal{H}^- + h_- \frac{\partial \mathcal{H}^-}{\partial t} \right) \right]. \quad (7.21)
$$

Proof. The identity can be established by a straightforward computation. $\square$

We refer to this identity as the *difference Hamiltonian identity*. It permits one to state the following result.

THEOREM 7.4. *The difference Hamiltonian equations* (7.6) *invariant with respect to the symmetry operator* (7.14) *possess the first integral*

$$
\mathcal{J} = \eta^i p_i - \xi \left(\mathcal{H}^- + h_- \frac{\partial \mathcal{H}^-}{\partial t} \right)
$$

if and only if the Hamiltonian function is invariant with respect to the same symmetry on the solutions of Eqs. (7.6).

Proof. This result is a consequence of identity (7.21). The invariance of the discrete Hamiltonian equations is needed to guarantee the invariance of the mesh defined by these equations. $\square$

Remark. Theorem 7.4 can be generalized to the case of divergence invariance of the Hamiltonian action, i.e., to the case in which

$$
\zeta_i^+ \underset{+h}{D}(q^i) + p_i^+ \underset{+h}{D}(\eta^i) - X(\mathcal{H}) - \mathcal{H} \underset{+h}{D}(\xi) = \underset{+h}{D}(V), \quad (7.22)
$$

where $V = V(t, \mathbf{q}, \mathbf{p})$. If this condition holds on the solutions of the discrete Hamiltonian equations (7.6), then one has the first integral

$$
\mathcal{J} = \eta^i p_i - \xi \left(\mathcal{H}^- + h_- \frac{\partial \mathcal{H}^-}{\partial t} \right) - V.
$$

Remark. For discrete Hamiltonian equations with Hamiltonian functions $\mathcal{H} = \mathcal{H}(h_+, \mathbf{q}, \mathbf{p}^+)$ invariant with respect to time translations, the energy is conserved,

$$\mathcal{E} = \mathcal{H}^- + h_- \frac{\partial \mathcal{H}^-}{\partial h_-} = \mathcal{H} + h_+ \frac{\partial \mathcal{H}}{\partial h_+}.$$

Note that $\mathcal{H}$ is not the discrete energy; it has the meaning of generating function of the discrete Hamiltonian flow.

Remark. In a similar way, one can consider the identity

$$
\zeta_i \underset{+h}{D}(q^i) + p_i \underset{+h}{D}(\eta^i) - X(\mathcal{H}) - \mathcal{H} \underset{+h}{D}(\xi)
$$
$$
\equiv -\xi \left(\frac{\partial \mathcal{H}}{\partial t} + \frac{h_-}{h_+} \frac{\partial \mathcal{H}^-}{\partial t} - \underset{+h}{D}(\mathcal{H}^-) \right) - \eta^i_+ \left(\underset{+h}{D}(p_i) + \frac{\partial \mathcal{H}}{\partial q^i_+} \right)
$$
$$
+ \zeta_i \left(\underset{+h}{D}(q^i) - \frac{\partial \mathcal{H}}{\partial p_i} \right) + \underset{+h}{D} \left[\eta^i p_i - \xi \left(\mathcal{H}^- + h_- \frac{\partial \mathcal{H}^-}{\partial t} \right) \right],
$$

which permits stating Noether's theorem for the second version of the difference Hamiltonian equations (7.7).

7.6. Invariance of Difference Hamiltonian Equations

An application of the discrete variational operators (7.11)–(7.13) to the expression

$$
\zeta^+_i (q^i_+ - q^i) + p^+_i (\eta^i_+ - \eta^i) - X(\mathcal{H}) h_+ - \mathcal{H}(\xi_+ - \xi)
$$
$$
\equiv \left(\zeta^+_i \underset{+h}{D}(q^i) + p^+_i \underset{+h}{D}(\eta^i) - X(\mathcal{H}) - \mathcal{H} \underset{+h}{D}(\xi) \right) h_+
$$

obtained by action of the symmetry operator X on the elementary action gives the following result.

LEMMA 7.5. *For any smooth function* $\mathcal{H} = \mathcal{H}(t, t_+, \mathbf{q}, \mathbf{p}_+)$, *one has the identities*

$$
\frac{\delta}{\delta p_j} \left(\zeta^+_i (q^i_+ - q^i) + p^+_i (\eta^i_+ - \eta^i) - X(\mathcal{H}) h_+ - \mathcal{H}(\xi_+ - \xi) \right)
$$
$$
= X \left(\frac{\delta \mathcal{H}}{\delta p_j} \right) + \frac{\partial \zeta_i}{\partial p_j} \frac{\delta \mathcal{H}}{\delta p_i} + \frac{\partial \eta^i}{\partial p_j} \frac{\delta \mathcal{H}}{\delta q^i} + \frac{\partial \xi}{\partial p_j} \frac{\delta \mathcal{H}}{\delta t}, \qquad j = 1, \dots, n,
$$
$$
\frac{\delta}{\delta q^j} \left(\zeta^+_i (q^i_+ - q^i) + p^+_i (\eta^i_+ - \eta^i) - X(\mathcal{H}) h_+ - \mathcal{H}(\xi_+ - \xi) \right)
$$
$$
= X \left(\frac{\delta \mathcal{H}}{\delta q^j} \right) + \frac{\partial \zeta_i}{\partial q^j} \frac{\delta \mathcal{H}}{\delta p_i} + \frac{\partial \eta^i}{\partial q^j} \frac{\delta \mathcal{H}}{\delta q^i} + \frac{\partial \xi}{\partial q^j} \frac{\delta \mathcal{H}}{\delta t}, \qquad j = 1, \dots, n,
$$

$$\frac{\delta}{\delta t}\left(\zeta_i^+(q_+^i - q^i) + p_i^+(\eta_+^i - \eta^i) - X(\mathcal{H})h_+ - \mathcal{H}(\xi_+ - \xi)\right)$$
$$= X\left(\frac{\delta\mathcal{H}}{\delta t}\right) + \frac{\partial\zeta_i}{\partial t}\frac{\delta\mathcal{H}}{\delta p_i} + \frac{\partial\eta^i}{\partial t}\frac{\delta\mathcal{H}}{\delta q^i} + \frac{\partial\xi}{\partial t}\frac{\delta\mathcal{H}}{\delta t}.$$

Using the lemma, we can relate the invariance of the discrete Hamiltonian equations to that of the Hamiltonian.

THEOREM 7.6. *If the discrete Hamiltonian $\mathcal{H}$ is invariant with respect to the operator (7.14), then so are the discrete Hamiltonian equations (7.6).*

Proof. If the difference Hamiltonian $\mathcal{H}$ is invariant, then the left-hand sides of the identities in Lemma 7.5 are zero. It follows that the variational equations (7.9) are invariant. Consequently, so are the difference Hamiltonian equations, which are equivalent to these variational equations. □

Remark. If the difference Hamiltonian $\mathcal{H}$ is divergence invariant, then so are the discrete Hamiltonian equations (7.6). This follows from the fact that total finite differences belong to the kernel of the discrete variational operators.

By using the identities in Lemma 7.5, we can refine the result of Theorem 7.6 and state a necessary and sufficient condition for the difference Hamiltonian equations to be invariant. This explicitly shows the distinction between the invariance of Hamiltonians and the invariance of Hamiltonian equations.

THEOREM 7.7. *The difference Hamiltonian equations (7.6) are invariant with respect to a symmetry (7.14) if and only if the following conditions are true (on the solutions of the discrete Hamiltonian equations):*

$$\frac{\delta}{\delta p_j}\left(\zeta_i^+(q_+^i - q^i) + p_i^+(\eta_+^i - \eta^i) - X(\mathcal{H})h_+ - \mathcal{H}(\xi_+ - \xi)\right)\big|_{(7.6)} = 0,$$
$$j = 1,\ldots,n,$$

$$\frac{\delta}{\delta q^j}\left(\zeta_i^+(q_+^i - q^i) + p_i^+(\eta_+^i - \eta^i) - X(\mathcal{H})h_+ - \mathcal{H}(\xi_+ - \xi)\right)\big|_{(7.6)} = 0,$$
$$j = 1,\ldots,n,$$

$$\frac{\delta}{\delta t}\left(\zeta_i^+(q_+^i - q^i) + p_i^+(\eta_+^i - \eta^i) - X(\mathcal{H})h_+ - \mathcal{H}(\xi_+ - \xi)\right)\big|_{(7.6)} = 0.$$

Proof. We use the fact that the discrete Hamiltonian equations (7.6) are equivalent to the variational equations (7.10). Now the claim follows from the identities in Lemma 7.5. □

7.7. Examples

In this section, we present applications of the theoretical results presented above to a number of differential equations and their discrete counterparts.

7.7.1. Discrete harmonic oscillator

Consider the one-dimensional harmonic oscillator

$$\ddot{u} + u = 0. \tag{7.23}$$

The symmetry group admitted by this equation and the corresponding first integrals can be found, for example, in [94].

Now consider the one-dimensional harmonic oscillator in the Hamiltonian form

$$\dot{q} = p, \qquad \dot{p} = -q. \tag{7.24}$$

These equations are generated by the Hamiltonian function

$$H(t, q, p) = \frac{1}{2}(q^2 + p^2).$$

Consider the discretization

$$\frac{q_+ - q}{h_+} = \frac{p + p_+}{2}, \qquad \frac{p_+ - p}{h_+} = -\frac{q + q_+}{2} \tag{7.25}$$

of Eqs. (7.24) on the uniform mesh $h_+ = h_-$ by the midpoint rule. This discretization can be rewritten as the system of equations

$$\underset{+h}{D}(q) = \frac{4}{4 - h_+^2}\left(p_+ + \frac{h_+}{2}q\right), \qquad \underset{+h}{D}(p) = -\frac{4}{4 - h_+^2}\left(q + \frac{h_+}{2}p_+\right), \tag{7.26}$$

$$h_+ = h_- = h.$$

It can be shown that this system is generated by the discrete Hamiltonian function

$$\mathcal{H}(t, t_+, q, p_+) = \frac{2}{4 - h_+^2}(q^2 + p_+^2 + h_+ q p_+).$$

Indeed, the first and second equations in (7.6) are exactly the same as in (7.26). The last equation in (7.6) acquires the form

$$-\frac{2(4 + h_+^2)}{(4 - h_+^2)^2}(q^2 + p_+^2) - \frac{16h_+}{(4 - h_+^2)^2}q p_+$$

$$+ \frac{2(4 + h_-^2)}{(4 - h_-^2)^2}(q_-^2 + p^2) + \frac{16h_-}{(4 - h_-^2)^2}q_- p = 0.$$

Using the first and second equations, we can rewrite it as

$$\left(-\frac{2}{4+h_+^2}+\frac{2}{4+h_-^2}\right)(q^2+p^2)=0.$$

Therefore, this equation can be taken in the equivalent form

$$h_+=h_-$$

provided that $q^2+p^2\neq 0$.

The system of difference equations (7.26) in particular admits the symmetries

$$X_1=\sin(\omega t)\frac{\partial}{\partial q}+\cos(\omega t)\frac{\partial}{\partial p},\qquad X_2=\cos(\omega t)\frac{\partial}{\partial q}-\sin(\omega t)\frac{\partial}{\partial p},$$

$$X_3=\frac{\partial}{\partial t},\qquad X_4=q\frac{\partial}{\partial q}+p\frac{\partial}{\partial p},\qquad X_5=p\frac{\partial}{\partial q}-q\frac{\partial}{\partial p},$$

where

$$\omega=\frac{\arctan(h/2)}{h/2}.$$

For the symmetry operators X_1 and X_2, we have the divergence invariance conditions

$$\zeta_+ \underset{+h}{D}(q)+p_+ \underset{+h}{D}(\eta)-X(\mathcal{H})-\mathcal{H}\underset{+h}{D}(\xi)=\underset{+h}{D}(V)$$

satisfied on the solutions of Eqs. (7.26) with the functions $V_1=q\cos(\omega t)$ and $V_2=-q\sin(\omega t)$, respectively. Therefore, we obtain the corresponding two first integrals

$$\mathcal{J}_1=p\sin(\omega t)-q\cos(\omega t),\qquad \mathcal{J}_2=p\cos(\omega t)+q\sin(\omega t).$$

The symmetry operator X_3 satisfies the invariance condition

$$\zeta_+ \underset{+h}{D}(q)+p_+ \underset{+h}{D}(\eta)-X(\mathcal{H})-\mathcal{H}\underset{+h}{D}(\xi)=0.$$

Thus, we obtain the first integral

$$\mathcal{J}_3=-\frac{4}{4-h_-^2}\left(\frac{4+h_-^2}{4-h_-^2}\frac{q^2+p^2}{2}+\frac{4h_-}{4-h_-^2}q_-p\right).$$

Using the first and second equations in (7.26), we can simplify it as

$$\mathcal{J}_3=-\frac{4}{4+h_-^2}\frac{q^2+p^2}{2}.$$

Since the first integrals $\mathcal{J}_1$ and $\mathcal{J}_2$ give the conservation law

$$\mathcal{J}_1^2+\mathcal{J}_2^2=q^2+p^2=\text{const},$$

we can equivalently take the third integral in the form $\tilde{\mathcal{J}}_3 = h_-$, which permits using a regular lattice.

The three first integrals $\mathcal{J}_1$, $\mathcal{J}_2$, $\tilde{\mathcal{J}}_3$ are sufficient for the integration of system (7.25). We obtain the solution

$$q = \mathcal{J}_2 \sin(\omega t) - \mathcal{J}_1 \cos(\omega t), \qquad p = \mathcal{J}_1 \sin(\omega t) + \mathcal{J}_2 \cos(\omega t) \qquad (7.27)$$

on the lattice

$$t_i = t_0 + ih, \qquad i = 0, \pm 1, \pm 2, \ldots, \qquad h = \tilde{\mathcal{J}}_3. \qquad (7.28)$$

7.7.2. Modified discrete harmonic oscillator (exact scheme)

The solutions of the discrete harmonic oscillator in the Lagrangian case (7.23) and in the Hamiltonian case (7.27), (7.28) follow the same trajectory as the solution of the continuous harmonic oscillator but at a different velocity. These discretization errors can be corrected by time reparametrization. Hence we obtain an *exact discretization of the harmonic oscillator*, i.e., a discretization that gives the solution of the underlying ordinary differential equation.

The discrete harmonic oscillator admits reparametrization. Consider the harmonic oscillator (7.24) discretized as

$$\frac{q_| - q}{h_+} = \Omega \frac{p + p_|}{2}, \qquad \frac{p_+ - p}{h_+} = -\Omega \frac{q + q_+}{2}, \qquad h_+ = h_- = h, \qquad (7.29)$$

where

$$\Omega = \frac{\tan(h/2)}{h/2}.$$

By analogy with Sec. 7.7.1, it can be shown that this discrete model of the harmonic oscillator is generated by the discrete Hamiltonian

$$\mathcal{H}(t, t_+, q, p_+) = \frac{2\Omega}{4 - \Omega^2 h_+^2}(q^2 + p_+^2 + \Omega^2 h_+ q p_+).$$

The system of difference equations (7.29) admits the symmetries

$$X_1 = \sin t \frac{\partial}{\partial q} + \cos t \frac{\partial}{\partial p}, \qquad X_2 = \cos t \frac{\partial}{\partial q} - \sin t \frac{\partial}{\partial p},$$

$$X_3 = \frac{\partial}{\partial t}, \qquad X_4 = q \frac{\partial}{\partial q} + p \frac{\partial}{\partial p}, \qquad X_5 = p \frac{\partial}{\partial q} - q \frac{\partial}{\partial p}.$$

For the symmetries X_1 and X_2, which satisfy the divergence invariance condition (7.22) with the functions $V_1 = q \cos t$ and $V_2 = -q \sin t$, we obtain two first integrals,

$$\mathcal{J}_1 = p \sin t - q \cos t, \qquad \mathcal{J}_2 = p \cos t + q \sin t.$$

The operator X_3 satisfies the invariance condition (7.20) and gives the first integral $\mathcal{J}_3$, which (by analogy with Sec. 7.7.1) can be taken in the equivalent form $\widetilde{\mathcal{J}}_3 = h_-$. The scheme (7.29) gives the exact solution of the harmonic oscillator, which can be found with the help of the first integrals $\mathcal{J}_1$, $\mathcal{J}_2$, and $\widetilde{\mathcal{J}}_3$ as

$$q = \mathcal{J}_2 \sin t - \mathcal{J}_1 \cos t, \qquad p = \mathcal{J}_1 \sin t + \mathcal{J}_2 \cos t.$$

This discrete solution is given on the lattice

$$t_i = t_0 + ih, \qquad i = 0, \pm 1, \pm 2, \ldots, \qquad h = \widetilde{\mathcal{J}}_3.$$

Exact schemes for the two- and four-dimensional harmonic oscillators were used in [80] to construct exact schemes for the two- and three-dimensional Kepler motion, respectively.

7.7.3. Nonlinear motion

The equations

$$\dot{q} = \frac{4}{p^2}, \qquad \dot{p} = 1$$

are generated by the Hamiltonian

$$H = -\frac{4}{p} - q.$$

Consider the discretization

$$\frac{q_+ - q}{h_+} = \frac{4}{(p_+ - h_+/2)(p + h_+/2)}, \qquad \frac{p_+ - p}{h_+} = 1 \qquad (7.30)$$

on the lattice

$$\frac{h_+}{p_+ - h_+/2} = \frac{h_-}{p - h_-/2}. \qquad (7.31)$$

This scheme is invariant with respect to the Lie group operators

$$X_1 = \frac{\partial}{\partial t}, \qquad X_2 = \frac{\partial}{\partial q}, \qquad X_3 = t\frac{\partial}{\partial t} - q\frac{\partial}{\partial q} + p\frac{\partial}{\partial p}.$$

The difference equations (7.30) can be rewritten as

$$\frac{q_+ - q}{h_+} = \frac{4}{(p_+ - h_+/2)^2}, \qquad \frac{p_+ - p}{h_+} = 1.$$

These equations are generated by the discrete Hamiltonian function

$$\mathcal{H}(t, t_+, q, p_+) = -\frac{4}{p_+ - h_+/2} - q.$$

The last discrete Hamiltonian equation in (7.6) is

$$-\frac{4p_+}{(p_+ - h_+/2)^2} - q + \frac{4p}{(p - h_-/2)^2} + q_- = 0.$$

This equation leads to the lattice equation (7.31) on the solutions of (7.30).

The Hamiltonian function is invariant with respect to the symmetry operators X_1 and X_3. For the symmetry X_2, we have divergence invariance with $V_2 = t$. Therefore, these symmetries give three first integrals,

$$\mathcal{I}_1 = \frac{4p_+}{(p_+ - h_+/2)^2} + q, \qquad \mathcal{I}_2 = p_+ - t_+,$$

$$\mathcal{I}_3 = -q_+ p_+ + t_+ \left(\frac{4p_+}{(p_+ - h_+/2)^2} + q \right).$$

Note that

$$4 - \mathcal{I}_1 \mathcal{I}_2 - \mathcal{I}_3 = \left(\frac{h_+}{p_+ - h_+/2} \right)^2$$

on the solutions of the difference equations (7.30), which justifies the lattice (7.31). By setting

$$\mathcal{I}_1 = A, \qquad \mathcal{I}_2 = B, \qquad \frac{h_+}{p_+ - h_+/2} = \varepsilon,$$

we find the solution of the discrete model in the form

$$q = A - \frac{4}{t + B} \left(1 - \frac{\varepsilon^2}{4} \right), \qquad p = t + B.$$

The integration of the lattice equation can be found in [48].

Chapter 8

Discrete Representation of Ordinary Differential Equations with Symmetries

In this chapter, we consider the relationships between the objects under study, namely, between differential and difference equations and transformation groups admitted by them.

Using the Taylor series, one can readily write out the differential representation of a given difference equation. This is a formal power series whose sum truncated at a certain term is the so-called *differential approximation to the difference equation*. The differential representation (an infinite-order differential equation) formally admits the same transformation group as the original difference equation, but the differential approximation, which is a finite-order differential equation, may preserve the symmetry of neither the original equation nor the difference equation.

Using the Lagrange formula and formal Newton series expansions, one can construct a *discrete representation of the differential equation*, which preserves the group of the original differential equation. But the truncated partial sums of such series, i.e., the *difference approximations to the differential equation*, need not preserve the admissible group in general.

In this section, we consider all these objects only for second-order ordinary differential equations and the corresponding schemes on a uniform mesh. For first-order ordinary differential equations whose symmetry is known, one can readily write out not only the invariant scheme but also the exact scheme; see Sec. 3.1. In Sec. 3.2, we succeeded in writing invariant schemes and meshes on a three-point stencil for the second-order ordinary differential equations. In Chapters 6 and 7, we constructed difference schemes that additionally have difference first integrals. The solutions of such schemes are very close to the solutions of ordinary differential equations not only in the approximation order but also in the form of the curves, which differ from the exact curves (solutions) only by an insignificant dilation or shrinking. It turns out that in the set of parametric families of invariant difference schemes for second-order ordinary differential equations one can single out schemes that have no approximation error, i.e., the *discrete representation of differential equations*, or exact difference schemes.

8.1. The Discrete Representation of ODE as a Series

1. Each second-order ordinary difference equation

$$\underset{h}{F}(x, x^+, x^-, v, v^+, v^-) = 0$$

on the uniform mesh

$$h^+ = h^- = h$$

can be represented in the "continuous" space of sequences $\widetilde{Z} = (x, u, u_1, u_2, \ldots)$ by means of the Taylor group with infinitesimal operator D.

   Let us illustrate this by an example.

EXAMPLE. Let the following ordinary differential equation be given in $\widetilde{Z}$:

$$u'' = u^2, \tag{8.1}$$

and let the following finite-difference equation on a uniform mesh approximating (8.1) up to the second order in h be given in $\underset{h}{Z}$:

$$\underset{h}{v_{x\bar{x}}} = v^2. \tag{8.2}$$

To represent (8.2) in $\widetilde{Z}$, one has to use the Taylor formula for $\underset{h}{v_{x\bar{x}}}$:

$$\underset{h}{v_{x\bar{x}}} = \underset{-h}{D} \underset{+h}{D}(v) = \sum_{s=1}^{\infty} \frac{(-h)^{s-1}}{s!} \sum_{k=1}^{\infty} \frac{h^{k-1}}{k!} D^{s+k}(v).$$

Thus, Eq. (8.2) in $\widetilde{Z}$ is a formal power series in h:

$$\sum_{s=1}^{\infty} \sum_{k=1}^{\infty} \frac{(-h)^{s-1} h^{k-1}}{s! k!} v_{s+k} = v^2, \tag{8.3}$$

where v_m is the mth derivative with respect to x.

   The representation (8.3) allows us to consider the approximate object, namely, the *differential approximation of the difference equation*. For example, omitting in (8.3) the terms of higher order than h^2, we obtain the *first differential approximation to (8.2)*:

$$v_2 + \frac{h^2}{12} v_4 = v^2.$$

Taking the terms of the next order into account, we similarly obtain the second, third, etc. differential approximations to the difference equation.

   In $\widetilde{Z}$, the differential approximation of any finite order occupies an intermediate position *in the functional-analytic sense* between the difference equation (in continuous representation) and the differential equation. But the fact that the differential approximation is close to the difference equation and to the differential equation in the sense of approximation does not guarantee the same closeness in the algebraic aspect, in particular, concerning the closeness of their group properties.

2. Thus, so far we have three objects in $\widetilde{Z}$ and $\underset{h}{Z}$: the differential equation (system), the difference equation, and the continuous representation of the difference equation. For symmetry reasons, it is necessary to have the fourth object, namely, an exact expression for the differential equation in the mesh space $\underset{h}{Z}$.

Earlier, for the Taylor group operator

$$D = \frac{\partial}{\partial x} + u_1 \frac{\partial}{\partial u} + u_2 \frac{\partial}{\partial u_1} + \cdots$$

we obtained the representation

$$D^{\pm} = \frac{\partial}{\partial x} + \underset{\pm h}{\widetilde{D}}(u) \frac{\partial}{\partial u} + \underset{\pm h}{\widetilde{D}}(\underset{h}{u_s}) \frac{\partial}{\partial \underset{h}{u_s}} + \cdots$$

in the mesh space, where $\underset{\pm h}{\widetilde{D}} = \sum_{n=1}^{\infty} \frac{(\mp h)^{n-1}}{n} \underset{\pm h}{D^n}$. The correspondence between the differential and finite-difference variables is given by the Lagrange formula

$$D \Longleftrightarrow \begin{cases} \sum\limits_{n=1}^{\infty} \frac{(-h)^{n-1}}{n} \underset{+h}{D^n}, \\ \sum\limits_{n=1}^{\infty} \frac{(+h)^{n-1}}{n} \underset{-h}{D^n} \end{cases} \tag{8.4}$$

Formula (8.4) permits one to represent the finite-difference variables in $\widetilde{Z}$ and hence to obtain a representation of any other difference equation.

EXAMPLE. We use the Lagrange formula (8.4) to rewrite Eq. (8.1), which was considered above on the uniform mesh, in the mesh space $\underset{h}{Z}$ as

$$\sum_{s=1}^{\infty} \frac{(\mp h)^{s-1}}{s} \underset{\pm h}{D^s} \sum_{n=1}^{\infty} \frac{(\mp h)^{n-1}}{n} \underset{\pm h}{D^n}(u) = u^2. \tag{8.5}$$

Formula (8.4) gives a nonunique representation for the differential variables: either the right half-line, or the left half-line, or the entire line of the independent variable x is used. With increasing order of the derivative u_s, the number of its discrete representations also increases. For example, the representation (8.5) generally means the following four representations:

$$\text{I.} \quad \sum_{s,n=1}^{\infty} \frac{(-h)^{s+n-2}}{s} \underset{+h}{D^{s+n}}(u) = u^2.$$

$$\text{II.} \quad \sum_{s,n=1}^{\infty} \frac{h^{s+n-2}}{s} \underset{-h}{D^{s+n}}(u) = u^2.$$

$$\text{III.} \quad \sum_{s=1}^{\infty} \sum_{n=1}^{\infty} \frac{(-h)^{s-1}h^{n-1}}{sn} \underset{+h}{D^s} \underset{-h}{D^n}(u) = u^2.$$

$$\text{IV.} \quad \sum_{s=1}^{\infty} \sum_{n=1}^{\infty} \frac{h^{s-1}(-h)^{n-1}}{sn} \underset{-h}{D^s} \underset{+h}{D^n}(u) = u^2.$$

$$\tag{8.6}$$

The formal series (8.6) represents the differential equation (8.1) in difference form; these series will be called the "exact difference scheme" for (8.1) or the *discrete representation of the differential equation* in series form.

In this one-dimensional case of a uniform mesh, the operators $\underset{+h}{D}$ and $\underset{-h}{D}$ commute, and representations III and IV in (8.6) coincide. Representations I and II are taken to each other by the discrete reflection group $x \to -x$, which acts as follows: $h \to -h$, $\underset{+h}{S} \to \underset{-h}{S}$, and $\underset{+h}{D} \to \underset{-h}{D}$. Representations III and IV are invariant under reflection.

Thus, there is a significant difference between representations I (II) and III (IV). To write out the differential equation in these representations, one uses a half-line or the entire line of the independent variable, respectively. From the algebraic point of view, this is the question of whether the reflection group is admissible. In this respect, representations III and IV are preferable, because the original ordinary differential equation admits the reflection group.

In more detail, from Eqs. (8.6) we obtain, to within $O(h^2)$,

$$\underset{h}{u_{xx}} - h \underset{h}{u_{xxx}} + O(h^2) = u^2,$$

$$\underset{h}{u_{\bar{x}\bar{x}}} + h \underset{h}{u_{\bar{x}\bar{x}\bar{x}}} + O(h^2) = u^2,$$

$$\underset{h}{u_{\bar{x}x}} + \frac{h}{2}(\underset{h}{u_{\bar{x}\bar{x}x}} - \underset{h}{u_{\bar{x}xx}}) + O(h^2) = u^2,$$

$$\underset{h}{u_{x\bar{x}}} + \frac{h}{2}(\underset{h}{u_{\bar{x}\bar{x}x}} - \underset{h}{u_{\bar{x}xx}}) + O(h^2) = u^2.$$

The finite-difference equations

$$\underset{h}{u_{xx}} - h \underset{h}{u_{xxx}} = u^2, \qquad \underset{h}{u_{\bar{x}\bar{x}}} + h \underset{h}{u_{\bar{x}\bar{x}\bar{x}}} = u^2, \qquad \underset{h}{u_{x\bar{x}}} - \frac{h^2}{2} \underset{h}{u_{x\bar{x}x\bar{x}}} = u^2 \qquad (8.7)$$

can be called the *first difference approximation* to the corresponding *differential equation* in the mesh space $\underset{h}{Z}$. In particular, it follows from (8.7) that the presence of the reflection group in representation III (IV) ensures the second order of approximation. The difference approximation (of any order) to some differential equation is a finite-order finite-difference equation and can be used to construct an approximate difference model of the differential equation. Clearly, the situation in the example considered above is of general character.

Note that we generally use distinct symbols (x, u) and (y, v) for the dependent and independent variables in the notation of the spaces $\widetilde{Z}$ and $\underset{h}{Z}$; for simplicity, we have assumed that $x = y$ and $u = v$ in the examples above. Moreover, the space $\widetilde{Z}$ should be supplemented with a nonlocal variable h, because the continuous representation of a finite-difference equation contains the mesh spacing h.

Differential-difference equations, which are often used to analyze difference schemes, occupy an independent position in the above scheme.

3. We use the same example to show how the group admitted by the original ordinary differential equation acts on the four models considered above.

One can consider the product

$$\underset{h}{\widetilde{Z}} = (x, u, u_1, u_2, \ldots, h; y, v, \underset{h}{v_1}, \underset{h}{v_2}, \ldots, h)$$

of the spaces $\widetilde{Z}$ and $\underset{h}{Z}$ and treat the transition from $\widetilde{Z}$ to $\underset{h}{Z}$ and vice versa as a *change of variables* in $\underset{h}{\widetilde{Z}}$:

$$y = f(x, u), \qquad v = g(x, u). \tag{8.8}$$

In a more general case, this change of variables (8.8) may be nonlocal.

The equation

$$u'' = u^2$$

in $\widetilde{Z}$ admits the group G_2 with generators

$$X_1 = \frac{\partial}{\partial u}, \qquad X_2 = x\frac{\partial}{\partial x} - 2u\frac{\partial}{\partial u}. \tag{8.9}$$

Let us pass to a similar group in $\underset{h}{Z}$ by using the well-known formulas of change of variables in the infinitesimal operator. We use the simplest change of the form (8.8), namely, the identity transformation

$$y = x, \qquad v = u$$

in the subspace (x, u, h, y, v, h). We obtain the following expressions for the coefficients of the operators X_1 and X_2 given by (8.9):

$$\bar{X}_i = X_i(f(x, u))\frac{\partial}{\partial y} + X_i(g(x, u))\frac{\partial}{\partial v}, \qquad i = 1, 2.$$

It follows that

$$\bar{X}_1 = \frac{\partial}{\partial y}, \qquad \bar{X}_2 = y\frac{\partial}{\partial y} - 2v\frac{\partial}{\partial v}.$$

Now it only remains to prolong $\bar{X}_2$ to finite-difference variables. (Note that $\bar{X}_1$ has no prolongation.) Denoting

$$\underset{h}{v_1} = \underset{+h}{D}(v), \qquad \underset{h}{v_2} = \underset{-h+h}{D\,D}(v), \ldots$$

and using the prolongation formulas obtained above, we have

$$\bar{X}_1 = \frac{\partial}{\partial y},$$

$$\bar{X}_2 = y\frac{\partial}{\partial y} - 2v\frac{\partial}{\partial v} - 3\underset{h}{v_1}\frac{\partial}{\partial \underset{h}{v_1}} - 4\underset{h}{v_2}\frac{\partial}{\partial \underset{h}{v_2}} - (n+2)\underset{h}{v_n}\frac{\partial}{\partial \underset{h}{v_n}} + \cdots + h\frac{\partial}{\partial h}. \tag{8.10}$$

The group $\bar{G}_2$ with operators (8.10) acts in the mesh space $\underset{h}{Z}$ and is similar to the group G_2 with operators (8.9).

Let us verify that the discrete representation (8.6) of a differential equation in $\underset{h}{Z}$ admits the operators (8.10). We rewrite the representation (8.6) in the same notation:

$$\sum_{s,n=1}^{\infty} \frac{(-h)^{s+n-2}}{sn} \underset{h}{v}_{n+s} - v^2 = 0. \tag{8.11}$$

Clearly, (8.11) admits the operator $\bar{X}_1$. The action of $\bar{X}_2$ on (8.11) gives

$$\sum_{s,n=1}^{\infty} \frac{(s+n-2)(-h)^{s+n-2}}{sn} \underset{h}{v}_{n+s} - \sum_{s,n=1}^{\infty} \frac{(n+s+2)(-h)^{s+n-2}}{sn} \underset{h}{v}_{n+s} + 4v^2 = 0,$$

or, after simple transformations,

$$-4 \left(\sum_{s,n=1}^{\infty} \frac{(-h)^{s+n-2}}{sn} \underset{h}{v}_{n+s} - v^2 \right) = 0,$$

whence it follows that the infinite-order difference equation (8.11) is invariant under $\bar{G}_2$.

In a similar way, we can verify that the other representations in (8.6) are invariant under the operators (8.10) as well.

In $\underset{h}{Z}$, consider the equation

$$\underset{h}{v}_{x\bar{x}} - v^2 = 0, \tag{8.12}$$

which is constructed with the guaranteed property of invariance under the group $\bar{G}_2$ with operators (8.10). As follows from (8.7), Eq. (8.12) is the "zero" difference approximation to representation III (IV) in (8.6); i.e., it differs by $O(h^2)$ from the infinite-order difference equation of the form III (IV) in (8.6). In this sense, the difference equation (8.12) is the most natural approximation to (8.6).

It is clear that any difference equation approximating III (IV) in (8.6) to within $O(h^2)$ is not necessarily invariant under $\bar{G}_r$.

4. Now consider the representation (8.12) in $\widetilde{Z}$,

$$\sum_{s=1}^{\infty} \frac{(-h)^{s-1}}{s!} \sum_{n=1}^{\infty} \frac{h^{n-1}}{n!} D^{s+n}(u) - u^2 = 0. \tag{8.13}$$

The group G_2, similar to $\bar{G}_2$, can be prolonged to the variables u_s and h,

$$X_1 = \frac{\partial}{\partial x},$$

$$X_2 = x\frac{\partial}{\partial x} - 2u\frac{\partial}{\partial u} - 3u_1\frac{\partial}{\partial u_1} - 4u_2\frac{\partial}{\partial u_2} - \cdots - (n+2)u_n\frac{\partial}{\partial u_n} + h\frac{\partial}{\partial h}.$$

The action of X_1 on (8.13) is zero, and the action of X_2 on (8.13) gives

$$-\sum_{s=1}^{\infty}\sum_{n=1}^{\infty}\frac{(-h)^{s-1}h^{n-1}}{s!n!}(s+n+2)u_{s+n}$$

$$+\sum_{s=1}^{\infty}\sum_{n=1}^{\infty}\frac{(-h)^{s-1}h^{s-1}}{s!n!}(s+n-2)u_{s+n}+4u^2=0,$$

which implies that

$$-4\left[\sum_{s=1}^{\infty}\sum_{n=1}^{\infty}\frac{(-h)^{s-1}h^{n-1}}{s!n!}u_{s+n}-u^2\right]\Bigg|_{(8.13)}=0.$$

It is obvious that (8.13) admits the group G_2 similar to $\bar{G}_2$. The invariance of the uniform mesh is also obvious.

Thus, one and the same group acts in $\underset{h}{\widetilde{Z}}$, and its representations in $\widetilde{Z}$ and $\underset{h}{Z}$ differ by a similarity transformation and can be prolonged to the differential variables (u_s) and the difference variables (v_n) by different prolongation formulas. The differential equation and its discrete representation in $\underset{h}{Z}$ admit similar groups G_r and $\bar{G}_r$. In general, an arbitrary difference scheme close in the approximation sense to a given discrete representation need not admit a given group $\bar{G}_r$. Apparently approximate models serving as differential approximations to difference schemes and difference approximations to differential equations should be considered from the standpoint of approximate groups [7]. In general, these approximate models need not inherit the groups G_r and $\bar{G}_r$ of the exact models.

8.2. Three-Point Exact Schemes for Nonlinear ODE

In this section, we consider two examples in which exact difference schemes can be represented in finite rather than series form. From the set of parametric families of invariant schemes obtained in Sec. 6.9, we single out exact schemes that have zero approximation error.

8.2.1. Let us construct an exact scheme for the ordinary differential equation

$$u'' = u^{-3}, \tag{8.14}$$

which was considered in Sec. 6.9, where we constructed a one-parameter family of invariant schemes with three first integrals.

Recall that Eq. (8.14) admits the three-parameter point transformation group generated by the operators

$$X_1 = \frac{\partial}{\partial x}, \qquad X_2 = 2x\frac{\partial}{\partial x} + u\frac{\partial}{\partial u}, \qquad X_3 = x^2\frac{\partial}{\partial x} + xu\frac{\partial}{\partial u}. \tag{8.15}$$

Equation (8.14) can be viewed as the Euler equation for the invariant functional with Lagrange function $(\frac{1}{u^2} - u_x^2)$. By Noether's theorem, Eq. (8.14) has three first integrals

$$J_1 = u_x^2 + \frac{1}{u^2} = A^0, \qquad J_2 = 2\frac{x}{u^2} - 2(u - u_x x)u_x = 2B^0,$$

$$J_3 = \frac{x^2}{u^2} + (u - xu_x)^2 = C^0.$$

The general solution of the ordinary differential equation has the form

$$A_0 u^2 = (A_0 x + B_0)^2 + 1.$$

Earlier, we obtained the one-parameter family of invariant meshes

$$\frac{h^+}{uu^+} = \frac{h^-}{uu^-} = \varepsilon, \qquad \varepsilon = \text{const}, \quad 0 < \epsilon \ll 1, \tag{8.16}$$

and the difference equation

$$\frac{\frac{u_x}{h} - \frac{u_{\bar{x}}}{h}}{h^-} = \frac{1}{u^2 u^-} \tag{8.17}$$

approximating the original equation (8.14) to the second order.

The exact solution of the invariant scheme (8.16)–(8.17)

$$A_0 u^2 = (A_0 x + B_0)^2 + 1 - \frac{\varepsilon^2}{4}. \tag{8.18}$$

is uniformly close to the exact solution of the original ordinary differential equation.

The exact scheme (if any) should admit the same transformation group as the ordinary differential equation (8.14). Since it is invariant, such scheme and mesh should be represented in terms of difference invariants. In particular, we can use the mesh (8.16) constructed from difference invariants. (Any other invariant mesh can be used as well.)

Let us construct an exact scheme starting from the parameter-dependent difference Lagrangian

$$\mathcal{L} = \frac{\delta}{uu^+} - \left(\frac{u^+ - u}{h^+}\right)^2,$$

where the parameter $\delta = \text{const}$ is as yet undefined.

The variational procedure on the same invariant mesh results in the following intersection of quasi-extremals (the global extremal):

$$\frac{h^+}{uu^+} = \frac{h^-}{uu^-} = \varepsilon, \qquad \varepsilon = \text{const}, \quad 0 < \varepsilon \ll 1, \tag{8.19}$$

$$\frac{\frac{u_x}{h} - \frac{u_{\bar{x}}}{h}}{h^-} = \frac{\delta}{u^2 u^-}. \tag{8.20}$$

Using the difference analog of Noether's theorem, we obtain the following first integrals:

1. $\quad u_x^2 + \dfrac{\delta}{u u^+} = A^0 = \text{const},$

2. $\quad \dfrac{2x + h^+}{2} u_x^2 + \delta \dfrac{2x + h^+}{2 u u^+} - \dfrac{u + u^+}{2} u_x = B^0 = \text{const},$

3. $\quad \delta \dfrac{x(x + h^+)}{u u^+} + \left(\dfrac{u + u^+}{2} - \dfrac{2x + h^+}{2} u_x \right)^2 = C^0 = \text{const}.$

The two-parameter family of schemes (8.19)–(8.20) with constants ε and δ contains the approximate scheme (8.16)–(8.17). To find the value of δ corresponding to the exact scheme, we substitute the exact solution (8.18) into the scheme (8.19)–(8.20) at three arbitrary points of some particular solution. This determines the constant δ:

$$\delta = 2 \frac{1 - \sqrt{1 - \varepsilon^2}}{\varepsilon^2}. \tag{8.21}$$

Since the action of the group G_3 with operators (8.15) takes every solution of the ordinary differential equation to every other solution and since the scheme (8.19)–(8.20) is invariant, it follows that the resulting scheme with constant (8.21) gives the *entire family of exact solutions*. Note that the scheme (8.19)–(8.20) still contains the arbitrary parameter ε.

Thus, the scheme (8.19)–(8.20) with constant (8.21) is an exact scheme for the ordinary differential equation (8.14); i.e., the family of solutions (8.18) of the ordinary differential equation (8.14) identically satisfies this scheme. Of course, the exact scheme (8.19)–(8.20) determines a set of points on the exact curve rather than the entire smooth curve. The density of these points on the curve depends on the parameter ε and can be arbitrary.

It is important to note that the first integrals (8.14) are difference (i.e., nonlocal) integrals and cannot be obtained from Noether's classical theorem for ordinary differential equations. But they hold both for the exact scheme and for the original ordinary differential equation.

Remark. It turns out that the invariant approximate scheme and the exact scheme are related by a similarity transformation. More precisely, the dilation

$$\widetilde{x} = x \cdot \alpha^2 \sqrt{1 - \frac{\varepsilon^2}{4}}, \qquad \widetilde{u} = u \cdot \alpha$$

of x, or u, or their combination, where $\alpha \neq 0$ is an arbitrary constant, relates the invariant scheme (8.16)–(8.17) to the exact scheme (8.19)–(8.20). By transforming the ODE (8.14), we can find a *differential equation for which the approximate invariant scheme* (8.16)–(8.17) *is exact*. This equation has the form

$$u'' = \frac{1}{u^3} \left(1 - \frac{\varepsilon^2}{4} \right) \tag{8.22}$$

Table 8.1: Differential equations and invariant difference models

Invariant ODEs	Invariant difference models
$u'' = \dfrac{1}{u^3}$	$\dfrac{\frac{u_x}{h} - \frac{u_{\bar{x}}}{h}}{h^-} = \dfrac{\delta}{u^2 u^-}, \qquad \delta = 2\dfrac{1 - \sqrt{1 - \varepsilon^2}}{\varepsilon^2}, \qquad \dfrac{h^+}{uu^+} = \dfrac{h^-}{uu^-} = \varepsilon$
$u'' = \dfrac{1 - \varepsilon^2/4}{u^3}$	$\dfrac{\frac{u_x}{h} - \frac{u_{\bar{x}}}{h}}{h^-} = \dfrac{1}{u^2 u^-}, \qquad \dfrac{h^+}{uu^+} = \dfrac{h^-}{uu^-} = \varepsilon$

for each ε.

Thus, the approximate invariant scheme (8.16)–(8.17) is exact for the approximate differential equations (8.22).

All four objects are shown in Table 8.1. The ordinary differential equation is on the left, and its discrete representation (exact scheme) is on the right. Each scheme approximates the ordinary differential equation in the other row.

8.2.2. Consider the operator algebra

$$X_1 = \frac{\partial}{\partial x}, \qquad X_2 = \frac{\partial}{\partial u}, \qquad X_3 = x\frac{\partial}{\partial x} + (x + u)\frac{\partial}{\partial u}.$$

The corresponding invariant ordinary differential equation

$$u'' = \exp(-u') \tag{8.23}$$

and invariant schemes were considered in Sec. 6.9. The Lagrangian

$$L = \exp(u') + u \tag{8.24}$$

admits the operators X_1 and X_2 as variational symmetries,

$$X_1 L + L D(\xi_1) = 0, \qquad X_2 L + L D(\xi_2) = 1 = D(x).$$

For the second operator X_2, one can find another Lagrangian,

$$L_2 = xu' - \exp(u'),$$

which ensures the exact (nondivergence) variational symmetry

$$X_2 L_2 + L_2 D(\xi_2) = 0$$

for X_2. One can show that there does not exist any Lagrangian providing variational symmetries for all three operators, but it suffices to have two symmetries to integrate Eq. (8.23). By Noether's theorem, the Lagrangian (8.24) permits easily computing the two first integrals

$$J_1 = \exp(u')(1 - u') + u = A^0, \qquad J_2 = \exp(u') - x = B^0,$$

from which one can readily find the general solution of Eq. (8.23) by eliminating u':

$$u = (B^0 + x)(\ln(B^0 + x) - 1) + A^0,$$

where A^0 and B^0 are arbitrary constants.

Recall that in Sec. 6.9 we constructed a conservative invariant model, which we now rewrite as

$$\frac{\alpha_1}{h_+}\left(\exp\left(y_x\right) - \exp\left(y_{\bar{x}}\right)\right) = 1, \quad h^+ = \varepsilon \exp(1 + \varepsilon^2)(1 + \varepsilon)^{-1 - 1/\varepsilon} \exp(y_x),$$

$$(8.25)$$

where the constant

$$\alpha_1 = \exp(1 + \varepsilon^2)(1 + \varepsilon)^{-1/\varepsilon}$$

ensures the second-order approximation. This scheme is completely invariant, has two first integrals (see Sec. 6.9), and is integrable. By using the integrals and by performing algebraic transformations, we find the general solution

$$u = \left(B^h + x\right)\ln\left(B^h + x\right) - (1 + \varepsilon^2)(x + B^h) + A^h. \qquad (8.26)$$

On this solution, the mesh equation in (8.25) is equivalent to the equation

$$h_+ = \varepsilon(x + B^h). \qquad (8.27)$$

Such a mesh is an integral of the two-point invariant equation

$$h_+ = (1 + \varepsilon)h_-.$$

The scheme (8.25) and its general solution contain the small parameter ε, which characterizes the mesh scale. This parameter can be found, for example, from the initial data (x_0, u_0, x_1, u_1) for system (8.25).

The general solution (8.26), (8.27) provides a uniform second-order approximation to the general solution of the original ordinary differential equation (8.23). We show that the family of models of the form (8.25) with a constant α_1 contains an exact scheme whose solution coincides with the solution of the ordinary differential equation at the mesh points. The mesh can be arbitrarily dense on the x-axis. The exact scheme should admit the same transformation group as the original equation; therefore, it should be expressed in terms of difference invariants.

In particular, this means that one can use the same mesh as in the approximate invariant scheme (8.25).

To find the constants that single out the exact scheme from the set of approximate schemes, we use the same idea as in the preceding example. Note that the scheme (8.25) with indeterminate constant α_1 is written out in the invariant form

$$J_0 = \frac{1}{h_+} \left(\exp(u_x) - \exp(u_{\bar{x}}) \right) = \frac{1}{\alpha}, \qquad \frac{h_+}{h_-} = (1 + \varepsilon), \qquad (8.28)$$

where J_0 is an invariant of the group. To find the value of J_0 on the exact solution, it suffices to calculate this value at three points of an arbitrary particular solution. For example, for such a solution we take

$$u = x \ln x - x \qquad (8.29)$$

on the mesh

$$h_+ = \varepsilon x. \qquad (8.30)$$

We take three points $x_1 = 1$, $x_2 = (1 + \varepsilon)$, and $x_3 = (1 + \varepsilon)^2$ on the x-axis according to the mesh (8.30). By substituting the corresponding three points of the particular solution into (8.28), we obtain the value

$$J_0 = e^{-1}(1 + \varepsilon)^{1/\varepsilon},$$

of the invariant J_0, which implies the value

$$\alpha = e(1 + \varepsilon)^{-1/\varepsilon}$$

corresponding to the exact scheme.

Thus, if the exact solution is known at three points, then we can not only reconstruct the entire curve of the particular solution passing through these three points but also write out the scheme (8.28), which, for the constant presented above, gives the entire set of solutions. Note that in general we need not have any analytic expression for the particular solution (8.29); we should only know three points of the exact solution.

The exact scheme, which is a special case of approximate invariant schemes, admits a variational statement as well. Just the same procedure as in the case of approximate invariant schemes gives the following exact integrals:

$$e(1 + \varepsilon)^{-1/\varepsilon} \exp(u_x) = x + h^+ + B,$$

$$\left(1 + \frac{1}{\varepsilon} \right) e(1 + \varepsilon)^{-(1+1/\varepsilon)} \ln(1 + \varepsilon) \exp(u_x) (u_x - 1) \qquad (8.31)$$

$$= u + \left(\left(1 + \frac{1}{\varepsilon} \right) \ln(1 + \varepsilon) - 1 \right) \frac{h^+}{\varepsilon} u_x + A.$$

From the integrals (8.31), we find the general solution

$$u = (x + B)\ln(x + B) - (x + B) + A,$$

which does not contain a small parameter and coincides with the exact solution of the ordinary differential equation (8.23).

The exact difference model has the same set of solutions as the original ordinary differential equation. Just as Eq. (8.23), it has two first integrals (8.31), and these difference integrals also hold for the differential equation (8.23).

The above examples show that approximate invariant schemes contain exact schemes as a subset and have the same algebraic structure. Exact schemes for other second-order ordinary differential equations with two or more symmetries can be constructed in a completely similar way. A more complicated example (an exact scheme for the Kepler problem) can be found in [80].

The existence of exact schemes whose solutions coincide at the points of an arbitrarily dense mesh with the corresponding values of the solution of the differential equation, gives rise to a peculiar *mathematical dualism*. The same physical processes can be described either by ordinary differential equations whose solutions are continuous curves or by discrete equations providing points on these curves. We believe that this dualism deserves attention of theoretical physicists interested in the construction of mathematical models.

Bibliography

[1] M. J. Ablowitz and B. M. Herbst. On homoclinic structure and numerically induced chaos for the nonlinear Schrödinger equation. *SIAM J. Appl. Math.*, **50**, No. 2, 1990, 339–351.

[2] M. J. Ablowitz, C. Schober, and B. M. Herbst. Numerical chaos, roundoff errors, and homoclinic manifolds. *Phys. Rev. Lett.*, **71**, No. 17, Oct 1993, 2683–2686.

[3] C. D. Ahlbrandt. Equivalence of discrete Euler equations and discrete Hamiltonian systems. *J. Math. Anal. Appl.*, **180**, No. 2, 1993, 498–517.

[4] G. D. Akrivis, V. A. Dougalis, and O. A. Karakashian. On fully discrete Galerkin methods of second-order temporal accuracy for the nonlinear Schrödinger equation. *Numer. Math.*, **59**, No. 1, 1991, 31–53.

[5] R. L. Anderson and N. H. Ibragimov. *Lie-Bäcklund Transformations in Applications*, volume 1 of *SIAM Studies in Applied Mathematics*. SIAM, Philadelphia, 1979.

[6] V. I. Arnold. *Mathematical Methods of Classical Mechanics*, volume 60 of *Graduate Texts in Mathematics*. Springer-Verlag, New York, second edition, 1989. Translated from the Russian by K. Vogtmann and A. Weinstein.

[7] V. A. Baikov, R. K. Gazizov, and N. H. Ibragimov. Approximate symmetries. *Mat. Sb. (N.S.)*, **136(178)**, No. 4, 1988, 435–450.

[8] M. I. Bakirova and V. A. Dorodnitsyn. An invariant difference model for the equation $u_t = u_{xx} + \delta u \ln u$. *Differ. Uravn.*, **30**, No. 10, 1994, 1697–1702. Translated in *Differ. Equ.*, **30**, 1994, No. 10, 1995, 1565–1570.

[9] M. I. Bakirova, V. A. Dorodnitsyn, and R. V. Kozlov. Symmetry-preserving difference schemes for some heat transfer equations. *J. Phys. A*, **30**, No. 23, 1997, 8139–8155.

[10] M. I. Bakirova et al. Invariant solutions of the heat equation that describe the directed propagation of combustion and spiral waves in a nonlinear medium. *Dokl. Akad. Nauk SSSR*, **299**, No. 2, 1988, 346–350.

[11] E. Bessel-Hagen. Über die Erhaltungssätze der Elektrodynamik. *Math. Ann.*, **84**, No. 3–4, 1921, 258–276.

[12] G. Birkhoff. *Hydrodynamics: A Study in Logic, Fact and Similitude*. Princeton Univ. Press, Princeton, NJ, revised edition, 1960.

[13] G. W. Bluman and S. C. Anco. *Symmetry and Integration Methods for Differential Equations*, volume 154 of *Applied Mathematical Sciences*. Springer-Verlag, New York, 2002.

[14] G. W. Bluman and J. D. Cole. *Similarity Methods for Differential Equations*, volume 13 of *Applied Mathematical Sciences*. Springer-Verlag, New York, 1974.

[15] G. W. Bluman and S. Kumei. *Symmetries and Differential Equations*, volume 81 of *Applied Mathematical Sciences*. Springer-Verlag, New York, 1989.

[16] A. Bobylev and V. Dorodnitsyn. Symmetries of evolution equations with non-local operators and applications to the Boltzmann equation. *Discrete Contin. Dyn. Syst.*, **24**, No. 1, 2009, 35–57.

[17] A. Bourlioux, C. Cyr-Gagnon, and P. Winternitz. Difference schemes with point symmetries and their numerical tests. *J. Phys. A*, **39**, No. 22, 2006, 6877–6896.

[18] A. Bourlioux, R. Rebelo, and P. Winternitz. Symmetry preserving discretization of $SL(2, \mathbb{R})$ invariant equations. *J. Nonlinear Math. Phys.*, **15**, Suppl. 3, 2008, 362–372.

[19] C. Budd and V. Dorodnitsyn. Symmetry-adapted moving mesh schemes for the nonlinear Schrödinger equation. *J. Phys. A*, **34**, No. 48, 2001, 10387–10400.

[20] C. J. Budd, S. Chen, and R. D. Russell. New self-similar solutions of the nonlinear Schrödinger equation with moving mesh computations. *J. Comput. Phys.*, **152**, No. 2, 1999, 756–789.

[21] C. J. Budd and G. J. Collins. An invariant moving mesh scheme for the nonlinear diffusion equation. In *Proceedings of the ICMS Conference on Grid Adaptation in Computational PDEs: Theory and Applications* (Edinburgh, 1996). *Appl. Numer. Math.*, **26**, No. 1-2, 1998, 23–39.

[22] C. J. Budd, G. J. Collins, W. Z. Huang, and R. D. Russell. Self-similar numerical solutions of the porous-medium equation using moving mesh methods. *R. Soc. Lond. Philos. Trans. Ser. A Math. Phys. Eng. Sci.*, **357**, No. 1754, 1999, 1047–1077.

[23] C. J. Budd, W. Huang, and R. D. Russell. Moving mesh methods for problems with blow-up. *SIAM J. Sci. Comput.*, **17**, No. 2, 1996, 305–327.

[24] G. B. Byrnes, R. Sahadevan, and G. R. W. Quispel. Factorizable Lie symmetries and the linearization of difference equations. *Nonlinearity*, **8**, No. 3, 1995, 443–459.

[25] E. Cartan. *Leçons sur les invariants intégraux. Cours professé à la Faculté des Sciences de Paris.* Hermann, Paris, 1922.

[26] A. K. Common, E. Hessameddini, and M. Musette. The Pinney equation and its discretization. *J. Phys. A*, **29**, No. 19, 1996, 6343–6352.

[27] A. K. Common and M. Musette. Two discretisations of the Ermakov–Pinney equation. *Phys. Lett. A*, **235**, No. 6, 1997, 574–580.

[28] V. A. Dorodnitsyn. Invariant solutions of the nonlinear heat equation with a source. *Zh. Vychisl. Mat. Mat. Fiz.*, **22**, No. 6, 1982, 1393–1400.

[29] V. A. Dorodnitsyn. Transformation groups in mesh spaces. In *Current Problems in Mathematics. Newest Results, Vol. 34 (Russian)*, Itogi Nauki i Tekhniki, 1989, pages 149–191. Akad. Nauk SSSR Vsesoyuz. Inst. Nauchn. i Tekhn. Inform., Moscow. Translated in *J. Soviet Math.*, **55**, 1991, No. 1, 1490–1517.

[30] V. A. Dorodnitsyn. A finite-difference analogue of Noether's theorem. *Dokl. Akad. Nauk*, **328**, No. 6, 1993, 678–682. Translated in *Phys. Dokl.*, **38**, 1993, No. 2, 66–68.

[31] V. A. Dorodnitsyn. Finite difference models entirely inheriting symmetry of original differential equations. In *Modern Group Analysis: Advanced Analytical and Computational Methods in Mathematical Physics (Acireale, 1992)*, 1993, pages 191–201. Kluwer Acad. Publ., Dordrecht.

[32] V. A. Dorodnitsyn. Finite difference models entirely inheriting continuous symmetry of original differential equations. *Internat. J. Modern Phys. C*, **5**, No. 4, 1994, 723–734.

[33] V. A. Dorodnitsyn. Invariant discrete model for the Korteweg–de Vries equation. Preprint CRM-2187, Centre de recherches mathématiques, Université de Montréal, 1994.

[34] V. A. Dorodnitsyn. Some new invariant difference equations on evolutionary grids. In *Proceedings of* 14*th IMACS World Congress of Computational and Applied Mathematics*, volume 1, 1994, pages 143–146.

[35] V. Dorodnitsyn. Continuous symmetries of finite-difference evolution equations and grids. In *Symmetries and Integrability of Difference Equations* (*Estérel, PQ,* 1994), volume 9 of *CRM Proc. Lecture Notes*, 1996, pages 103–112. Amer. Math. Soc., Providence, RI.

[36] V. A. Dorodnitsyn. Conservation laws for difference equations. In N. Ibragimov, K. Razi Nagvi, and E. Straume, editors, *Modern Group Analysis* VII, *Developments in Theory, Computations and Application*, 1999, pages 91–97. MARS Publishers, Symmetry Foundation, Trondheim, Norway.

[37] V. A. Dorodnitsyn. *Lie Group Properties of Difference Equations*. MAKS-Press, Moscow, 2000. Revised edition: Moscow, Fizmatgiz, 2001.

[38] V. A. Dorodnitsyn. Invariant difference model for nonlinear Schrödinger equation with conservation of Lagrangian structure. In N. H. Ibragimov, editor, *Proceedings of the International Conference MOGRAN* 2000, 2001, pages 49–52. USATU Publishers, Ufa, Russia.

[39] V. Dorodnitsyn. Noether-type theorems for difference equations. *Appl. Numer. Math.*, **39**, No. 3-4, 2001, 307–321. Special issue: Themes in geometric integration.

[40] V. A. Dorodnitsyn. On the linearization of second-order differential and difference equations. *SIGMA*, **2**, 2006, 065. http://arxiv.org/abs/nlin/0608038.

[41] V. A. Dorodnitsyn, G. G. Elenin, and S. P. Kurdyumov. Exact solutions of certain problems for a quasilinear equation of parabolic type. In *Computational Mathematics* (Warsaw, 1980), volume 13 of *Banach Center Publ.*, 1984, pages 113–123. PWN, Warsaw.

[42] V. Dorodnitsyn and R. Kozlov. A heat transfer with a source: the complete set of invariant difference schemes. *J. Nonlinear Math. Phys.*, **10**, No. 1, 2003, 16–50.

[43] V. A. Dorodnitsyn and R. Kozlov. First integrals of difference Hamiltonian equations, *J. Phys. A*, **45**, 2009, 454007.

[44] V. A. Dorodnitsyn and R. Kozlov. Invariance and first integrals of continuous and discrete Hamiltonian equations. *J. Engineering Math.*, **66**, No. 1–3, 2010, 253–270.

[45] V. A. Dorodnitsyn and R. Kozlov. Lagrangian and Hamiltonian formalism for discrete equations: symmetries and first integrals. (To appear in Cambridge Univ. Press, Cambridge, UK).

[46] V. Dorodnitsyn, R. Kozlov, and P. Winternitz. Lie group classification of second-order ordinary difference equations. *J. Math. Phys.*, **41**, No. 1, 2000, 480–504.

[47] V. Dorodnitsyn, R. Kozlov, and P. Winternitz. On Lie group classification of second-order ordinary difference equations. In *Proceedings of the Workshop on Nonlinearity, Integrability and All That: Twenty Years after NEEDS '79 (Gallipoli, 1999)*, 2000, pages 250–257. World Sci. Publ., River Edge, NJ.

[48] V. Dorodnitsyn, R. Kozlov, and P. Winternitz. Symmetries, Lagrangian formalism and integration of second order ordinary difference equations. *J. Nonlinear Math. Phys.*, **10**, No. suppl. 2, 2003, 41–56.

[49] V. Dorodnitsyn, R. Kozlov, and P. Winternitz. Continuous symmetries of Lagrangians and exact solutions of discrete equations. *J. Math. Phys.*, **45**, No. 1, 2004, 336–359.

[50] V. Dorodnitsyn and P. Winternitz. Lie point symmetry preserving discretizations for variable coefficient Korteweg–de Vries equations. *Nonlinear Dynam.*, **22**, No. 1, 2000, 49–59.

[51] R. D. Driver. *Ordinary and Delay Differential Equations*. Springer-Verlag, New York, 1977. *Applied Mathematical Sciences, Vol. 20*.

[52] N. A. Elnatanov and J. Schiff. The Hamilton–Jacobi difference equation. *Funct. Differ. Equ.*, **3**, No. 3-4, 1996, 279–286 (1997).

[53] L. H. Erbe and P. X. Yan. Disconjugacy for linear Hamiltonian difference systems. *J. Math. Anal. Appl.*, **167**, No. 2, 1992, 355–367.

[54] M. Fels and P. J. Olver. Moving frames and coframes. In *Algebraic Methods in Physics* (Montréal, QC, 1997), CRM Ser. Math. Phys., 2001, pages 47–64. Springer-Verlag, New York.

[55] R. Floreanini and L. Vinet. Quantum algebras and q-special functions. *Ann. Physics*, **221**, No. 1, 1993, 53–70.

[56] R. Floreanini and L. Vinet. Symmetries of the q-difference heat equation. *Lett. Math. Phys.*, **32**, No. 1, 1994, 37–44. Preprint CRM-1919, Centre de recherches mathématiques, Université de Montréal, 1993.

[57] R. Floreanini and L. Vinet. Lie symmetries of finite-difference equations. *J. Math. Phys.*, **36**, No. 12, 1995, 7024–7042.

[58] V. A. Galaktionov et al. A quasilinear equation of heat conduction with a source: peaking, localization, symmetry, exact solutions, asymptotic behavior, structures. In *Current Problems in Mathematics. Newest Results,* Vol. 28 (Russian), Itogi Nauki i Tekhniki, 1986, pages 95–205, 316. Akad. Nauk SSSR Vsesoyuz. Inst. Nauchn. i Tekhn. Inform., Moscow. Translated in *J. Soviet Math.*, **41**, 1988, No. 5, 1222–1292.

[59] V. A. Galaktionov and S. R. Svirshchevskii. *Exact Solutions and Invariant Subspaces of Nonlinear Partial Differential Equations in Mechanics and Physics. Chapman & Hall/CRC Applied Mathematics and Nonlinear Science Series*. Chapman & Hall/CRC, Boca Raton, FL, 2007.

[60] I. M. Gelfand and S. V. Fomin. *Calculus of Variations*. Prentice-Hall Inc., Englewood Cliffs, NJ, revised English edition, 1963.

[61] A. O. Gel'fond. *Calculus of Finite Differences*. Nauka, Moscow, third edition, 1967.

[62] D. Gómez-Ullate, S. Lafortune, and P. Winternitz. Symmetries of discrete dynamical systems involving two species. *J. Math. Phys.*, **40**, No. 6, 1999, 2782–2804.

[63] A. González-López, N. Kamran, and P. J. Olver. Lie algebras of vector fields in the real plane. *Proc. London Math. Soc.*, **64**, No. 2, 1992, 339–368.

[64] E. Hairer, C. Lubich, and G. Wanner. *Geometric Numerical Integration*, volume 31 of *Springer Series in Computational Mathematics*. Springer-Verlag, Berlin, second edition, 2006.

[65] R. Hernández Heredero, D. Levi, M. A. Rodríguez, and P. Winternitz. Lie algebra contractions and symmetries of the Toda hierarchy. *J. Phys. A*, **33**, No. 28, 2000, 5025–5040.

[66] R. Hernández Heredero, D. Levi, and P. Winternitz. Symmetries of the discrete Burgers equation. *J. Phys. A*, **32**, No. 14, 1999, 2685–2695.

[67] R. Hernández Heredero, D. Levi, and P. Winternitz. Symmetry preserving discretization of the Burgers equation. In *SIDE III—Symmetries and Integrability of Difference Equations* (Sabaudia, 1998), volume 25 of *CRM Proc. Lecture Notes*, 2000, pages 197–208. Amer. Math. Soc., Providence, RI.

[68] P. E. Hydon. Symmetries and first integrals of ordinary difference equations. *R. Soc. Lond. Proc. Ser. A Math. Phys. Eng. Sci.*, **456**, No. 2004, 2000, 2835–2855.

[69] P. E. Hydon. *Symmetry Methods for Differential Equations. A Beginner's Guide. Cambridge Texts in Applied Mathematics.* Cambridge University Press, Cambridge, 2000.

[70] P. E. Hydon. Conservation laws of partial difference equations with two independent variables. *J. Phys. A*, **34**, No. 48, 2001, 10347–10355.

[71] I. I. Ibragimov and M. V. Keldysh. On the interpolation of entire functions. *Mat. Sb.*, **20**(**62**), 1947, 283–290.

[72] N. H. Ibragimov. Invariant variational problems and the conservation laws (remarks on E. Noether's theorem). *Teoret. Mat. Fiz.*, **1**, No. 3, 1969, 350–359.

[73] N. H. Ibragimov. *Transformation Groups Applied to Mathematical Physics. Mathematics and Its Applications (Soviet Series).* D. Reidel Publishing Co., Dordrecht, 1985. Translated from the Russian.

[74] N. H. Ibragimov, editor. *CRC Handbook of Lie Group Analysis of Differential Equations. Vol. 1.* W. F. Ames et al., CRC Press, Boca Raton, FL, 1994.

[75] N. H. Ibragimov, editor. *CRC Handbook of Lie Group Analysis of Differential Equations. Vol. 2.* A. V. Aksenov et al., CRC Press, Boca Raton, FL, 1995.

[76] N. H. Ibragimov, editor. *CRC Handbook of Lie Group Analysis of Differential Equations. Vol. 3.* R. L. Anderson et al., CRC Press, Boca Raton, FL, 1996.

[77] N. H. Ibragimov. *A Practical Course in Differential Equations and Mathematical Modelling.* ALGA Publications, Karlskrona, 2006.

[78] V. Kolmanovskii and A. Myshkis. *Applied Theory of Functional-Differential Equations*, volume 85 of *Mathematics and Its Applications (Soviet Series).* Kluwer Acad. Publ., Dordrecht, 1992.

[79] V. M. Kostin. Certain invariant solutions of equations of the Korteweg-de Vries type. *Zh. Prikl. Mekh. Tekhn. Fiz.*, 1969, No. 4, 69–73.

[80] R. Kozlov. Conservative discretizations of the Kepler motion. *J. Phys. A*, **40**, No. 17, 2007, 4529–4539.

[81] J.-L. Lagrange. Sur une nouvelle espèce de calcul relatif à la différentiation et à l'intégration des quantités variables. *Nouveaux mémoires de l'Académie royale des sciences et belles-lettres de Berlin*, 1772. Œuvres de Lagrange, tome 3, Gauthier-Villars, Paris, 1867–92, pp. 441–476.

[82] J.-L. Lagrange. Mémoire sur la méthode d'interpolation. *Nouveaux mémoires de l'Académie royale des sciences et belles-lettres de Berlin*, 1792–93. Œuvres de Lagrange, tome 5, Gauthier-Villars, Paris, 1867–92, pp. 663–684.

[83] S. Lall and M. West. Discrete variational Hamiltonian mechanics. *J. Phys. A*, **39**, No. 19, 2006, 5509–5519.

[84] B. Leimkuhler and S. Reich. *Simulating Hamiltonian Dynamics*, volume 14 of *Cambridge Monographs on Applied and Computational Mathematics*. Cambridge University Press, Cambridge, 2004.

[85] D. Levi, S. Tremblay, and P. Winternitz. Lie point symmetries of difference equations and lattices. *J. Phys. A*, **33**, No. 47, 2000, 8507–8523.

[86] D. Levi and P. Winternitz. Continuous symmetries of discrete equations. *Phys. Lett. A*, **152**, No. 7, 1991, 335–338.

[87] D. Levi and P. Winternitz. Symmetries and conditional symmetries of differential-difference equations. *J. Math. Phys.*, **34**, No. 8, 1993, 3713–3730.

[88] D. Levi and P. Winternitz. Symmetries of discrete dynamical systems. *J. Math. Phys.*, **37**, No. 11, 1996, 5551–5576.

[89] D. Levi and P. Winternitz. Continuous symmetries of difference equations. *J. Phys. A*, **39**, No. 2, 2006, R1–R63.

[90] S. Lie. *Vorlesungen über Differentialgleichungen mit bekannten infinitesimalen Transformationen*. B. G. Teubner, Leipzig, 1891. Reprinted as *Differentialgleichungen*, Chelsea, New York, 1967.

[91] S. Lie. *Theorie der Transformationsgruppen, dritter (und letzter) Abschnitt. Unter Mitwirkung von Fr. Engel bearbeitet*. B. G. Teubner, Leipzig, 1893.

[92] S. Lie. *Vorlesungen über continuirliche Gruppen mit geometrischen und anderen Anwendungen*. B. G. Teubner, Leipzig, 1893.

[93] S. Lie. Klassifikation und Integration von gewöhnlichen Differentialgleichungen zwischen x, y die eine Gruppe von Transformationen gestatten. I–IV. In *Gesammelte Abhandlungen*, volume 5, 1924. B. G. Teubner, Leipzig.

[94] M. Lutzky. Symmetry groups and conserved quantities for the harmonic oscillator. *J. Phys. A*, **11**, No. 2, 1978, 249–258.

[95] S. Maeda. Canonical structure and symmetries for discrete systems. *Math. Japon.*, **25**, No. 4, 1980, 405–420.

[96] S. Maeda. Extension of discrete Noether theorem. *Math. Japon.*, **26**, No. 1, 1981, 85–90.

[97] S. Maeda. The similarity method for difference equations. *IMA J. Appl. Math.*, **38**, No. 2, 1987, 129–134.

[98] J. E. Marsden and T. S. Ratiu. *Introduction to Mechanics and Symmetry. A Basic Exposition of Classical Mechanical Systems*, volume 17 of *Texts in Applied Mathematics*. Springer-Verlag, New York, second edition, 1999.

[99] L. Martina, S. Lafortune, and P. Winternitz. Point symmetries of generalized Toda field theories. II. Symmetry reduction. *J. Phys. A*, **33**, No. 36, 2000, 6431–6446.

[100] R. McLachlan. Symplectic integration of Hamiltonian wave equations. *Numer. Math.*, **66**, No. 4, 1994, 465–492.

[101] S. V. Meleshko. *Methods for Constructing Exact Solutions of Partial Differential Equations. Mathematical and Analytical Techniques with Applications to Engineering*. Springer, New York, 2005.

[102] S. V. Meleshko and S. Moyo. On the complete group classification of the reaction-diffusion equation with a delay. *J. Math. Anal. Appl.*, **338**, No. 1, 2008, 448–466.

[103] A. D. Myshkis. *Linear Differential Equations with Retarded Argument*. Nauka, Moscow, second edition, 1972.

[104] E. Noether. Invariante Variationsprobleme. *Gött. Nachr.*, 1918, 235–257.

[105] N. E. Nörlund. *Differenzenrechnung*. Springer-Verlag, Berlin, 1924.

[106] P. J. Olver. Evolution equations possessing infinitely many symmetries. *J. Math. Phys.*, **18**, No. 6, 1977, 1212–1215. Reprinted in: *Solitons and Particles*, C. Rebbi and G. Soliani, eds., World Scientific, Singapore, 1984, pp. 235–238.

[107] P. J. Olver. *Applications of Lie Groups to Differential Equations*, volume 107 of *Graduate Texts in Mathematics*. Springer-Verlag, New York, second edition, 1993.

[108] P. J. Olver. Geometric foundations of numerical algorithms and symmetry. *Appl. Algebra Engrg. Comm. Comput.*, **11**, No. 5, 2001, 417–436. Special issue "Computational geometry for differential equations."

[109] P. J. Olver. An introduction to moving frames. In *Geometry, Integrability and Quantization. Papers from the 5th International Conference held in Varna, June 5–12, 2003*, I. M. Mladenov and A. C. Hirshfeld, eds., 2004, pages 67–80. Softex, Sofia.

[110] P. J. Olver and J. Pohjanpelto. Moving frames for Lie pseudo-groups. *Canad. J. Math.*, **60**, No. 6, 2008, 1336–1386.

[111] L. V. Ovsiannikov. *Group Analysis of Differential Equations*. Academic Press Inc. [Harcourt Brace Jovanovich Publishers], New York, 1982. Translated from the Russian by Y. Chapovsky. Translation edited by William F. Ames.

[112] L. V. Ovsyannikov. Group properties of the equation of non-linear heat conductivity. *Dokl. Akad. Nauk SSSR*, **125**, 1959, 592–495.

[113] L. V. Ovsyannikov. *Group Properties of Differential Equations*. Izdat. Sibirsk. Otdel. Akad. Nauk SSSR, Novosibirsk, 1962.

[114] L. V. Ovsyannikov. *Lectures on the Theory of Group Properties of Differential Equations*. Novosibirsk. Gos. Univ., Novosibirsk, 1966.

[115] L. V. Ovsyannikov. *Lectures on Basic Gas Dynamics*. Nauka, Moscow, 1981.

[116] L. V. Ovsyannikov. On the property of x-autonomy. *Dokl. Akad. Nauk*, **330**, No. 5, 1993, 559–561. Translated in *Russian Acad. Sci. Dokl. Math.*, **47**, 1993, No. 3, 581–584.

[117] P. Pue-on and S. V. Meleshko. Group classification of second-order delay ordinary differential equations. *Commun. Nonlinear Sci. Numer. Simul.*, **15**, No. 6, 2010, 1444–1453.

[118] G. R. W. Quispel, H. W. Capel, and R. Sahadevan. Continuous symmetries of differential-difference equations: The Kac–van Moerbeke equation and Painlevé reduction. *Phys. Lett. A*, **170**, No. 5, 1992, 379–383.

[119] G. R. W. Quispel and R. Sahadevan. Lie symmetries and the integration of difference equations. *Phys. Lett. A*, **184**, No. 1, 1993, 64–70.

[120] M. A. Rodríguez and P. Winternitz. Lie symmetries and exact solutions of first-order difference schemes. *J. Phys. A*, **37**, No. 23, 2004, 6129–6142.

[121] R. Sahadevan, G. B. Byrnes, and G. R. W. Quispel. Linearisation of difference equations using factorisable Lie symmetries. In *Symmetries and Integrability of Difference Equations* (Estérel, PQ, 1994), volume 9 of *CRM*

Proc. Lecture Notes, 1996, pages 337–343. Amer. Math. Soc., Providence, RI.

[122] A. A. Samarskii. *The Theory of Difference Schemes*, volume 240 of *Monographs and Textbooks in Pure and Applied Mathematics*. Marcel Dekker Inc., New York, 2001.

[123] A. A. Samarskii, V. A. Galaktionov, S. P. Kurdyumov, and A. P. Mikhailov. *Blow-Up in Quasilinear Parabolic Equations*, volume 19 of *de Gruyter Expositions in Mathematics*. Walter de Gruyter & Co., Berlin, 1995. Translated from the 1987 Russian original by Michael Grinfeld and revised by the authors.

[124] A. A. Samarskii, P. P. Matus, and P. N. Vabishchevich. *Difference Schemes with Operator Factors*, volume 546 of *Mathematics and Its Applications*. Kluwer Acad. Publ., Dordrecht, 2002.

[125] A. A. Samarskii and E. S. Nikolaev. *Numerical Methods for Grid Equations. Vol. II: Iterative methods*. Birkhäuser Verlag, Basel, 1989. Translated from the Russian and with a note by Stephen G. Nash.

[126] A. A. Samarskii and I. M. Sobol'. Examples of numerical calculation of temperature waves. *Zh. Vychisl. Mat. Mat. Fiz.*, **3**, 1963, 702–719.

[127] A. A. Samarskii and P. N. Vabishchevich. *Computational Heat Transfer*, volume 1, 2. Wiley, Chichester, 1995.

[128] J. M. Sanz-Serna and M. P. Calvo. *Numerical Hamiltonian Problems*, volume 7 of *Applied Mathematics and Mathematical Computation*. Chapman & Hall, London, 1994.

[129] Y. Shi. Symplectic structure of discrete Hamiltonian systems. *J. Math. Anal. Appl.*, **266**, No. 2, 2002, 472–478.

[130] H. Stephani. *Differential Equations. Their Solution Using Symmetries*. Cambridge University Press, Cambridge, 1989.

[131] C. Sulem and P.-L. Sulem. *The Nonlinear Schrödinger Equation. Self-Focusing and Wave Collapse*, volume 139 of *Applied Mathematical Sciences*. Springer-Verlag, New York, 1999.

[132] Y. B. Suris. *The Problem of Integrable Discretization: Hamiltonian Approach*, volume 219 of *Progress in Mathematics*. Birkhäuser Verlag, Basel, 2003.

[133] J. Tanthanuch and S. V. Meleshko. On definition of an admitted Lie group for functional differential equations. *Commun. Nonlinear Sci. Numer. Simul.*, **9**, No. 1, 2004, 117–125. Group analysis of nonlinear wave problems (Moscow, 2002).

[134] W. Thirring. *A Course in Mathematical Physics. Vol. I. Classical Dynamical Systems*. Springer-Verlag, New York, 1978. Translated from the German by Evans M. Harrell.

[135] A. N. Tikhonov and A. A. Samarskii. Convergence of difference schemes in the class of discontinuous coefficients. *Dokl. Akad. Nauk SSSR*, **124**, No. 2, 1959, 529–532.

[136] A. N. Tikhonov and A. A. Samarskii. *Equations of Mathematical Physics*. Translated by A. R. M. Robson and P. Basu; translation edited by D. M. Brink. A Pergamon Press Book. The Macmillan Co., New York, 1963.

[137] Y. Tourigny and J. M. Sanz-Serna. The numerical study of blowup with application to a nonlinear Schrödinger equation. *J. Comput. Phys.*, **102**, No. 2, 1992, 407–416.

[138] J. Wu. *Theory and Applications of Partial Functional-Differential Equations*, volume 119 of *Applied Mathematical Sciences*. Springer-Verlag, New York, 1996.

Index